High-Throughput Image Reconstruction and Analysis

Artech House Series
Bioinformatics & Biomedical Imaging

Series Editors
Stephen T. C. Wong, The Methodist Hospital and Weill Cornell Medical College
Guang-Zhong Yang, Imperial College

Advances in Diagnostic and Therapeutic Ultrasound Imaging, Jasjit S. Suri, Chirinjeev Kathuria, Ruey-Feng Chang, Filippo Molinari, and Aaron Fenster, editors

Biological Database Modeling, Jake Chen and Amandeep S. Sidhu, editors

Biomedical Informatics in Translational Research, Hai Hu, Michael Liebman, and Richard Mural

Genome Sequencing Technology and Algorithms, Sun Kim, Haixu Tang, and Elaine R. Mardis, editors

High-Throughput Image Reconstruction and Analysis, A. Ravishankar Rao and Guillermo A. Cecchi, editors

Life Science Automation Fundamentals and Applications, Mingjun Zhang, Bradley Nelson, and Robin Felder, editors

Microscopic Image Analysis for Life Science Applications, Jens Rittscher, Stephen T. C. Wong, and Raghu Machiraju, editors

Next Generation Artificial Vision Systems: Reverse Engineering the Human Visual System, Maria Petrou and Anil Bharath, editors

Systems Bioinformatics: An Engineering Case-Based Approach, Gil Alterovitz and Marco F. Ramoni, editors

Text Mining for Biology and Biomedicine, Sophia Ananiadou and John McNaught, editors

Translational Multimodality Optical Imaging, Fred S. Azar and Xavier Intes, editors

High-Throughput Image Reconstruction and Analysis

A. Ravishankar Rao
Guillermo A. Cecchi

Editors

ARTECH
HOUSE

BOSTON | LONDON
artechhouse.com

Library of Congress Cataloging-in-Publication Data
A catalog record for this book is available from the U.S. Library of Congress.

British Library Cataloguing in Publication Data
A catalogue record for this book is available from the British Library.

ISBN-13: 978-1-59693-295-1

685 Canton Street
Norwood, MA 02062

10 9 8 7 6 5 4 3 2 1

Contents

CHAPTER 1

Introduction

A. Ravishankar Rao and Guillermo A. Cecchi

Progress in biology is dependent on the ability to observe, measure, and model the behavior of organisms at multiple levels of abstraction, from the microscopic to the macroscopic. There has been a tremendous growth recently in the techniques to probe the structure and workings of cellular and organ-level mechanisms. Significant advances have been made in areas such as serial block face microscopy and knife-edge microscopy, which allow microstructure information to be gathered at unprecedented levels of both detail and scope. At larger spatial scales, it is now possible to image human whole-brain activity using functional magnetic resonance imaging (fMRI). At the same time, advances have also been made in gathering temporal image data streams from microscopic samples with the use of fluorescent and multiphoton imaging techniques. The increasing spatial and temporal resolution available, combined with advanced sectioning techniques are providing extremely content-rich data to biologists and put unprecedented power in their hands. The complexity of this data is a reflection of the complexity of biological systems, which contain 3D structures with multiple components, interacting with intracellular and extracellular variables.

This is also a game-changing development, since scientists are no longer limited to carrying out experiments to test a single hypothesis at a time. They are now able to vary multiple parameters simultaneously and observe several phenomena of relevance using multispectral techniques. This can be combined with recent advances in data mining techniques to determine relationships and correlations amongst the many variables of interest. This allows a significantly larger parameter space to be explored. The large amount of data being made available improves the capability of scientists to generate hypotheses and conduct experiments. As an example, assays can be performed where multiple cell populations can be simultaneously studied with varying ambient conditions such as temperature and intrinsic conditions such as concentrations of injected drugs. Such assays would be very useful to pharmaceutical companies interested in exploring the parameter space for drug development and discovery.

However, there is a mounting problem being faced by practitioners and researchers today, which is the computational bottleneck: the data storage and processing needs are growing exponentially. It may take several hours or even days to process the collected data, and the resulting throughput time may be unacceptable to support desired workflows in laboratories. Unless the computational issues are addressed immediately, scientists will be overwhelmed with the data collected and will not have adequate tools available to process and extract meaning from the data. A related problem is that of building models from the collected data, which is a useful technique to test the understanding of the phenomena of interest. As the data expose interactions at finer spatial and time scales, the variables that are modeled also increase in number and complexity. This increases the computational burden on the modeling effort as well.

Fortunately, there is a solution to this problem, and that is offered by the area of high-performance computing (HPC). There has been a long history of using HPC to model problems in physics, but its use in the biological sciences has been very recent and rather limited. In general, HPC has not been used much in bio-imaging applications due to the difficulty in porting code to parallel machines. However, the landscape is rapidly changing due to the increased availability of HPC platforms, improvements in parallel programming environments (such as the emergence of the message passing interface as a standard), and the availability of toolkits to perform parallel data mining.

HPC has significant potential to be applied to problems in biology, and particularly to imaging applications. The high computational demands of simulation and modeling complex systems can also be addressed through HPC. So a single HPC architecture can support multiple computational requirements, ranging from analyzing data to building and simulating models.

The purpose of this book is to explore how the coupling of HPC with automated imaging techniques can impact and further our understanding of complex biological systems. The proposed solution involves algorithms to perform automated or semi-automated image analysis and data mining coupled with HPC. This will provide biologists with a powerful tool to probe and understand interactions in complex biological systems. This book will familiarize readers with game changing techniques in bio-imaging, analyze the impact they will have, and provide recommendations on how to best harness this emerging technology.

The intended audience of the book includes cell biologists, users of the Open Microscopy Environment (OME), microscope manufacturers, developers of high-performance computing systems, creators of high-resolution image atlases of multiple organs (e.g., brain/heart/liver), and anyone interested in emerging challenges in microscopy.

This book presents a state-of-the art survey of leading edge micrsoscopy techniques, and the upcoming challenges they face. The contributions in the book describe recent developments in high-performance computing systems and their relevance to solving problems in microscopy. Overall, the book provides an integrated view of a new capability in science, where high-performance computing is used advantageously to build and analyze complex models gathered from large biological datasets.

The book is broadly outlined as follows. We initially examine emerging biological applications where microscopy is playing a crucial role in understanding the structure of organs and cells. We also observe that there is growing interest in simulating cellular processes, and applying them to study emergent phenomena such as cancer growth. The large volumes of data mandate the use of parallel processing techniques. The prevailing technology to utilize parallel computation is explained. We then examine several specific applications that make use of parallel computation. We have tried to illuminate the problem from many perspectives, including the creation of databases, data mining, and vizualization. In order to provid a well rounded perspective, we have included recent research that is not necessarily tied to microscopy, but whose ideas and techniques are quite general and very relevant.

1.1 Part I: Emerging Technologies to Understand Biological Systems

In Part I, we examine new technologies to investigate biological systems. These enable unprecendented levels of observation in spatial and temporal detail.

1.1.1 Knife-Edge Scanning Microscopy: High-Throughput Imaging and Analysis of Massive Volumes of Biological Microstructures

In Chapter 2, Choe, Abbott, Han, Huang, Keyser, Kwon, Mayerich, Melek, and McCormick describe a unique apparatus that combines sectioning and imaging of biological tissue. This apparatus is suitable for acquiring images of entire organs, such as a plastic-embedded mouse brain, at a resolution of approximately 0.6 μ/pixel. This results in a total data size of approximately 20 terabytes. The authors have developed vector tracking methods to process these images to extract meaningful structures such as vascular or neural filaments. They exploit parallel algorithms in order to address the high computation required. Finally, they employ special techniques from the field of visualization, such as streamlines, in order to be able to render the large data sets effectively.

1.1.2 4D Imaging of Multicomponent Biological Systems

In Chapter 3, Palaniappan, Bunyak, Nath, and Goffeney investigate time-varying imaging of biological entities such as cells in motion. This provides important information about temporal characteristics, which can be used to understand phenomena such as cell growth and development. This chapter describes the application of graphical processing units (GPUs) to overcome the computational bottleneck.

1.1.3 Utilizing Parallel Processing in Computational Biology Applications

In Chapter 4, Wagner and Jordan discuss an important problem in the field of computational biology: the modeling of tumor growth. A mathematical model is formulated that captures essential processes governing tumor state, including creation, migration, and death, and the interactions of tumor cells with a tissue environment, including oxygen consumption and enzyme production. The behavior of the model is probed through a discrete-time simulation. The challenge is to be able to apply the model to a biologically realistic number of tumor cells, in order to be clinically relevant. This turns out to be of the order of a billion tumor cells. The authors show that this is computationally tractable through parallelization techniques. Their techniques, as discussed in the section on domain decomposition, are quite general and can be applied to other problems, such as the tracking of moving cells in 4D microscopy applications. Other applications, such as contour tracking in 3D image stacks, can also benefit from the domain decomposition techniques described in this chapter, as it concerns changes in the process that owns a cell's computations as the cell migrates. A similar situation occurs when the contour of a cell shifts as one traces it along successive sections.

1.2 Part II: Understanding and Utilizing Parallel Processing Techniques

Having examined the growing need for computation in biological applications, and in microscopy in particular, we realize that parallel processing is inevitable. We take a closer look at how computations can be parallelized. In spite of major advances in computing, the effective parallelization of programs has not been fully automated, and requires significant input from the end user. This has slowed the adoption of parallel computing by the communities that could benefit from it. However, this need not be the case, and several tasks in image processing can benefit from relatively straightforward parallelization. An important task in image analysis involves image segmentation and classification, which requires the use of machine learning techniques.

1.2.1 Introduction to High-Performance Computing Using MPI and OpenMP

In Chapter 5, Garg provides a very concise and readable account of the challenges involved in creating systems that support parallel computation. Taking the viewpoint of a designer of parallel systems, Garg shows that there is a trade-off between operating at a high level of abstraction when specifying a parallel program and the optimum performance it can achieve. For the best performance, a knowledge of the underlying architecture is required. Nevertheless, the task of specifying parallel programs has been simplified by the emergence of standards such as the message passing interface (MPI) and OpenMP. Garg illustrates how eight basic MPI functions are sufficient to get started with MPI programming and quickly harness the utility of parallel computing. The MPI standard contains several more functions which are useful for advanced users, but the eight that have been highlighted are sufficient in the beginning. This should lower the barrier of entry for practitioners of image processing and computational biology to start utilizing parallel computation.

1.2.2 Parallel Feature Extraction

In Chapter 6, Rao and Cecchi address the issue of processing large image datasets that arise in microscopy applications. They focus on several essential algorithms that are applied to a wide variety of images, including filtering operations, 3D connected component analysis, and morphological operations. They examine the use of appropriate domain decomposition techniques, and show how specific MPI features can be used to achieve the most advantageous decompositions. They illustrate the use of these parallel techniques to process serial slice data of neural tissue, by extracting contours of neural structures.

1.2.3 Machine Learning Techniques for Large Data

In Chapter 7, Yom-Tov provides a broad overview of machine learning techniques that are especially suited for large data sets as arising in computational biology or microscopy applications. This chapter highlights recent advances in machine learning, which include efficient parallel techniques for feature selection, clustering and classification. Since support-vector machines are a popularly used classifier, this chapter summarizes research devoted to their efficient, parallel implementation.

1.3 Part III: Specific Applications of Parallel Computing

In this collection of chapters, we examine specific applications of parallel computing to solve biological imaging problems.

1.3.1 Scalable Image Registration and 3D Reconstruction at Microscopic Resolution

In Chapter 8, Cooper, Huang, Ujaldon, and Ruiz provide a detailed exposition of the problem of image registration and 3D reconstruction. Image registration is a fundamental operation with wide applicability in biomedical imaging. Several recent advances in imaging of cellular and tissue structure have resulted in the acquisition of very large image sizes. Typical image sizes being handled by the authors are in the 23K×62K pixel range, which leads to cumulative image data in the terabyte range. Though image registration has been well studied in the past, their application to such large image datasets needs a new approach. Conventional techniques would take several weeks to proces this data, which is unacceptable. The main contribution of this chapter is to demonstrate how the image registration problem can be advantageously decomposed to run efficiently in a parallel environment. There are several steps in registration that lend themselves to parallelization, such as the identification of candidates for intensity feature matching. The authors exploit a cluster of 32 CPUs as well as their associated GPUs to achieve a significant speedup of 49x over a serial implementation. This is sufficient to bring the computation time from the order of weeks to a few hours, rendering the problem tractable.

1.3.2 Data Analysis Pipeline for High-Content Screening in Drug Discovery

In Chapter 9, Tao, Young, and Feng examine the importance of high content screening, a technique which seeks to measure the phenotypic characteristics of cells as a function of immediate environmental influences such as drug toxicity. Though there are mutliple challenges in the gathering of high content screening data, including image acquisition and analysis, the authors focus on the key problem of data mining. There are currently no vendor solutions that provide tools to adequately mine the vast amounts of data collected, which range in the tens of terabytes. Furthermore, the number of parameters can be huge, complicating the search for relevant models and the subsequent interpretation of the experimental data. The authors present their solution to this problem, which involves statistical and machine learning techniques including decision trees and support vector machine classifiers. The ideas for large scale data mining presented by Yom-Tov in Chapter 7 would be very relevant to the problems being faced in high-content screening.

1.3.3 Information About Color and Orientation in the Primate Visual Cortex

In Chapter 10, Xiao and Kaplan apply imaging techniques to investigate the representation of visual information in the primate cortex. Their research seeks to understand whether information about different visual attributes such as color and orientation are processed in segregated compartments in the cortex, or are processed by the same location. This has been an open question in the literature, with studies supporting both hypotheses. Xiao and Kaplan designed a technique to measure the cortical response to stimuli that differed only in orientation or color. By using support vector machines to train a classifier, they were able to show that certain areas of the cortex could predict the input stimulus with a high degree of accuracy, indicating they were coding the stimulus. Though the problem can be thus formulated in a mathematically elegant fashion, it is very computationally demanding, as separate classifiers need to be designed for every local region of the cortex. They overcame this by parallelizing the computation, bringing down the time from a few years to a few hours. Overcoming the computational bottleneck allows scientists the ability to think about and formulate their problem in the most appropriate manner.

1.3.4 High-Throughput Analysis of Microdissected Tissue Samples

In Chapter 11, Rodriguez-Canales, Hanson, Tangrea, Mukherjee, Erickson, Albert, Majumdar, Bonner, Pohida, Emmert-Buck, and Chuaqui describe exciting new developments in applying molecular techniques to tissue samples for studying diseases. A technique that is being increasingly used is tissue microdissection, which requires advances in multiple fields, including histopathology, image analysis, and microscopy instrumentation. This technique opens new directions of investigation in the biological sciences. This chapter provides important background to understand innovative techniques that are being applied to probe and isolate relevant microscopic cellular features from biological samples. The advantages and limitations of these techniques are presented, which should be valuable to practitioners interested in applying image processing techniques to this domain. An emerging application is the use of fluorescent labeling to identify different cell populations. This poses several challenges in image analysis, such as segmentation.

1.3.5 Applications of High-Performance Computing to Functional Magnetic Resonance Imaging (fMRI) Data

In Chapter 12, Garg examines the analysis of fMRI images, an area of considerable interest in neuroscience, as it provides data on a whole, behaving brain in human subjects. Due to the large number of spatial voxels and long temporal durations used in typical experiments, the size of the data gathered can be enormous, especially if multiple subjects are considered. As a result, researchers have focused on utilizing less computationally demanding techniques such as the general linear model. However, more sophisticated modeling of the temporal aspects of the signals gathered leads to a richer interpretation of the data, including the establishment of networks of related voxels, and directed graphs that represent causal relationships between voxels. This constitutes a paradigm shift in the analysis of fMRI images, and enables a wider variety of experimental protocols to be adopted. The price to be paid for this increased capability is the increased computation involved. Garg shows how the effective parallelization of these techniques renders tractable the more sophisticated mathematical modeling techniques such as Granger causality analysis.

1.4 Part IV: Postprocessing

Once the microscope images have been acquired and analyzed, they need to be stored in databases that facilitate their retrieval. The original images and the result of processing also need to be visualized effectively, in order to provide the best insight to the viewer.

1.4.1 Bisque: A Scalable Biological Image Database and Analysis Framework

In Chapter 13, Singh and Kvilekval describe how the availability of large image databases and search technology has enabled a new level of hypothesis creation and testing, thus transforming the field of biology. The emerging field of bioimage informatics addresses a key concern, which is effective management and mining of bioimage databases. The authors describe an image database solution they have developed, called Bisque. This addresses several critical issues such as appropriate data modeling, data integration from multiple sources, and scalability. The bulk of existing search technology relies on textual keyword matching. Bisque goes beyond this, and enables matching of images based on features derived automatically from them. This solution is made scalable by using a services oriented architecture, where services based on Web protocols link the user to the data.

1.4.2 High-Performance Computing Applications for Visualization of Large Microscopy Images

In Chapter 14, Singh, Lin, Schulze, Peltier, Martone, and Ellisman examine the problem of effective display, navigation, and manipulation of large microscopy images. The creation of large multiresolution image datasets of a single biological entity such as an organ opens up the possibility of visually inspecting the entire dataset in a seamless manner. This allows scientists the ability to integrate information across multiple scales, leading to better insight and understanding. However, significant technological challenges need to be overcome before this vision can be turned into reality. The large image sizes prevent them from being loaded onto the memory of a single processor. Furthermore, for effective interactivity with the data, computation needs to be done in near real-time. The visual display of the large image sizes requires the use of multitiled displays. Due to all these considerations, the use of high-performance computing becomes a necessity. The authors describe a solution to this problem using a scalable cluster-based approach.

1.5 Conclusion

Though this book covers a large number of relevant and interesting topics, the field is vast, and there wil be many aspects missing. One of them is the description of a single integrated system devoted to providing a comprehensive solution to image processing tasks carried out on a parallel platform. We realize that to make this happen, a concerted effort is required that combines the expertise of researchers across academia, industry, and possibly governmental agencies. Though readers may not find their specific problem addressed and solved in this book, it is hoped that the techniques and methods described herein will enable the reader to make significant progress in the right direction.

Acknowledgments

The editors wish to thank all the chapter authors for their valuable contributions which enable the reader to gain insight into potential problems and their solutions. We appreciate the feedback and encouragement from the anonymous reviewers which resulted in an improved scope and direction for the book. It was a pleasure to work with the cooperative and understanding editorial staff at Artech House. Finally, we are grateful to the management at the Computational Biology Center, IBM Research for their support and encouragement in taking on this project.

PART I

Emerging Technologies to Understand Biological Systems

CHAPTER 2

Knife-Edge Scanning Microscopy: High-Throughput Imaging and Analysis of Massive Volumes of Biological Microstructures

Yoonsuck Choe, Louise C. Abbott, Donhyeop Han, Pei-San Huang, John Keyser, Jaerock Kwon, David Mayerich, Zeki Melek, and Bruce H. McCormick

Recent advances in physical-sectioning microscopy have enabled high-throughput imaging of massive volumes of biological microstructure at a very high resolution. The Knife-Edge Scanning Microscope (KESM) we have developed is one of the few that combines serial sectioning and imaging in an integrated process. The KESM is capable of imaging biological tissue (about 1 cm^3) at 300 nm $\times$ 300 nm $\times$ 500 nm resolution within 100 hours, generating data at a rate of 180 Mbps. The resulting data per organ (e.g., a mouse brain) can easily exceed tens of terabytes. High-performance computing methods are required at every stage in the lifetime of the generated data set: (1) distributed storage and retrieval, (2) image processing to remove noise and cutting artifacts, (3) image and texture segmentation, (4) three-dimensional tracking and reconstruction of microstructures, and (5) interactive visualization. In this chapter, we will review the capabilities and latest results from the KESM (Section 2.2) and discuss the computational challenges arising from the massive amounts of data, along with a survey of our ongoing efforts to address these challenges (Sections 2.3 to 2.5). We expect the integration of high-throughput imaging and high-performance computing to lead to major breakthroughs in scientific discovery in biological sciences.

2.1 Background

In this section, we will provide a brief review of high-throughput imaging and analysis methods, to provide a proper context for the work we will discuss in the remainder of the chapter.

2.1.1 High-Throughput, Physical-Sectioning Imaging

Currently, the standard approach for microscopic imaging of a volume of tissue is confocal microscopy [1]. The basic idea is to change the depth of focus (focal plane), and use a pinhole aperture to detect photons originating only from the target depth. This is called *optical sectioning* where virtual, not actual, sections are obtained. In conventional optical sectioning, the main limiting factor is not in the resolution along the x-y plane ($\sim$ 250 nm) but in that of the z direction ($\sim$ 700 nm) [2]. Although the resolution and imaging depth can be improved using more advanced schemes such as multiphoton microscopy [3], optical sectioning techniques are limited to the tissue surface and have limited z-axis resolution. Slow imaging speed is another issue, for both confocal and multiphoton, such that even in enhanced two-photon microscopy, the data rate is less than 8 MB/s (512 $\times$ 484 at 30 frames/s reported in [4], and about 1 frame/s in [3]).

Physical sectioning combined with microscopy is one alternative to overcome the above issues, since z-axis resolution depends only on how thin the tissue can be sectioned (it can go down to $\sim$ 30 nm using a vibrating microtome), and there is virtually no depth limit on the tissue thickness.

The five most notable approaches in this direction are listed here:

1. All-optical histology [5];
2. Knife-edge scanning microscopy (KESM) [6–10];
3. Array tomography [11];
4. Serial-block-face scanning electron microscopy (SBF-SEM) [12];
5. Automatic tape-collecting lathe ultramicrotome (ATLUM) [13].

All-Optical Histology

There are some efforts to eliminate the depth limit in optical sectioning microscopy. For example, all-optical histology combines multiphoton microscopy with tissue ablation to allow deeper imaging [5]. Imaging is performed using standard optical sectioning. Next, femtosecond laser pulses are used to ablate $\sim$150-μm-thick sections on the top of the tissue block, exposing new tissue to be imaged. The main advantage of all-optical histology is that it overcomes the tissue thickness limit in confocal and multiphoton microscopy by ablating thick chunks of tissue. However, since multiphoton microscope is used for imaging, it suffers from the same vertical resolution limit and slow imaging speed.

Knife-Edge Scanning Microscopy

The KESM (U.S. patent #6,744,572) has been designed at Texas A&M University (TAMU) in recent years. The instrument, shown in Figure 2.1, is capable of scanning a complete mouse brain ($\sim$310 mm^3) at 300-nm sampling resolution within 100 hours when scanning in full production mode. The basic idea is to simultaneously cut and image thin sections of embedded biological tissue. We will discuss KESM in more technical detail in the following section (Section 2.2).

Array Tomography

Array tomography uses an ultramicrotome to manually section embedded tissue. Adhesives on the edge of the embedded tissue block allow successive sections to stick to each other. As sequential sections are cut, this forms an array of ultrathin sections. The resulting tissue array is placed on a glass slide and can be repeatedly washed, stained, and imaged using fluorescence microscopes and scanning electron microscopes. Volume data are obtained by combining together the imaged sections. The main advantages of array tomography is that it enables the imaging of multiple

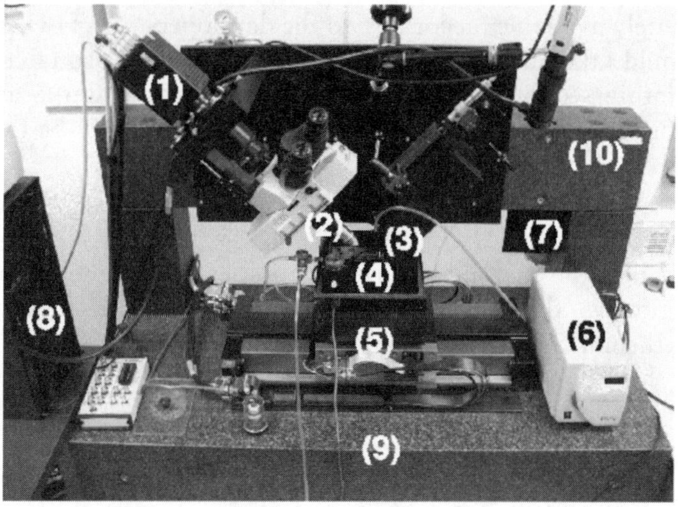

Figure 2.1 The Knife-Edge Scanning Microscope (KESM). A photo of the KESM is shown with its major components marked: (1) high-speed line-scan camera, (2) microscope objective, (3) diamond knife assembly and light collimator, (4) specimen tank (for water immersion imaging), (5) three-axis precision air-bearing stage, (6) white-light microscope illuminator, (7) water pump (in the back) for the removal of sectioned tissue, (8) PC server for stage control and image acquisition, (9) granite base, and (10) granite bridge.

molecular markers from the same slide, thus providing perfect registration across different data modalities.

Serial Block-Face Scanning Electron Microscopy

Serial block-face scanning electron microscopy (SBF-SEM), is capable of imaging tissue volumes of approximately 500 μ in linear dimension at a resolution on the order of 10 nm $\times$ 10 nm $\times$ 50 nm (exact values range from 12.7 nm to 13.4 nm for x and y; and 55 nm to 63 nm for z) [12]. The main advantage of SBF-SEM is its extremely high resolution, where detailed cellular and subcellular ultrastructure can be imaged. However, the volume is limited to cubes of several hundred μm, and scanning time can be prohibitive if sufficient signal-to-noise is to be achieved (a 200-μm cube at the above resolution can take up to one year). Improved staining methods to increase contrast can (theoretically) improve the speed 400-fold. SBF-SEM development is in a mature stage, where production models are now available from a commercial vendor (GATAN, Inc.).

Automatic Tape-Collecting Lathe Ultramicrotome

Automatic tape-collecting lathe ultramicrotome (ATLUM) is the latest development in serial sectioning microscopy [13]. ATLUM is capable of collecting a continuous ribbon sized 1 mm $\times$ 5 mm $\times$ 50 nm that get automatically pasted on a continuously running tape. The tape is then cut and organized into a ultrathin section library. The typical volume ATLUM can handle is about 10 mm^3, which is roughly a cube with a 2.15-mm linear dimension. One distinction in ATLUM is that imaging (typically with SEM) is done only when needed, thus a digitized library is not immediately available (according to the developers of ATLUM, imaging such a volume would take hundreds of years). Thus, a more reasonable approach is to have the ultrathin section library containing the chips to serve as the database itself. In the future, robotic random access SEM imaging will be used to do on-demand imaging of objects of interest.

Summary and Comparison

Even though the methods above are unified under the common theme of physical sectioning, the resolution and the typical volume they can handle all differ, putting them in a relatively complementary role with respect to each other. In other words, these methods cannot be ranked on an absolute scale since there are relative advantages and disadvantages to each method. Table 2.1 provides a summary comparison of these microscopy methods.

Finally, we wish to emphasize that no matter what physical volume these methods deal with (ranging from 100^3 μm^3 up to 1 cm^3), the resulting volume data can exceed several TBs, thus posing serious challenges to computational analysis, and high-performance computing methods could help address these challenges.

2.1.2 Volumetric Data Analysis Methods

Once the volume data are obtained from physical sectioning microscopy, the next task is to extract objects of interest from the data set. This is a nontrivial task due to distortions, noise, and artifacts resulting from the cutting process (often due to vibrations known as *chatter*). Furthermore, the difficulty in automation for this kind of data is dramatically increased by the density and the huge number of objects in the microscopy data. This is unlike volume data from medical imaging (magnetic resonance imaging or computer-aided tomography) where object count and density are both low, for which various automated methods exist (e.g., National Library of Medicine's Insight Toolkit, http://www.itk.org/).

In this section, we will review three main approaches taken in 3D reconstruction.

Image Registration, Segmentation, and Reconstruction

A traditional approach for 3D reconstruction is to consider the volume data as a stack of images, while processing one image at a time. This is often necessary because the z-axis resolution is so much lower than the lateral resolution for most

Table 2.1 Summary Comparison

Method	*Resol. (x,y)*	*Resol. (z)*	*Volume*	*Modality*	*Time*
All-optical hist.	0.5 μm	1 μm	1 cm^3	Fluorescence	~900 hours
KESM	0.3–0.6 μm	0.5–1 μm	1 cm^3	Bright field, fluorescence*	~100 hours
Array tomography	~0.2 μm	0.05–0.2 μm	~100^3 μm^3	Fluorescence, EM	N/A
SBF-SEM	~0.01 μm	~0.03 μm	~500^3 μm^3	EM	N/A
ATLUM	~0.01 μm	0.05 μm	~2.15^3 mm^3	EM	N/A

*Expected in the near future.

imaging modalities. First, depending on the imaging modality, alignment (registration) of successive images in the image stack constituting the volume data may be necessary. Next, the foreground objects and the background need to be segmented. Finally, the segmented objects in individual images need to be stitched together (reconstruction) to form a geometric representation of objects such as neurons and vasculatures.

The process is often times manual, and computer algorithms only play a supportive role. For example, see the RECONSTRUCT package [14], and Neuron_Morpho [15]. There are some efforts to automate the entire process [16]. One disadvantage of the segment-then-reconstruct approach is that it is hard to incorporate the 3D context in the process when there is ambiguity (foreground or background? connect or not?) since the approach operates primarily in 2D. Finally, manual algorithms are far too time-consuming for the dense, high-frequency structures found in high-throughput microscopy.

Convolutional Neural Networks

Jain et al. recently developed an alternative approach for 3D reconstruction, using convolutional networks [17]. Convolutional networks is a special type of artificial neural network that can be trained using supervised learning algorithms such as backpropagation [18]. The input is a cubic volume of data, and the output is the segmentation of the input volume. With an $n \times n \times n$ voxel volume, connecting to another $n \times n \times n$ voxel volume in the hidden layers, the number of connection weights (tunable parameters) can become prohibitive (for full connectivity, we need n^6). Convolutional networks avoid such an issue by *sharing* the connection weights. So, for one connection bundle connecting two $n \times n \times n$ voxel volumes only n^3 connection weights would be needed, instead of n^6. In some sense, these n^3 connections work as a filter, performing a convolution operation in 3D. Thus, the hierarchy in the neural network works as a series of 3D convolutions and combinations of the results.

For the actual training, small cubic samples from the raw data volume are plugged into the input, while manually labeled target values are plugged into the output. A gradient-based backpropagation algorithm is the used to adjust the connection weights. Thus, the learning process is basically converging to the most effective 3D convolution kernels. One disadvantage of this approach is that, even though the convolution strategy drastically reduces the number of tunable parameters, the training can be very slow, often lasting for weeks on modern parallel computers. The slowness is partly due to the computationally demanding 3D convolution operation, and partly to the gradient-descent learning algorithm. The advantage of the convolutional network approach is that fairly high levels of accuracy can be reached based on a small amount of manually labeled ground truth.

Vector Tracking

The reconstruction methods discussed above (also see [19] for a review of traditional methods such as such as voxel-based skeletonization) have their own advantages, but in many cases they are not efficient enough computationally to be applied to large volume datasets, especially those with fiber-like structures (see [20] for a review). Part of the problem is that each and every voxel has to be visited (multiple times) during processing. One way to overcome this problem is to visit only a subset of relevant voxels.

One of the latest approaches addressing the above issues is the vector tracking algorithm [20, 21], where only local information around the object of interest is examined, thus avoiding visiting every voxel in the volume data set. The basic idea is to begin with a seed point, estimate the trajectory of a fiber by means of template matching. Subsequently, small steps are taken along the fiber, continually adjusting the current position based on the estimated trajectory. Our group at Texas A&M generalized the approach for more robust tracking [22], which we will describe in detail in Sections 2.3 and 2.4 (also see [23], which uses level sets).

2.2 Knife-Edge Scanning Microscopy

The instrument comprises four major subsystems: (1) precision positioning stage (Aerotech), (2) microscope/knife assembly (Micro Star Technology), (3) image capture system (Dalsa), and (4) cluster computer (Dell). The specimen, a whole mouse brain, is embedded in a plastic block and mounted atop a three-axis precision stage. A custom diamond knife, rigidly mounted to a massive granite bridge overhanging the three-axis stage, cuts consecutive thin serial sections from the block. Unlike block face scanning, the KESM concurrently cuts and images (under water) the tissue ribbon as it advances over the leading edge of the diamond knife. A white light source illuminates the rear of the diamond knife, providing illumination at the leading edge of the diamond knife with a strip of intense illumination reflected from the beveled knife-edge, as illustrated in Figure 2.2. Thus, the diamond knife performs two distinct functions: as an optical prism in the collimation system, and as the tool for physically cutting thin serial sections. The microscope objective, aligned perpendicular to the top facet of the knife, images the transmitted light. A high-sensitivity line-scan camera repeatedly samples the newly cut thin section at the knife-edge, prior to subsequent major deformation of the tissue ribbon after imaging. The imaged stripe is a 20-μm-wide band located at the bevel at the very tip of the diamond knife, spanning the entire width of the knife. Finally, the digital video signal is passed through image acquisition boards and stored for subsequent analysis in a small dedicated computing server. The current server is a dual processor PC (3.2 GHz/2MB Cache, Xeon) with 6 GB of memory, built-in 1-TB storage, connected to an archival RAID attachment. The process of sectioning and imaging is fully automated with minimal human intervention. Figure 2.3 shows a screen-shot of the stage controller/imaging application developed and maintained in our lab.

A quick calculation puts us in context, regarding the massiveness of the data that KESM can produce. Consider the acquisition of volume data representing a plastic-embedded mouse brain (15 mm Anterior-Posterior, 12 mm Medial-Lateral, 6 mm Dorsal-Ventral). A 40X objective has a field of view (knife width) of 0.625 mm. Sixteen strips (each 0.625 mm wide by 15 mm long) are cut for each z-axis section (like plowing a field). For a (z-axis) block height of 6 mm, 12,000 sections must be cut, each 0.5 μm thick. The integrated tissue ribbon length (15 mm/strip $\times$ 16 strips/section $\times$ 12,000 sections/mouse brain) is 2.9 km. The

(a)

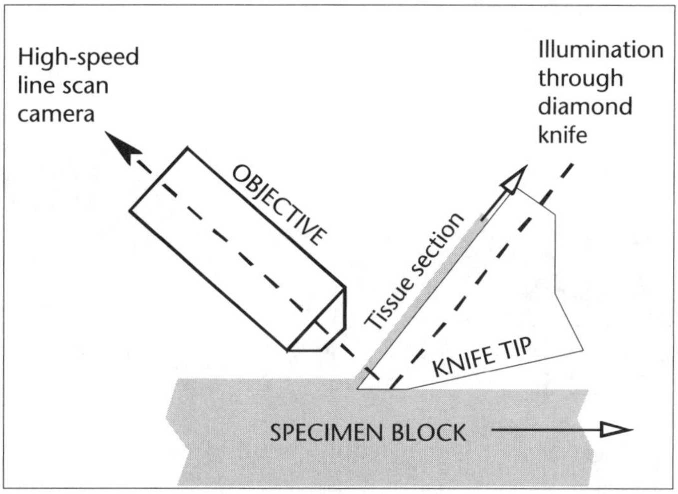

(b)

Figure 2.2 Tissue sectioning and imaging in KESM. (a) A close-up of the parts 2, 3, and 4 in Figure 2.1 is shown. To the left is the microscope objective, and to the right the diamond knife and light collimator. Submerged under water in the center is the plastic-embedded brain tissue held in a specimen ring. (b) The principle of operation of KESM is illustrated. The objective and the knife is held in place, while the specimen affixed on the positioning stage moves (arrow with solid line) at the resolution of 20 nm and travel speed of 1–5, and gets scraped against the diamond knife (5 mm wide for 10X objective), generating a thin section flowing over the knife (arrow with solid line). Line-scan imaging is done near the very tip of the knife where the distortion is minimal (maximum 106 μm from the tip of the knife). Illumination if provided through the diamond knife (arrow with dashed line indicates the light path). Note that the size of the knife is exaggerated. (Adapted from [9].)

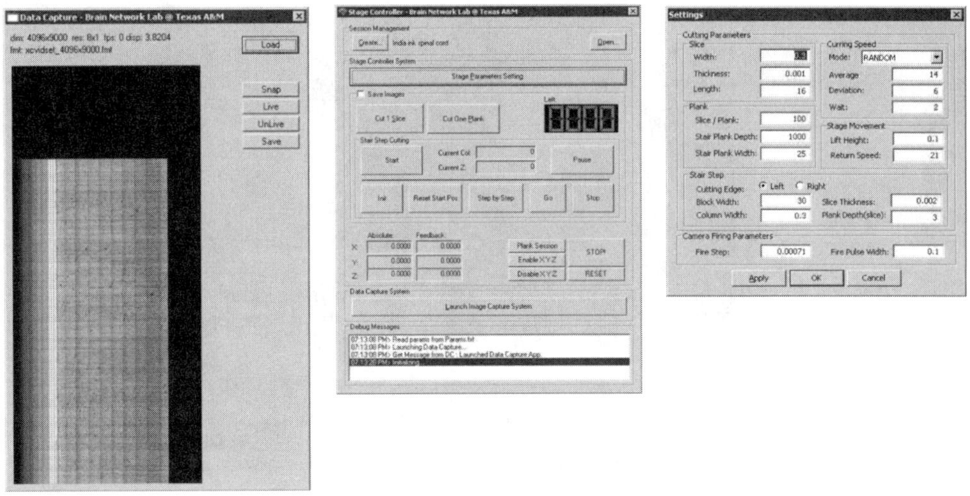

Figure 2.3 KESM automation software. A screenshot of the automated stage control and image capture application is shown.

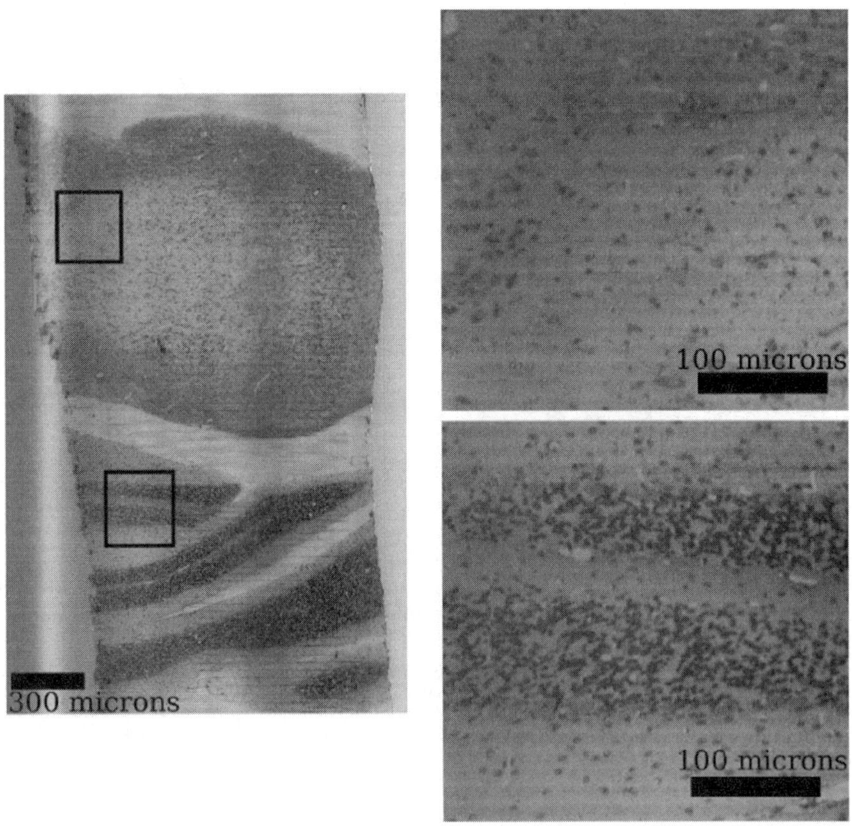

Figure 2.4 Nissl data from KESM. Coronal section of mouse brain stem is shown, with part of the cerebellum visible to the bottom half. Close-up of the inserts are shown to the right. The pixel resolution of the images is 0.6 μm/pixel, with a section thickness of 1 μm. (Adapted from [9].)

tissue ribbon is line-sampled at 300-nm resolution, near the Nyquist rate for an ideal optical resolution of (0.77)(532 nm) = (0.80 NA) = 512 nm. Based on this, the total data size (assuming one byte per voxel) comes out to 20 TB (at half the resolution in each dimension, it would be ~2.5 TB). The tissue ribbon can be sampled at 11 mm/s by line sampling at 44 kHz (180 Mbps), the camera maximum (Dalsa CT-F3-4096 pixels). Sampling the 2.9-km tissue ribbon requires 265,000s = 73 hr. Because mice brains are not cubical, stage return takes time, and soon we add 50% overhead, resulting in ~100 hr.

Figures 2.4 and 2.5 show typical data that can be obtained using the KESM [9]. Nissl staining dyes the RNA in the cytoplasm of all neurons and the DNA in cell bodies in all cells. However, the dendritic arbors and axons remain unstained.

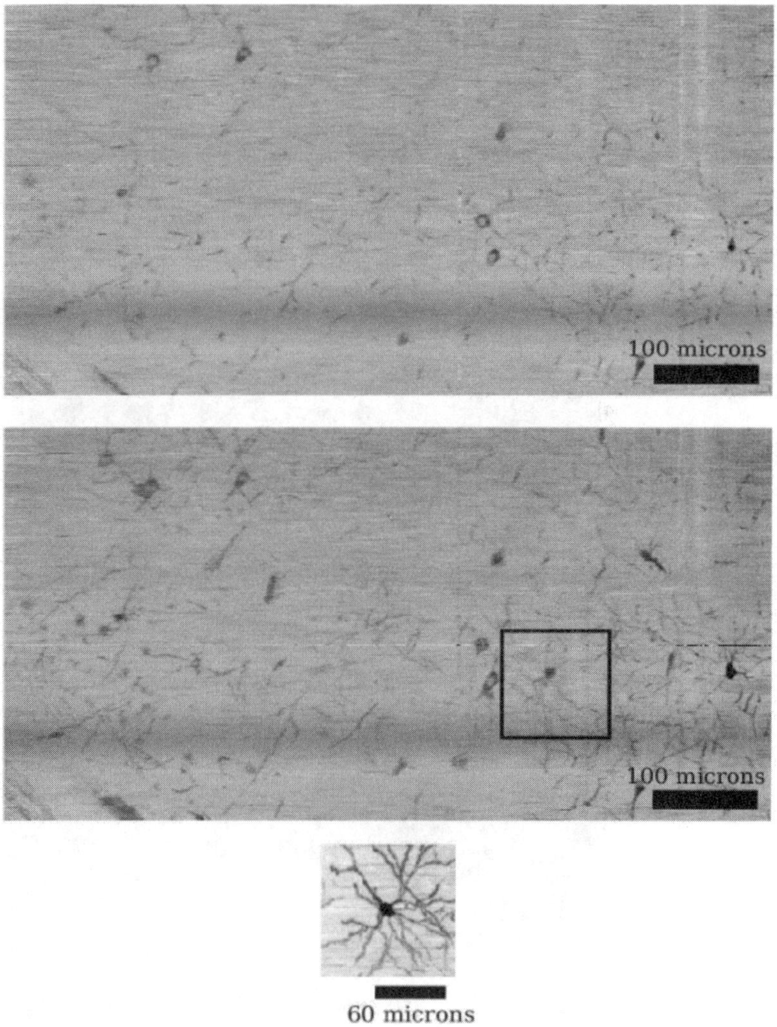

Figure 2.5 Golgi data from KESM. The stack of images generated by KESM can be viewed from the side of the stack (resectioning). A single resectioned plane is shown at the top. Golgi stain results in sparse data, so oftentimes it is easier to see familiar structures by overlaying multiple planes (middle). A single neuron can be observed when approximately 300 planes are overlayed (bottom). (Adapted from [9].)

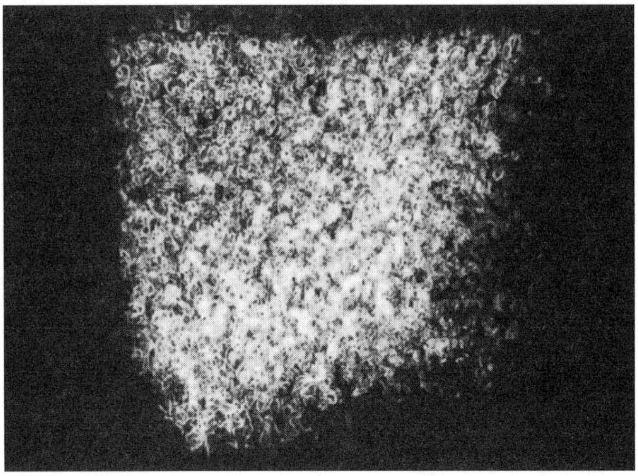

(a)

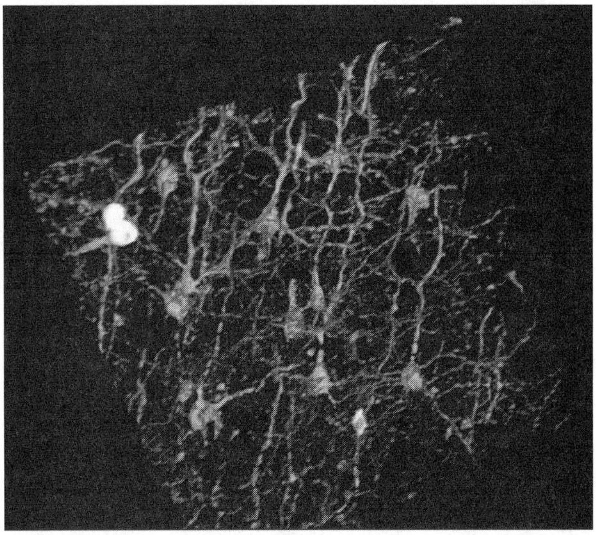

(b)

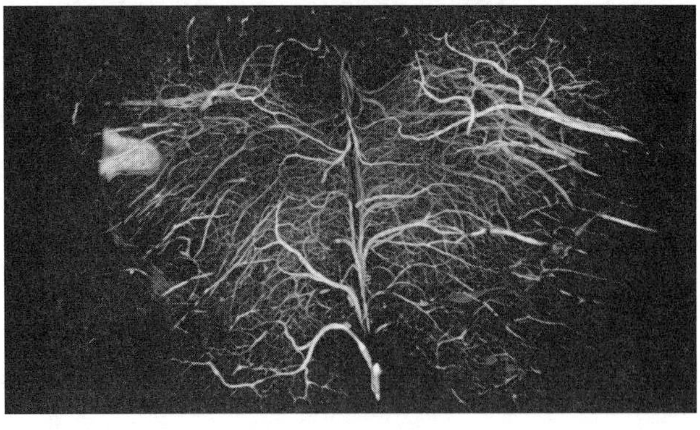

(c)

Figure 2.6 Volume visualization of KESM data. Volume visualizations of KESM data using Amira [24] are shown. (a) Nissl-stained mouse cerebral cortex ($\sim 300^3$ μm^3 cube). (b) Golgi-stained mouse cerebral cortex ($\sim 300^3$ μm ^{3} cube). (c) India-ink-stained vasculature in mouse spinal cord (1.8 mm $\times$1.8 mm $\times$ 1.2 mm, adapted from [46, 47]). Voxel size $\sim$ 0.6 μm $\times$ 0.7 μm $\times$ 1 μm.

Thus, Nissl staining allows us to reconstruct the distribution of all cell bodies in the mouse brain, and in particular their distribution within the six layers of the cerebral cortex. Golgi staining, in contrast, reveals the entire structure of neurons, as it stains just 1% of the neurons in the tissue. Individual neurons can be seen clearly, permitting reconstruction. India ink enables high-contrast staining of the entire vascular network. Figure 2.6 shows a rendering of the data volume using Amira [24].

2.3 Tracing in 2D

As we have seen in the previous section, the amount of data generated by high-throughput microscopy instruments is staggering. Simply iterating through the entire data set can take a long time. Thus, traditional image processing and 3D reconstruction algorithms are not suitable for this kind of data, since they require intense computation on every pixel/voxel in the data volume. Can et al. [21] and Haris et al. [25] developed template-based methods to overcome this issue.

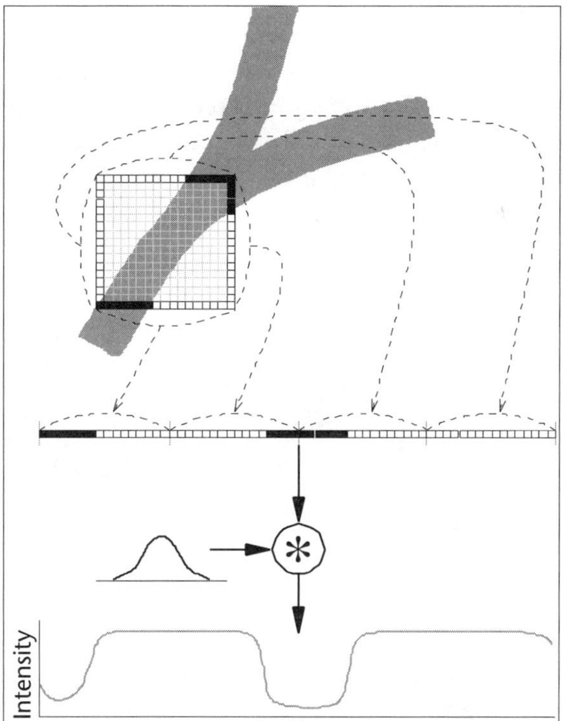

Figure 2.7 Identifying cross-section of fiber object and the moving window. An illustration of the method used to identify fiber segments (the "Y"-shaped object in the background) that overlap with the circumference of the moving-window template is shown (the blacksolid pixels). First, all the pixels are taken from the moving window's boundary, and convolution is done with a Gaussian filter to obtain an intensity profile (bottom). The valleys in the intensity profile are identified as fiber cross-sections. This process helps deal with inherent noise in biomedical images. Note that the actual algorithm extracts a "band" of pixels from the window border, and applies a small 2D Gaussian filter to extract the intensity profile.

In this section, we will present an extended template-based 2D tracking algorithm that can track fiber-like structures in biological volume data, such as neuronal processes or vascular networks. The basic idea is shown in Figures 2.7 and 2.8. A moving window template is generated around a seed point (the size of the window is proportional to the fiber width), and the pixels along the circumference of the moving window are sampled to construct an intensity profile. The intensity profile is convolved with a Gaussian filter, and based on this, *fiber cross-sections* (FCSs) are identified, where the border of the window overlaps the underlying object (Figure 2.7). The current position is then moved to the center of the FCS, and a cubic tangential trace spline (CTTS) is used to interpolate between the current and the next seed point (or center point) (Figure 2.8, left). CTTS is an interpolation method we developed by combining Lagrange interpolation [26] and B-splines [27], for fast and accurate interpolation. When there is a branch, linear interpolation is used instead of the CTTS (Figure 2.8, right). To make sure that the interpolation line is over actual data, the pixels directly below the spline are checked. All the other pixels in the moving window are safely ignored, greatly reducing the amount of data to be inspected.

Our algorithm (MW-CTTS, for moving window with CTTS) improves upon the previous approaches [21, 25] by adding robustness to noise, improved branch handling, and longer tracked distance from a single seed point. We quantitatively measured the noise-robustness of our algorithm using synthetic data with added noise. Two types of fibers were generated (line-type and curve-type), with varying fiber widths (20 to 50 pixels). We randomly generated 2,400 such synthetic inputs and ran our algorithm against Can et al.'s. Figure 2.9 shows the results. We

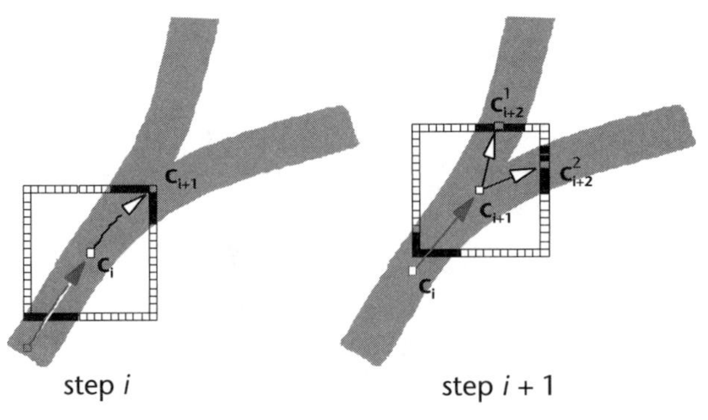

Figure 2.8 Fiber tracing with moving window templates. An illustration of the fiber tracing algorithm is shown. (Left) Given an initial seed point (or moving window center from the previous step) c_i, first we construct a moving window of an appropriate size and identify fiber cross-sections (FCS) in the advancing direction. By taking the center of the identified FCS, we can obtain the next center point c_{i+1} (gray pixel). In this example, there is no branch within the moving window. Once the target is found, a cubic tangential trace spline (CTTS) is used to connect the two center points c_i and c_{i+1}, and pixels underneath the interpolation line are checked to certify that the interpolation is over an existing fiber object. (Right) Starting from the moving window center estimated from the previous step (c_{i+1}), another moving window is constructed and the candidate FCSs identified. In this case we have two candidates, thus we generate two center points for the continued trace (c^1_{i+2} and c^2_{i+2}, marked as gray pixels). For cases like this including branches, we used linear inerpolation. Note that the full grid is not shown, to avoid clutter.

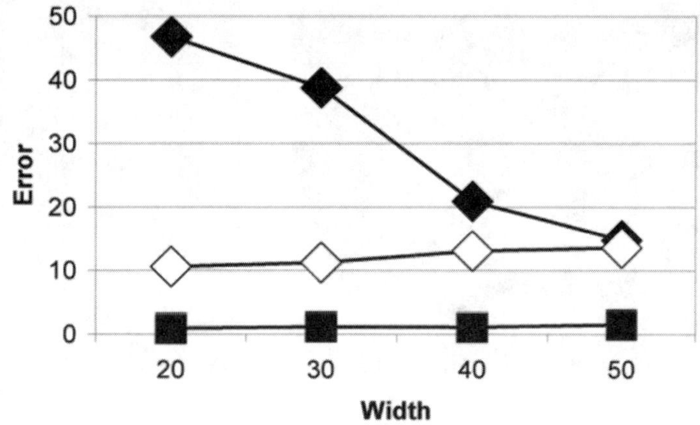

(a)

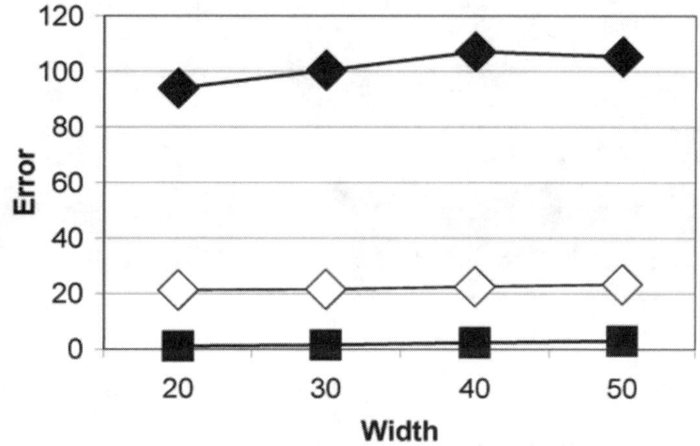

(b)

Figure 2.9 (a, b) Performance of MW-CTTS tracing algorithm on synthetic data. (a) Line-type and (b) curve-type input. The average performance of our MW-CTTS tracing algorithm on synthetic data (linear or curvy objects) are shown (squares), compared with that of Can et al. [21] (solid diamonds) and Haris et al. [25] (open diamonds). In both cases, our algorithm showed superior performance (near zero error).

also ran our algorithm and the two other algorithms on a real vascular data set [Figure 2.10(a)]. Can et al.'s method cannot handle branches [Figure 2.10(b)], and Haris et al.'s can show limited coverage due to incorrect branch handling [Figure 2.10(c)]. On the other hand, our algorithm shows the longest tracked distance, coupled with accurate traces [Figure 2.10(d)]. Our algorithm is also fast compared to the two other methods. Table 2.2 summarizes the total distance tracked and the speed of tracking.

The algorithm described above is efficient, requiring $O(k)$ complexity (in the number of pixels examined), where k is the length of the tracked fiber. The number of pixels needed to be inspected depends on: (1) the moving window's side length,

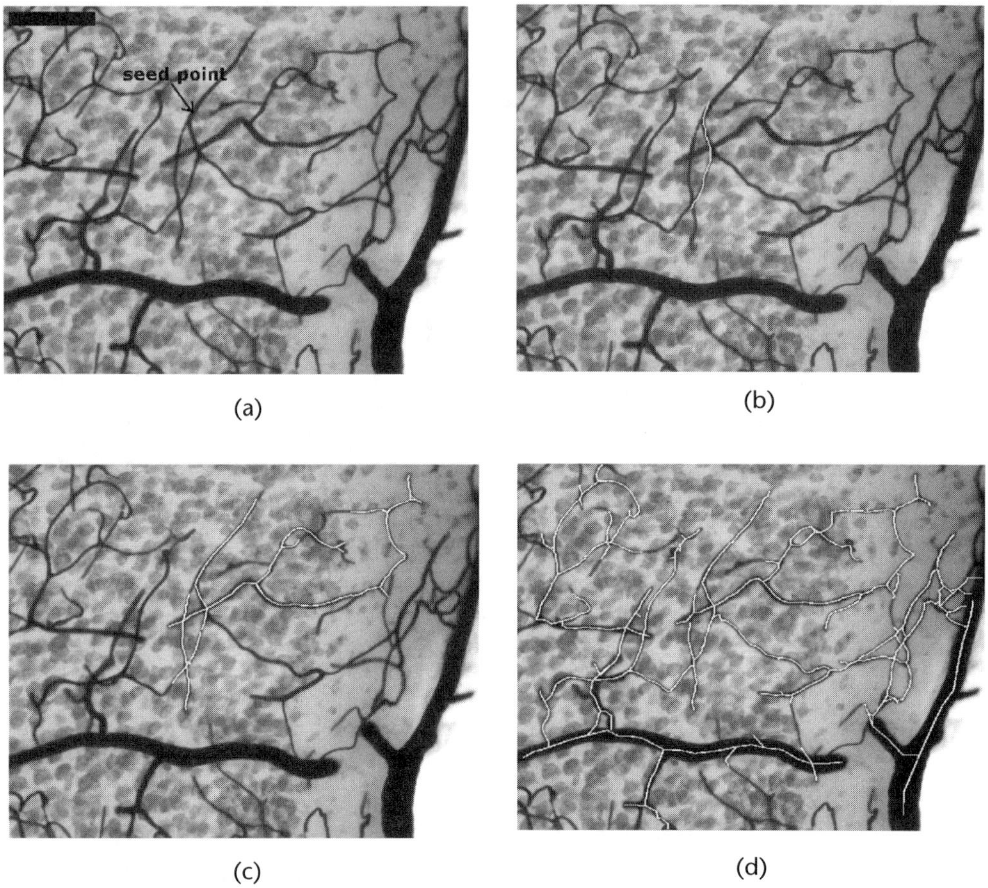

Figure 2.10 (a–d) Tracing results using MW-CTTS compared to other algorithms. Tracing results on a vascular image is shown for three different algorithms (scale bar = 20 μm, mouse cortex with vasculature stained with India ink and cell nuclei stained with Nissl, imaged using conventional light microscope). All three algorithms were initiated with the same single seed point in (a). Our MW-CTTS algorithm exhibits the most extensive trace.

(2) the fiber width, (3) the size of the Gaussian filter, and (4) total number of moving windows. With the fiber width n, moving window side length of $2n\varepsilon$ (ε is a small parameter between 1.1 and 1.5), a Gaussian filter size of 5×5, and the total number of moving windows of $\frac{k}{2n\varepsilon}$, the number of pixels come out to $(2n\varepsilon \times 4) \times (5 \times 5) \times \frac{k}{2n\varepsilon} = 100k$. That is, the complexity of the algorithm is $O(k)$. This calculation shows that our algorithm can scale up quite well to large volumes of data. It also helps that vascular and neural data are very sparse ($< 6\%$ for vascular and $<$ 1–2% for neural data).

In the next section, we will show how a similar approach can be extended into 3D.

Table 2.2 Performance Comparison Using One Seed Point on 6 Mouse Cerebellum Data

	Can et al.		*Haris et al.*		*MW-CTTS*	
	total dist.	*μsec/dist.*	*total dist.*	*μsec/dist.*	*total dist.*	*μsec/dist.*
#1	68 px	50.21177	831 px	8.791092	936 px	6.705957
#2	25 px	100.6608	1826 px	16.84223	1912 px	12.57990
#3	76 px	85.74671	1473 px	22.86413	7845 px	9.974812
#4	272 px	14.07918	2190 px	12.16080	3712 px	11.60712
#5	196 px	16.77046	1461 px	14.55146	4045 px	10.91478
#6	216 px	20.83727	1449 px	16.10389	7731 px	11.80340

The μm/pixel is 1.1. We performed the trace on a PC with Intel Pentium 4 (2.4 GHz) processor, 512MB of memory Windows under XP operating system. C/C++ programming language and OpenGL library were used. Note that MW-CTTS always has the largest number of traced distance within real-time.

2.4 Tracing in 3D

In this section, we discuss the extension of vector tracking techniques to three-dimensional datasets like those found in high-throughput microscopy. In general, 2D techniques have limited use since a single section contains a very limited amount of data. One method of dealing with this would be to combine several sections together, creating images similar to those in Figure 2.10. Two-dimensional tracking methods can then be used to trace filament data. However, this method causes a loss in resolution along the projected axis.

Instead, we employ tracking techniques that work completely in three dimensions. One of the most basic techniques for segmentation in 2D and 3D datasets is template matching [28, 29]. In this case, a template is created that has a structure similar to the objects to be segmented. We then run the template across the image (or volume), centering the template at each pixel (or voxel). We then compare the local region with the template. Template matching has many advantages, the most important of which is that it is robust in the presence of noise. In addition, the operation is highly parallel and different regions can be tested for correlation to the template simultaneously.

Although this works well for structures that have a fixed size and are rotationally invariant, segmentation of varying structures, such as blood vessels, can be time consuming. This is because the template must also be sampled at different sizes and orientations, turning what was a 3D spatial problem into a seven-dimensional problem; three spatial, three rotational, and one in scale-space. This is particularly problematic when dealing with large data sets. Additional problems arise when trying to discretize these other dimensions. While the spatial dimensions of the image are already discrete, only a finite number of template sizes and orientations can be tested.

We solve this problem by using the vector tracking methods described in Section 2.3 along with a heuristic-based approach to reduce the number of samples required to identify a vascular or neuronal filament. By starting from an initial seed point, we predict the path of the filament by taking samples of the cross-section along several directions (Figure 2.11). These samples are then compared with a template. The sample that provides the best response is selected as the trajectory of the filament. We then take a small step along this direction, at which point we take another series of samples (Figure 2.12). Every step on the way, we need to

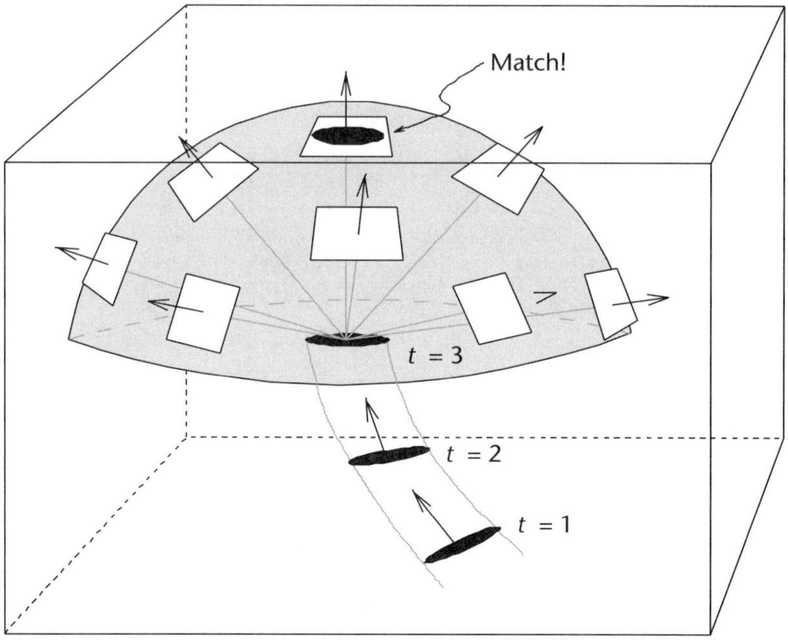

(a)

(b)

Figure 2.11 Fast fiber tracing. The basic idea behind our fast fiber tracing algorithm is shown. (a) Starting from a seed point ($t = 1$), the direction of the fiber is estimated ($t = 2$ to $t = 3$). The next point in the fiber trajectory is estimated using a template-based approach. Images are formed as interpolated slices (computed via graphics hardware) through the data volume—sampling in several directions around the current point. (Note that the distance between trace points is exaggerated.) (b) A typical array of slices from the tangential planes are shown. Only a small number of these resemble a circular cross-section of a fiber (upper-right corner). The best matching one is selected as the next point in the fiber trajectory. (c) An array of templates are shown. Using the graphics processing unit (GPU), comparison of (b) and (c) can be done extremely fast, through the texture rendering logic. (Adapted from [47].)

(c)

Figure 2.11 *(continued)*

make small adjustments such as axis correction and estimation of the fiber radius. Figure 2.13 shows tracing results on a vascular data set. These techniques can be used to limit sampling to the filament surface [30] and cross-section [31].

This kind of tracking allows us to reduce the required number of samples in several ways:

- Similar to vector tracking, samples are only taken local to the filament.
- By understanding the curvature of the filament, we can sample a limited number of orientations based on the orientation used in the previous step.

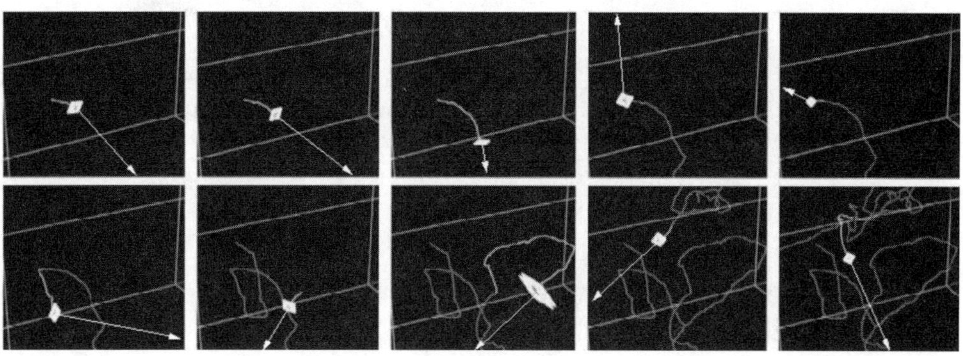

Figure 2.12 Tracing process. An example of using the tracing algorithm to track a sequence of fibers is shown. The small squares show the current trace center point, and the white arrows indicate the prediction direction. Gray trailing traces show the fibers that have been traced so far. (Adapted from [47].)

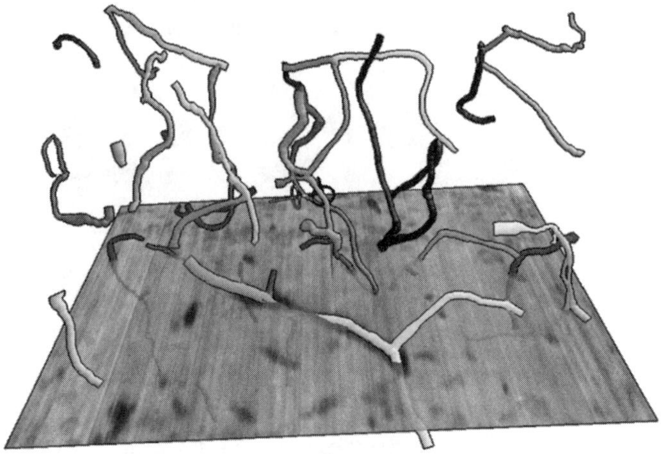

(a)

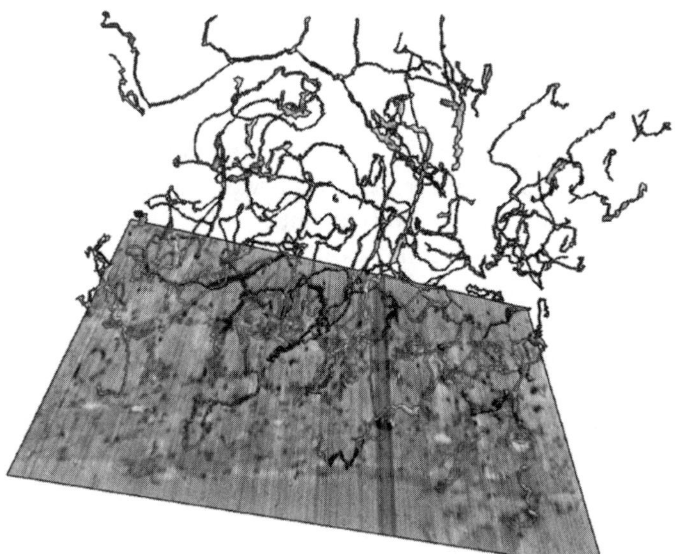

(b)

Figure 2.13 (a, b) Tracing results. The tracing results on a vascular data set are shown, with a slice of the actual data shown on the bottom. The fibers are colored with different shades of gray to represent independent traces. (Adapted from [47].)

- Since filaments gradually change size, we can limit the number of samples in the scale dimension to sizes local to the previous step.

Finally, we note that evaluating each sample is a highly parallel operation that maps well to modern graphics processing units (GPUs). By implementing these tracking methods completely on commodity GPUs, we are able to achieve even

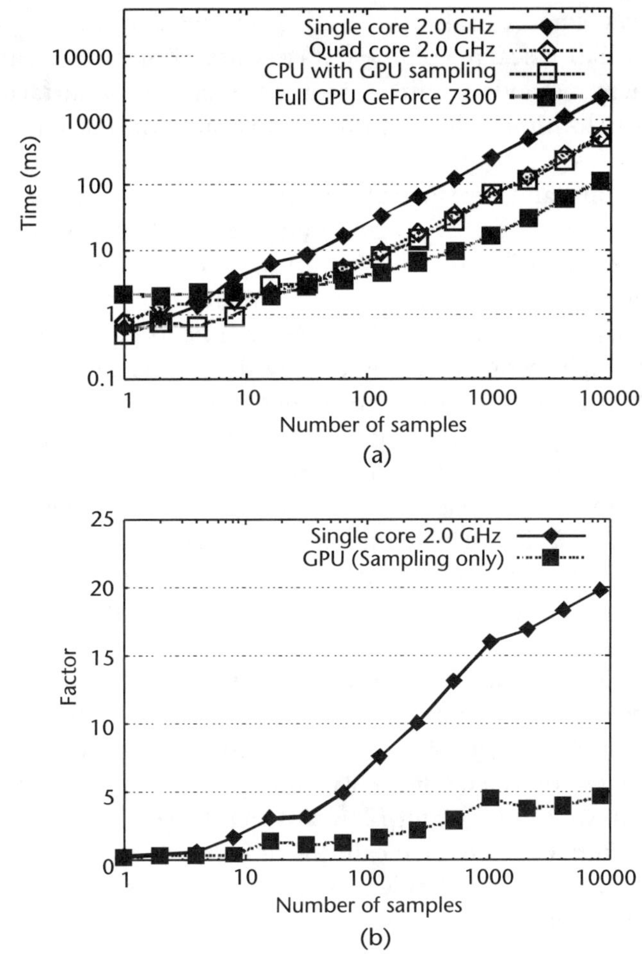

Figure 2.14 Fast fiber tracing using the graphics processing unit (GPU). Performance figures demonstrate the speedup obtained by using GPU computation. The GPU version results: (a) computation time taken to trace the same amount of data on different combinations of CPUs (single- or multicore) and GPUs are shown. The use of GPU gives an order-of-magnitude reduction in computation time. (b) The speedup achieved by using the full capacity of GPUs as compared to that of single-core CPU (diamond) or GPU sampling only (square) is shown, as a speed-up ratio (Factor = comparison time/full GPU time). The results show an almost 20-fold speedup compared to single-core CPU-based runs. (Adapted from [47].)

greater speedup over standard template matching schemes (Figure 2.14). See [31] for details.

2.5 Interactive Visualization

Both the size and density of structures in high-throughput datasets pose several problems in modeling and visualization. In this section, we discuss interactive visualization techniques that we use to explore dense bundles of filaments in volumetric networks such as neuronal trees and microvascular networks.

Direct volume visualization and isosurface construction are standard methods used to visualize scalar volume datasets like those created using KESM. Unfortunately, these methods require that a large dataset or surface be stored in memory during visualization, which is often impossible on the limited memory of current graphics cards.

Streamline and stream tube methods [32] provide a means of limiting the amount of active data by storing only the simplest representation of the filament data. These methods are often used for visualization in Diffusion Tensor MRI to display bundles fibers in white-matter regions of the brain. At the resolution of high-throughput microscopy, however, filaments rarely travel in coherent groups. Rendering high-throughput microscopy data using these techniques creates images that are cluttered and difficult to understand.

The techniques that we use are based on ideas demonstrated by streamlines. We first segment a filament network, as described in Section 2.4. We then render the filaments as a series of billboards that are always oriented towards the viewer (Figure 2.15). This eliminates the need to store large volumes, in the case of volume rendering, or polygon meshes, in the case of surface representations. Instead, only the series of line segments and radii are used to display the network (Figure 2.16).

This allows a user to interactively explore the filament data but does little to reduce the clutter produced by the dense network of filaments. Since we have the positional information for each filament, we can selectively visualize parts of the network based on the connectivity and trajectory of filaments [33]. By selecting an orientation, we can highlight filaments that travel in the selected direction, looking for trends in the dataset. We call this *orientation filtering* and use it to understand the flow of axons in a complex network as well as general trends in vascular flow through different types of tissue.

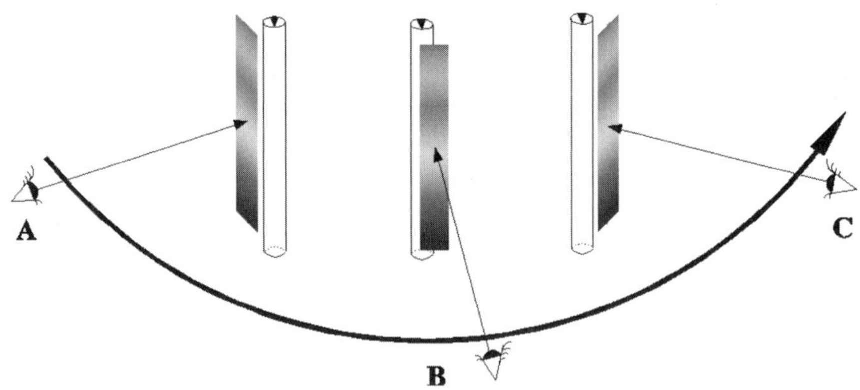

Figure 2.15 Self-orienting surfaces. The geometric data is replaced by a single geometric object (such as a surface). This object (surface) is oriented to always face the viewer. The object itself is also modified so that the appearance is correct from any angle. Note that the orientation of the data (cylinder) is not changing (small arrow head on top of the cylinders).

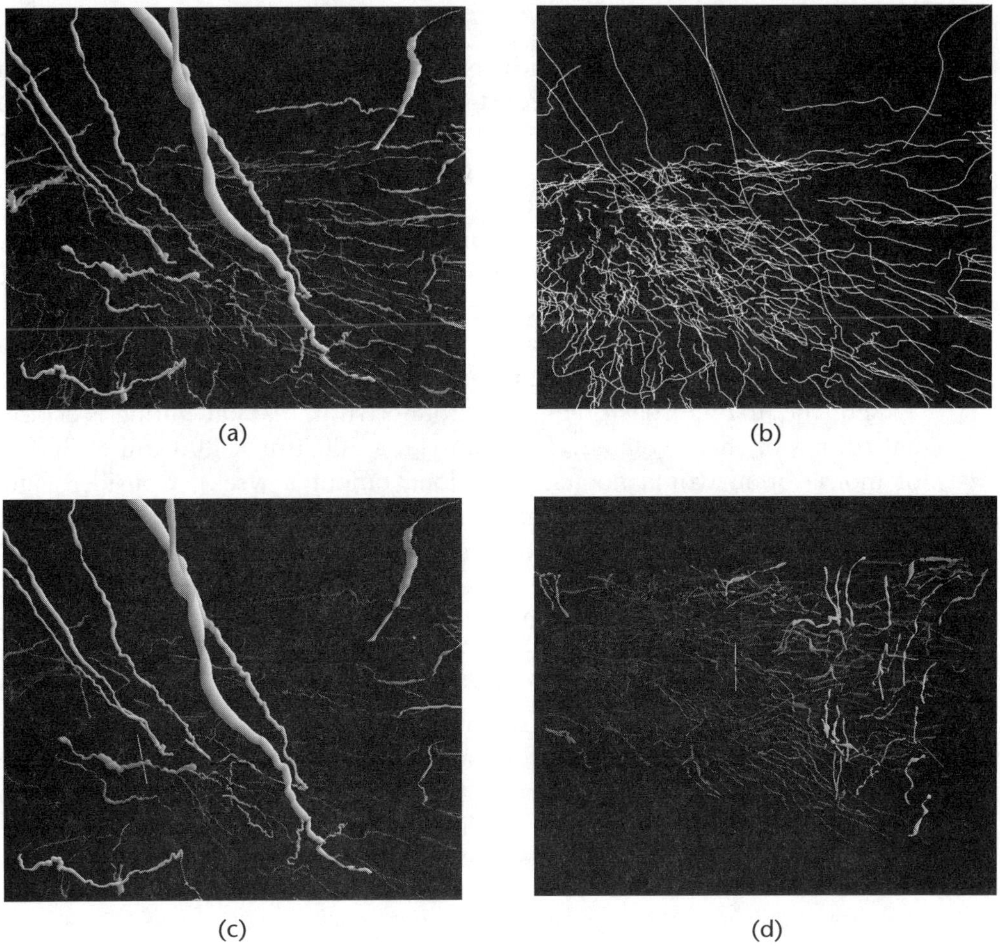

Figure 2.16 Interactive visualization of fibrous and thread-like data. (a) Self-orienting surfaces implemented on the GPU give high-quality renderings of neuron data viewed close up, and (b) allow large volumes to be viewed at interactive rates. (c) Orientation filtering allows us to interactively explore orientation of fibers, allowing us (d) to discover fiber bundle directions across larger datasets. (Adapted from [33].)

2.6 Discussion

The main contribution of our work is three-fold: (1) high-throughput serial-sectioning imaging of brain microstructures, (2) fast and accurate algorithms for 2D and 3D reconstruction of neural and vascular morphology, and (3) interactive real-time visualization of the reconstructed structures. These techniques are expected to provide critical neuroanatomical data at a submicron-level for whole small animal brains. There are several open issues that need to be addressed. In the following sections, we will discuss validation and editing, as well as how to exploit parallelism.

2.6.1 Validation and Editing

Quality control becomes a critical issue for any automated data processing procedure, and our parallel 3D reconstruction algorithm is no exception. In medical imaging, this process is called "validation" [34–36]. The challenge here is that obtaining full ground truth for validation could require performing a reconstruction of the entire dataset. Typically, two approaches are taken for validation: (1) use of manually labeled data as ground truth, and (2) use of digital phantoms (i.e., synthetically generated data mimicking structures observed in a known ground truth) (see [36] for a review). The two approaches have their own limitations, but they are complementary. We are currently looking into optimally combining these two approaches for accurate, efficient, high-throughput validation.

For the first approach requiring ground truth, manual editing is critical. Manual editing of small volumes, for example, a 300 μm × 300 μm × 300 μm cube of mouse brain would not take too long (about a week). Consider Figure 2.17 which illustrates the image acquisition and reconstruction process. Given the data volume (image stack I) obtained from KESM, domain experts can label the data to provide the ground truth $\hat{M}_e$ (note that the domain expert's labeling is in itself an estimation of the true ground truth M). Concurrently with manual labeling, the data volume I can be put through our segmentation and reconstruction algorithms, producing the estimated microstructure $\hat{M}$. Validation can then be done by comparing $\hat{M}_e$ (by the domain expert) and $\hat{M}$ (by the reconstruction algorithm). Measures like mutual information can be used to quantitatively measure the difference. Furthermore, we can train validation functions using machine learning techniques, and use the resulting validator for large-scale validation (see Chapter 7 by Yom-Tov on machine learning for large-scale datasets).

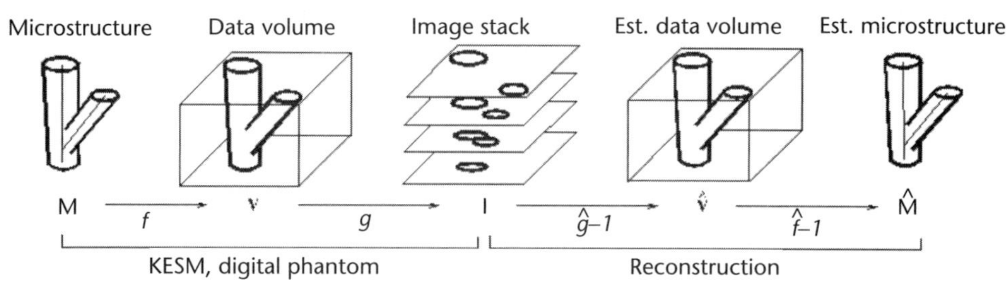

Figure 2.17 Microstructure-to-image mapping and reconstruction. The process by which a microstructure (real or synthetically generated) is turned into a stack of images in KESM and how they are reconstructed is shown. Modeling this process ($g \circ f$, composition of g and f) enables the generation of realistic synthetic image data (digital phantoms). On the other hand, the task of recovering the structural descriptions from the image data is basically the inverse: $\hat{f}^{-1} \circ \hat{g}^{-1}$, a composition of the segmentation ($\hat{g}^{-1}$) and the 3D reconstruction process ($\hat{f}^{-1}$). (The "^" symbol indicates that these functions are estimates.) Validation can take three approaches: (1) given I from KESM and $\hat{M}_e$ from human labeling, generate reconstruction $\hat{M}$ from I and compare it with $\hat{M}_e$; (2) given a digital phantom I with ground truth M, reconstruct $\hat{M}$ from I and compare it with M; or (3) Given I and its reconstruction $\hat{M}$, use the mapping $g \circ f$ to create $\hat{I}$ and show the difference image $\hat{I} - I$ that can be accepted or rejected.

For the second approach, digital phantoms can be used. Digital phantoms are synthetically generated, realistic data, produced from a known ground truth model (see, e.g., [37]). Given a known ground truth M, the 3D volume V is generated in a straight-forward manner, and then image formation, distortion, and noise models are used to produce the final product, the image stack I (Figure 2.17). The KESM imaging process can be modeled (as in [38]), and based on this, noisy digital phantoms (I) can be generated from a synthetic ground truth (M). The reconstruction algorithm can then be executed on this digital phantom, and the estimated microstructure $\hat{M}$ compared to the ground truth (M).

These two approaches will help perform large-scale validation of our automated algorithms.

2.6.2 Exploiting Parallelism

With the advent of high-throughput microscopy instruments like KESM, data acquisition is no longer a bottleneck: the reconstruction and computational analysis becomes a major bottleneck [16]. Fortunately, many tasks involved in reconstruction and analysis can be conducted locally, thus allowing straightforward parallelization (see, e.g., [39]). For example, the large data volume can be partitioned into small unit cubes that can fit in several gigabites of memory, and the reconstruction can be done within each cube on multiple computing nodes in parallel. The latest techniques developed for parallel feature extraction and 3D reconstruction using high-performance computing can be adapted easily for these tasks (see Chapter 6 by Rao and Cecchi and Chapter 8 by Cooper et al.).

Besides parallelizing by dividing our dataset into multiple regions, each to be handled by a single computing node, we can exploit the avaiable GPUs and multicores on modern CPUs to run multiple processes simultaneously tracking different fibers in the same unit cube. This will allow for larger memory blocks, and thus fewer artificial boundaries across which fiber tracking is inherently more difficult.

Even though these computations can be carried out locally, in the end, the results of these computations need to be merged, since processes from neurons (especially long axons) can project across the full span of the brain. After initial unit cubes are processed, we must merge data from adjacent unit cubes. It may also be necessary to modify our reconstruction in one unit cube based on the data obtained from the boundary of an adjacent unit cube. Some ideas on how a similar merger can be achieved is discussed in [40].

Another important observation is that the result from parallel reconstruction and merging need to be stored. As morphological information from adjacent unit cubes are combined, the fibers within each area must be merged. This can create potential problems, as the merged fiber-like data loses the locality it previously enjoyed. Furthermore, the data structure could become extremely complex (e.g., the entire vascular system could be one connected component); methods for representing such complex structures *hierarchically* will be needed.

Once all the reconstructions are done and stored so that a network of connected components emerges (e.g., neuronal or vascular networks), large-scale network analysis needs to be conducted to extract principles that connect network structure to function [41–45]. For such a task, parallel processing techniques developed for computational biology applications could be utilized (see Chapter 4 by Wagner).

Finally, various visualization techniques may also require parallel and high-performance computing capabilities. Machine learning techniques to visualize relevant features in the input (Chapter 10 by Xiao et al.), and the use of new high-performance computing paradigms (Chapter 14 by Singh et al.) show that visualization also needs to tap into high-performance computing.

2.7 Conclusion

Emerging high-throughput microscopy techniques such as the Knife-Edge Scanning Microscopy are starting to generate immense volumes of high-quality, high-resolution data from whole animal organs such as the mouse brain. Once these data are made available, the real bottleneck becomes computational analysis. Computational demand is very high at every stage of analysis, from image processing, 3D reconstruction, up to validation and network analysis (e.g., for neuronal and vascular networks). These challenges can be met effectively only through a combination of: (1) efficient and accurate algorithms and (2) high-performance computing. In this chapter, we surveyed high-throughput microscopy techniques with a focus on our own KESM, and the fast algorithms for morphological reconstruction. An extension of these methods, coupled with high-performance computing paradigms, can open the door to exciting new discoveries in biological sciences.

Acknowledgments

The results reported in this chapter have been supported in part by the National Science Foundation (MRI award #0079874 and ITR award #CCR-0220047), Texas Higher Education Coordinating Board (ATP award #000512-0146-2001), National Institute of Neurological Disorders and Stroke (Award #1R01-NS54252); and the Department of Computer Science, and the Office of the Vice President for Research at Texas A&M University. We dedicate this chapter to Bruce H. McCormick, our dear colleague and the main architect of the KESM, who passed away while this chapter was being written.

References

[1] K. Carlsson, P. Danielsson, R. Lenz, A. Liljeborg, and N. Aslund, "Three-dimensional microscopy using a confocal laser scanning microscope," *Optics Letter*, vol. 10, pp. 53–55, 1985.

[2] J. B. Pawley, *Handbook of Biological Confocal Microscopy*, New York: Plenum Press, 1995.

[3] W. Denk, J. H. Strickler, and W. W. Webb, "Two-photon laser scanning fluorescence microscopy," *Science*, vol. 248, pp. 73–76, 1990.

[4] G. Y. Fan, H. Fujisaki, A. Miyawaki, R.-K. Tsay, R. Y. Tsien, and M. H. Elisman, "Video-rate scanning two-photon excitation fluorescence microscopy and ratio imaging with cameleons," *Biophysical Journal*, vol. 76, pp. 2412–2420, 1999.

[5] P. S. Tsai, B. Friedman, A. I. Ifarraguerri, B. D. Thompson, V. Lev-Ram, C. B. Schaffer, Q. Xiong, R. Y. Tsien, J. A. Squier, and D. Kleinfeld, "All-optical histology using ultrashort laser pulses," *Neuron*, vol. 39, pp. 27–41, 2003.

[6] B. H. McCormick, "System and method for imaging an object," 2004. USPTO patent #US 6,744,572 (for Knife-Edge Scanning; 13 claims).

[7] B. H. McCormick, "The knife-edge scanning microscope," tech. rep., Department of Computer Science, Texas A&M University, 2003. http://research.cs.tamu.edu/bnl/.

[8] B. H. McCormick, L. C. Abbott, D. M. Mayerich, J. Keyser, J. Kwon, Z. Melek, and Y. Choe, "Full-scale submicron neuroanatomy of the mouse brain," in *Society for Neuroscience Abstracts*, Washington, DC: Society for Neuroscience, 2006. Program No. 694.5. Online.

[9] D. Mayerich, L. C. Abbott, and B. H. McCormick, "Knife-edge scanning microscopy for imaging and reconstruction of three-dimensional anatomical structures of the mouse brain," *Journal of Microscopy*, vol. 231, pp. 134–143, 2008.

[10] B. H. McCormick and D. M. Mayerich, "Three-dimensional imaging using Knife-Edge Scanning Microscope," *Microscopy and Microanalysis*, vol. 10 (Suppl. 2), pp. 1466–1467, 2004.

[11] K. Micheva and S. J. Smith, "Array tomography: A new tool for imaging the molecular architecture and ultrastructure of neural circuits," *Neuron*, vol. 55, pp. 25–36, 2007.

[12] W. Denk and H. Horstmann, "Serial block-face scanning electron microscopy to reconstruct three-dimensional tissue nanostructure," *PLoS Biology*, vol. 19, p. e329, 2004.

[13] K. Hayworth and J. W. Lichtman, Automatic Tape-Collecting Lathe Ultramicrotome (ATLUM), http://www.mcb.harvard.edu/lichtman/ATLUM/ATLUM_web.htm, 2007.

[14] J. C. Fiala and K. M. Harris, "Extending unbiased stereology of brain ultrastructure to three-dimensional volumes," *J. Am. Med. Inform. Assoc.*, vol. 8, pp. 1–16, 2001.

[15] K. M. Brown, D. E. Donohue, G. D'Alessandro, and G. A. Ascoli, "A cross-platform freeware tool for digital reconstruction of neuronal arborizations from image stacks," *Neuroinformatics*, vol. 3, pp. 343–359, 2007.

[16] D. Chklovskii, "From neuronal circuit reconstructions to principles of brain design," in *Proceedings of the 5th Computational and Systems Neuroscience Meeting (COSYNE 2008 Abstracts)*, p. 331, 2008.

[17] V. Jain, J. F. Murray, F. Roth, H. S. Seung, S. Turaga, K. Briggman, W. Denk, and M. Helmstaedter, "Using machine learning to automate volume reconstruction of neuronal shapes from nanoscale images," in *Society for Neuroscience Abstracts*, Washington, DC: Society for Neuroscience, 2007. Program No. 534.7. Online.

[18] Y. LeCun, B. Boser, J. S. Denker, D. Henderson, R. E. Howard, W. Hubbard, and L. D. Jackel, "Backpropagation applied to handwritten zip code recognition," *Neural Computation*, vol. 1, pp. 541–551, 1989.

[19] J. J. Capowski, ed., *Computer Techniques in Neuroanatomy*, Plenum, 1989.

[20] K. A. Al-Kofahi, S. Lasek, D. H. Szarowski, C. J. Pace, G. Nagy, J. N. Turner, and B. Roysam, "Rapid automated three-dimensional tracing of neurons from confocal image stacks," *IEEE Transactions on Information Technology in Biomedicine*, vol. 6, pp. 171–187, 2002.

[21] A. Can, H. Shen, J. N. Turner, H. L. Tanenbaum, and B. Roysam, "Rapid automated tracing and feature extraction from retinal fundus images using direct exploratory algorithms," *IEEE Transactions on Information Technology in Biomedicine*, vol. 3, pp. 125–138, 1999.

[22] D. M. Mayerich, L. C. Abbott, and B. H. McCormick, "Imaging and reconstruction of mouse brain vasculature and neighboring cells using knife-edge scanning microscopy," in *Society for Neuroscience Abstracts*, Washington, DC: Society for Neuroscience, 2006. Program No. 694.4. Online.

[23] B. Busse, S. J. Smith, C. A. Taylor, and R. Y. Arakaki, "Development of a semi-automated method for three-dimensional neural structure segmentation," in *Society for Neuroscience Abstracts*, Washington, DC: Society for Neuroscience, 2006. Program No. 834.13. Online.

[24] Zuse Institute Berlin (ZIB) and Mercury Computer Systems, Berlin, "Amira: Advanced 3D visualization and volume modeling," http://www.amiravis.com, 2006.

[25] K. Haris, S. Efstratiadis, N. Maglaveras, C. Pappas, J. Gourassas, and G. Louridas, "Model-based morphological segmentation and labeling of coronary angiograms," *IEEE Trans. Med. Imag.*, vol. 18, pp. 1003–1015, October 1999.

[26] R. W. Hamming, *Numerical Methods for Scientists and Engineers*, New York: McGraw-Hill, 1962.

[27] T. Pavlidis, *Algorithms for Graphics and Image Processing*, Computer Science Press, 1982.

[28] R. C. Gonzalez and R. E. Woods, *Digital Image Processing*, 2nd ed., Prentice Hall, 2002.

[29] Y. Sato, S. Nakajima, N. Shiraga, H. Atsumi, S. Yoshida, T. Koller, G. Gerig, and R. Kikinis, "Three-dimensional multi-scale line filter for segmentation and visualization of curvilinear structures in medical images," *Medical Image Analysis*, vol. 2, pp. 143–168, 1998.

[30] K. Al-Kofahi, S. Lasek, D. Szarowski, C. Pace, G. Nagy, J. Turner, and B. Roysam, "Rapid automated three-dimensional tracing of neurons from confocal image stacks," *IEEE Transactions on Information Technology in Biomedicine*, vol. 6, pp. 171–186, 2002.

[31] D. M. Mayerich, Z. Melek, and J. Keyser, "Fast filament tracking using graphics hardware," *Technical Report*, vol. TAMU-CS-TR-2007-11-3, 2007.

[32] C. Stoll, S. Gumhold, and H.-P. Seidel, "Visualization with stylized line primitives," *16th IEEE Visualization 2005*, p. 88, 2005.

[33] Z. Melek, D. Mayerich, C. Yuksel, and J. Keyser, "Visualization of fibrous and thread-like data," *IEEE Transactions on Visualization and Computer Graphics*, vol. 12, no. 5, pp. 1165–1172, 2006.

[34] T. Yoo, N. J. Ackerman, and M. Vannier, "Toward a common validation methodology for segmentation and registration algorithms," in *Lecture Notes In Computer Science, Vol. 1935; Proceedings of the Third International Conference on Medical Image Computing and Computer-Assisted Intervention*, (London), pp. 422–431, Springer, 2000.

[35] S. J. Warfield, K. H. Zou, and W. M. Wells, "Validation of image segmentation and expert quality with expectation-minimization algorithm," in *Lecture Notes in Computer Science, Vol. 2488; Proceedings of the 5th International Conference on Medical Image Computing and Computer-Assisted Intervention*, pp. 298–306, 2002.

[36] D. L. Pham, C. Xu, and J. L. Prince, "Current methods in medical image segmentation," *Annual Review of Biomedical Engineering*, vol. 2, pp. 315–337, 2000.

[37] R. A. Koene, "Large scale high resolution network generation: Producing known validation sets for serial reconstruction methods that use histological images of neural tissue," in *International Conference on Complex Systems*, 2007. [Presentation].

[38] J. S. Guntupalli, "Physical sectioning in 3D biological microscopy," Master's thesis, Department of Computer Science, Texas A&M University, 2007.

[39] A. R. Rao, G. A. Cecchi, and M. Magnasco, "High performance computing environment for multidimensional image analysis," *BMC Cell Biology*, vol. 8 (Suppl 1), p. S9, 2007.

[40] J. Kwon, D. Mayerich, Y. Choe, and B. H. McCormick, "Lateral sectioning for knife-edge scanning microscopy," in *Proceedings of the IEEE International Symposium on Biomedical Imaging*, 2008. In press.

[41] O. Sporns and G. Tononi, "Classes of network connectivity and dynamics," *Complexity*, vol. 7, pp. 28–38, 2002.

[42] M. Kaiser and C. C. Hilgetag, "Nonoptimal component placement, but short processing paths, due to long-distance projections in neural systems," *PLoS Computational Biology*, vol. 2, pp. 805–815, 2006.

[43] D. Chklovskii, T. Schikorski, and C. Stevens, "Wiring optimization in cortical circuits," *Neuron*, vol. 34, pp. 341–347, 2002.

[44] R. Milo, S. Shen-Orr, S. Itzkovitz, N. Kashtan, D. Chklovskii, and U. Alon, "Network motifs: Simple building blocks of complex networks," *Science*, vol. 298, pp. 824–827, 2002.

[45] A.-L. Barabási, *Linked*, Cambridge, MA: Perseus Publishing, 2002.

[46] D. M. Mayerich and J. Keyser, "Filament tracking and encoding for complex biological networks," *Proceedings of Solid Modeling*, 2008.

[47] D. M. Mayerich, Z. Melek, and J. Keyser, "Fast filament tracking using graphics hardware," Tech. Rep. TAMU-CS-TR-2007-11-3, Department of Computer Science, Texas A&M University, 2007.

CHAPTER 3

Parallel Processing Strategies for Cell Motility and Shape Analysis

K. Palaniappan, F. Bunyak, S. Nath, and J. Goffeney

3.1 Cell Detection

The first step in cell behavior analysis is cell detection. Cell detection returns an initial contour close to actual cell boundaries, that is later refined by the cell segmentation process. A wide range of techniques have been applied for detecting and segmenting biological objects of interest in video microscopy imagery including spatially adaptive thresholding, morphological watershed, mean shift, active contours, graph cuts, clustering, multidimensional classifiers like neural networks, and genetic algorithms. Various factors such as type of the cells, environmental conditions, and imaging characteristics such as zoom factor affect the appearance of cells, and thus choice of detection method. But in general, features used in cell detection can be grouped into three categories: intensity-based, texture-based, and spatio-temporal features. Intensity-based detection approaches (i.e., intensity thresholding [1, 2] or clustering [3]) are suitable for lower resolution or stained images where cells or nuclei appear semi-homogeneous and have distinct intensity patterns compared to the background. For high-resolution unstained images where interior of the cells appear highly heterogeneous or where interior and exterior intensity distributions are similar, features based on spatial texture (i.e., ridgeness measures in [4]) or features based on spatiotemporal features or motion are needed.

In the following section, we describe the flux tensor framework for accurate detection of moving objects (which in this context correspond to moving cells) in time lapse images that effectively handles accurate detection of nonhomogeneous cells in the presence of complex biological processes, background noise, and clutter.

3.1.1 Flux Tensor Framework

The classical spatiotemporal orientation or 3D grayscale structure tensor has been widely utilized for low-level motion detection, segmentation, and estimation [5], since it does not involve explicit feature tracking or correlation-based matching. The traditional structure tensor fails to disambiguate between stationary and moving features such as edges and junctions without an explicit (and expensive) eigendecomposition step at every pixel to estimate a dense image velocity or optical flow field. Flux tensor successfully discriminates between stationary and moving image structures more efficiently than structure tensors using only the trace.

3D Structure Tensors

Structure tensors are a matrix representation of partial derivative information. As they allow both orientation estimation and image structure analysis, they have many applications in image processing and computer vision. 2D structure tensors have been widely used in edge/corner detection and texture analysis; 3D structure tensors have been used in low-level motion estimation and segmentation [6, 7].

Under the constant illumination model, the optic-flow (OF) equation of a spatiotemporal image volume $\mathbf{I}(\mathbf{x})$ centered at location $\mathbf{x} = [x, y, t]$ is given by (3.1) [8] where, $\mathbf{v}(\mathbf{x}) = [v_x, v_y, v_t]$ is the optic-flow vector at $\mathbf{x}$,

$$\begin{aligned}\frac{d\mathbf{I}(\mathbf{x})}{dt} &= \frac{\partial \mathbf{I}(\mathbf{x})}{\partial x} v_x + \frac{\partial \mathbf{I}(\mathbf{x})}{\partial y} v_y + \frac{\partial \mathbf{I}(\mathbf{x})}{\partial t} v_t \\ &= \nabla \mathbf{I}^T(\mathbf{x}) \mathbf{v}(\mathbf{x}) = 0\end{aligned} \tag{3.1}$$

and $\mathbf{v}(\mathbf{x})$ is estimated by minimizing (3.1) over a local 3D image patch $\boldsymbol{\Omega}(\mathbf{x}, \mathbf{y})$, centered at $\mathbf{x}$. Note that v_t is not 1 since spatiotemporal orientation vectors will be computed. Using Lagrange multipliers, a corresponding error functional $e_{ls}(\mathbf{x})$ to minimize (3.1) using a least-squares error measure can be written as (3.2) where $W(\mathbf{x}, \mathbf{y})$ is a *spatially invariant* weighting function (e.g., Gaussian) that emphasizes the image gradients near the central pixel [7].

$$\begin{aligned}e_{ls}(\mathbf{x}) = &\int_{\boldsymbol{\Omega}(\mathbf{x},\mathbf{y})} (\nabla \mathbf{I}^T(\mathbf{y}) \mathbf{v}(\mathbf{x}))^2 \;\; W(\mathbf{x}, \mathbf{y}) \, d\mathbf{y} \\ &+ \lambda \left(1 - \mathbf{v}(\mathbf{x})^T \mathbf{v}(\mathbf{x})\right)\end{aligned} \tag{3.2}$$

Assuming a constant $\mathbf{v}(\mathbf{x})$ within the neighborhood $\boldsymbol{\Omega}(\mathbf{x}, \mathbf{y})$ and differentiating $e_{ls}(\mathbf{x})$ to find the minimum, leads to the standard eigenvalue problem (3.3) for solving $\hat{\mathbf{v}}(\mathbf{x})$ the best estimate of $\mathbf{v}(\mathbf{x})$,

$$\mathbf{J}(\mathbf{x}, \mathbf{W}) \hat{\mathbf{v}}(\mathbf{x}) = \lambda \hat{\mathbf{v}}(\mathbf{x}) \tag{3.3}$$

The 3D structure tensor matrix $\mathbf{J}(\mathbf{x}, \mathbf{W})$ for the spatiotemporal volume centered at $\mathbf{x}$ can be written in expanded matrix form, without the spatial filter $W(\mathbf{x}, \mathbf{y})$ and the positional terms shown for clarity, as

$$\mathbf{J} = \begin{bmatrix} \int_{\boldsymbol{\Omega}} \frac{\partial \mathbf{I}}{\partial x} \frac{\partial \mathbf{I}}{\partial x} d\mathbf{y} & \int_{\boldsymbol{\Omega}} \frac{\partial \mathbf{I}}{\partial x} \frac{\partial \mathbf{I}}{\partial y} d\mathbf{y} & \int_{\boldsymbol{\Omega}} \frac{\partial \mathbf{I}}{\partial x} \frac{\partial \mathbf{I}}{\partial t} d\mathbf{y} \\ \int_{\boldsymbol{\Omega}} \frac{\partial \mathbf{I}}{\partial y} \frac{\partial \mathbf{I}}{\partial x} d\mathbf{y} & \int_{\boldsymbol{\Omega}} \frac{\partial \mathbf{I}}{\partial y} \frac{\partial \mathbf{I}}{\partial y} d\mathbf{y} & \int_{\boldsymbol{\Omega}} \frac{\partial \mathbf{I}}{\partial y} \frac{\partial \mathbf{I}}{\partial t} d\mathbf{y} \\ \int_{\boldsymbol{\Omega}} \frac{\partial \mathbf{I}}{\partial t} \frac{\partial \mathbf{I}}{\partial x} d\mathbf{y} & \int_{\boldsymbol{\Omega}} \frac{\partial \mathbf{I}}{\partial t} \frac{\partial \mathbf{I}}{\partial y} d\mathbf{y} & \int_{\boldsymbol{\Omega}} \frac{\partial \mathbf{I}}{\partial t} \frac{\partial \mathbf{I}}{\partial t} d\mathbf{y} \end{bmatrix} \tag{3.4}$$

A typical approach in motion detection is to threshold **trace(J)** (3.5); but this results in ambiguities in distinguishing responses arising from stationary versus moving features (e.g., edges and junctions with and without motion), since **trace(J)** incorporates total gradient change information but fails to capture the nature of these gradient changes (i.e., spatial only versus temporal).

$$\mathbf{trace(J)} = \int_{\boldsymbol{\Omega}} \|\nabla I\|^2 dy \tag{3.5}$$

To resolve this ambiguity and to classify the video regions experiencing motion, the eigenvalues and the associated eigenvectors of **J** are usually analyzed [9, 10]. However, eigenvalue decompositions at every pixel are computationally expensive especially if real-time performance is required.

Flux Tensors

In order to reliably detect only the moving structures *without* performing expensive eigenvalue decompositions, the concept of the *flux tensor* is proposed. Flux tensor is the temporal variations of the optical flow field within the local 3D spatiotemporal volume.

Computing the second derivative of (3.1) with respect to t, (3.6) is obtained where, $\mathbf{a}(\mathbf{x}) = [a_x, a_y, a_t]$ is the acceleration of the image brightness located at $\mathbf{x}$.

$$\begin{aligned}\frac{\partial}{\partial t}\left(\frac{d\mathbf{I}(\mathbf{x})}{dt}\right) &= \frac{\partial^2 \mathbf{I}(\mathbf{x})}{\partial x \partial t} v_x + \frac{\partial^2 \mathbf{I}(\mathbf{x})}{\partial y \partial t} v_y + \frac{\partial^2 \mathbf{I}(\mathbf{x})}{\partial t^2} v_t \\ &\quad + \frac{\partial \mathbf{I}(\mathbf{x})}{\partial x} a_x + \frac{\partial \mathbf{I}(\mathbf{x})}{\partial y} a_y + \frac{\partial \mathbf{I}(\mathbf{x})}{\partial t} a_t\end{aligned} \tag{3.6}$$

which can be written in vector notation as

$$\frac{\partial}{\partial t}(\nabla \mathbf{I}^T(x)\mathbf{v}(\mathbf{x})) = \frac{\partial \nabla \mathbf{I}^T(\mathbf{x})}{\partial t}\mathbf{v}(\mathbf{x}) + \nabla \mathbf{I}^T(\mathbf{x})\ \mathbf{a}(\mathbf{x}) \tag{3.7}$$

Using the same approach for deriving the classic 3D structure, minimizing (3.6) assuming a constant velocity model and subject to the normalization constraint $\|\mathbf{v}(\mathbf{x})\| = 1$ leads to

$$\begin{aligned}e_{ls}^F(\mathbf{x}) &= \int_{\boldsymbol{\Omega}(\mathbf{x},\mathbf{y})} \left(\frac{\partial(\nabla \mathbf{I}^T(\mathbf{y}))}{\partial t}\mathbf{v}(\mathbf{x})\right)^2 W(\mathbf{x},\mathbf{y})\ dy \\ &\quad + \lambda\left(1 - \mathbf{v}(\mathbf{x})^T\mathbf{v}(\mathbf{x})\right)\end{aligned} \tag{3.8}$$

Assuming a constant velocity model in the neighborhood $\boldsymbol{\Omega}(\mathbf{x},\mathbf{y})$, results in the acceleration experienced by the brightness pattern in the neighborhood $\boldsymbol{\Omega}(\mathbf{x},\mathbf{y})$ to be zero at every pixel. As with its 3D structure tensor counterpart **J** in (3.4), the 3D flux tensor $\mathbf{J_F}$ using (3.8) can be written as

$$\mathbf{J_F}(\mathbf{x},\mathbf{W}) = \int_{\boldsymbol{\Omega}} W(\mathbf{x},\mathbf{y})\frac{\partial}{\partial t}\nabla \mathbf{I}(\mathbf{x}) \cdot \frac{\partial}{\partial t}\nabla \mathbf{I}^{\mathrm{T}}(\mathbf{x})dy \tag{3.9}$$

and in expanded matrix form as

$$\mathbf{J_F} = \begin{bmatrix} \int_{\Omega} \left\{ \frac{\partial^2 \mathbf{I}}{\partial x \partial t} \right\}^2 d\mathbf{y} & \int_{\Omega} \frac{\partial^2 \mathbf{I}}{\partial x \partial t} \frac{\partial^2 \mathbf{I}}{\partial y \partial t} d\mathbf{y} & \int_{\Omega} \frac{\partial^2 \mathbf{I}}{\partial x \partial t} \frac{\partial^2 \mathbf{I}}{\partial t^2} d\mathbf{y} \\ \int_{\Omega} \frac{\partial^2 \mathbf{I}}{\partial y \partial t} \frac{\partial^2 \mathbf{I}}{\partial x \partial t} d\mathbf{y} & \int_{\Omega} \left\{ \frac{\partial^2 \mathbf{I}}{\partial y \partial t} \right\}^2 d\mathbf{y} & \int_{\Omega} \frac{\partial^2 \mathbf{I}}{\partial y \partial t} \frac{\partial^2 \mathbf{I}}{\partial t^2} d\mathbf{y} \\ \int_{\Omega} \frac{\partial^2 \mathbf{I}}{\partial t^2} \frac{\partial^2 \mathbf{I}}{\partial x \partial t} d\mathbf{y} & \int_{\Omega} \frac{\partial^2 \mathbf{I}}{\partial t^2} \frac{\partial^2 \mathbf{I}}{\partial y \partial t} d\mathbf{y} & \int_{\Omega} \left\{ \frac{\partial^2 \mathbf{I}}{\partial t^2} \right\}^2 d\mathbf{y} \end{bmatrix} \tag{3.10}$$

As seen from (3.10), the elements of the flux tensor incorporate information about temporal gradient changes which leads to efficient discrimination between stationary and moving image features. Thus the trace of the flux tensor matrix, which can be compactly written and computed as

$$\mathbf{trace}(\mathbf{J_F}) = \int_{\Omega} \left\| \frac{\partial}{\partial t} \nabla \mathbf{I} \right\|^2 d\mathbf{y} \tag{3.11}$$

and can be directly used to classify moving and nonmoving regions without the need for expensive eigenvalue decompositions. If motion vectors are needed, then (3.8) can be minimized to get $\hat{\mathbf{v}}(\mathbf{x})$ using

$$\mathbf{J_F}(\mathbf{x}, \mathbf{W})\ \hat{\mathbf{v}}(\mathbf{x}) = \lambda \hat{\mathbf{v}}(\mathbf{x}) \tag{3.12}$$

In this approach the eigenvectors need to be calculated at just moving feature points.

3.1.2 Flux Tensor Implementation

To detect motion blobs, only the trace of flux tensor $\mathbf{trace}(\mathbf{J_F}) = \int_{\Omega(\mathbf{y})} \left\| \frac{\partial}{\partial t} \nabla \mathbf{I} \right\|^2 d\mathbf{y}$ needs to be computed. That requires computation of I_{xt}, I_{yt}, and I_{tt} and the integration of squares of I_{xt}, I_{yt}, I_{tt} over the area $\mathbf{\Omega}(\mathbf{y})$. The following notation is adopted for simplicity:

$$I_{xt} = \frac{\partial^2 I}{\partial x \partial t}, \quad I_{yt} = \frac{\partial^2 I}{\partial y \partial t}, \quad I_{tt} = \frac{\partial^2 I}{\partial t \partial t} \tag{3.13}$$

The calculation of the derivatives is implemented as convolutions with a filter kernel. By using separable filters, the convolutions are decomposed into a cascade of 1D convolutions. For numerical stability as well as noise reduction, a smoothing filter is applied to the dimensions that are not convolved with a derivative filter (e.g., calculation of I_{xt} requires smoothing in y-direction, and calculation of I_{yt} requires smoothing in x-direction). I_{tt} is the second derivative in temporal direction; the smoothing is applied in both spatial directions. As smoothing and derivative filters, optimized filter sets presented by Scharr et al. in [11, 12] are used.

The integration is also implemented as an averaging filter decomposed into three 1D filters. As a result, calculation of trace at each pixel location requires

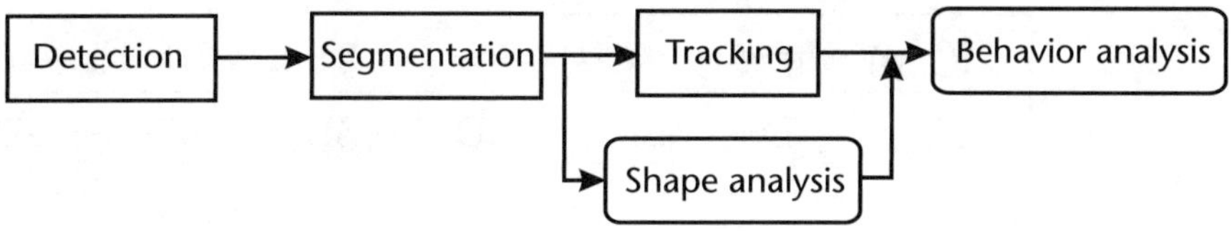

Figure 3.1 General framework for cell shape and behavior analysis.

three 1D convolutions for derivatives and three 1D convolutions for averages in the corresponding spatiotemporal cubes.

A brute-force implementation where spatial and temporal filters are applied for each pixel separately within a spatiotemporal neighborhood would be computationally very expensive since it would have to recalculate the convolutions for neighboring pixels. For an efficient implementation, the spatial (x and y) convolutions are separated from the temporal convolutions, and the 1D convolutions are applied to the whole frames one at a time. This minimizes the redundancy of computations and allows reuse of intermediate results.

The spatial convolutions required to calculate I_{xt}, I_{yt}, and I_{tt} are I_{xs}, I_{sy}, and I_{ss} where s represents the smoothing filter. Each frame of the input sequence is first convolved with two 1D filters, either a derivative filter in one direction and a smoothing filter in the other direction, or a smoothing filter in both directions. These intermediate results are stored as frames to be used in temporal convolutions, and pointers to these frames are stored in a first-in first-out (FIFO) buffer of size $n_{FIFO} = n_{Dt} + n_{At} - 1$ where n_{Dt} is the length of the temporal derivative filter

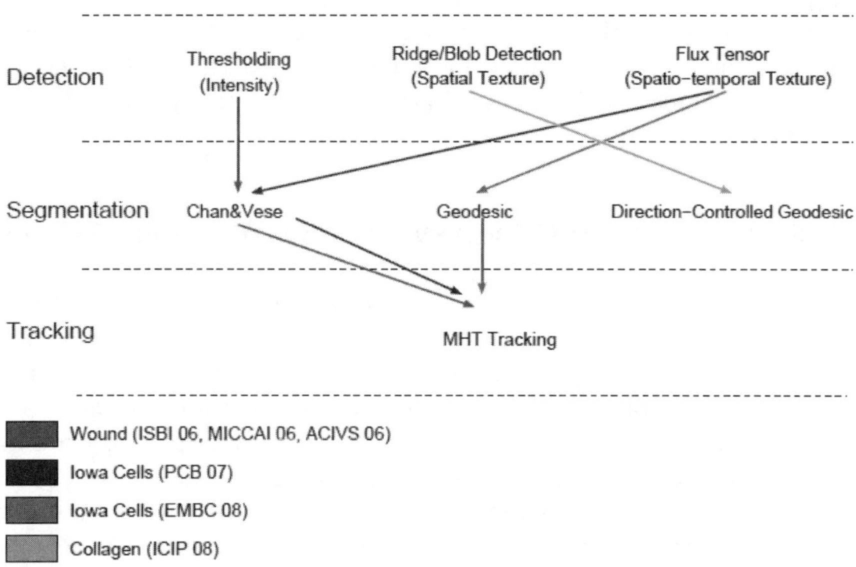

Figure 3.2 Cell shape and motility analysis approaches for different datasets.

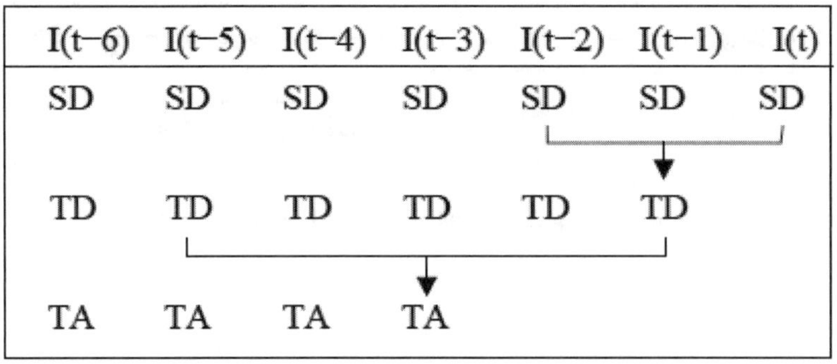

Figure 3.3 Steady state operation of the FIFO for temporal derivative filter of length $n_{Dt} = 3$ and temporal averaging filter of length $n_{At} = 5$. SD: spatial derivatives (I_{xs}, I_{sy}, I_{ss}); TD: temporal derivatives (I_{xt}, I_{yt}, I_{tt}); TA: temporal averages $(\mathbf{trace}(\mathbf{J_F}) = \int_{\Omega(\mathbf{y})}(I_{xt}^2 + I_{yt}^2 + I_{tt}^2)d\mathbf{y})$.

and n_{At} is the length of the temporal averaging filter. For each input frame, three frames I_{xs}, I_{sy}, and I_{ss} are calculated and stored. Once n_{Dt} frames are processed and stored, FIFO has enough frames for the calculation of the temporal derivatives I_{xt}, I_{yt}, and I_{tt}. Since averaging is distributive over addition, $I_{xt}^2 + I_{yt}^2 + I_{tt}^2$ is computed first and spatial averaging is applied to this result and stored in the FIFO structure to be used in the temporal part of averaging. Once flux tensor trace of n_{At} frames are computed, temporal averaging is applied. Motion mask FG_M is obtained by thresholding and post-processing averaged flux tensor trace. Post-processing include morphological operations to join fragmented objects and to fill holes.

Figure 3.3 shows the steady state operation of the FIFO for $n_{Dt} = 3$ and $n_{At} = 5$. Curent frame at time t is processed and the spatial results are stored at the last slot in the temporal filtering FIFO. At this point, the intermediate flux trace result can be produced for frame (t−1) and stored at the last slot in the temporal averaging FIFO, which in turn can be used the produce the flux trace for the frame (t−3).

3.2 Cell Segmentation Using Level Set-Based Active Contours

In recent years, PDE-based segmentation methods such as active contours have gained a considerable interest particularly in biomedical image analysis because of their advantages such as mathematical description of the underlying physics phenomena, subpixel precision, and direct extension to higher dimensions [13, 14]. Active contours evolve/deform a curve C, subject to constraints from a given image. Active contours are classified as parametric [15, 16] or geometric [17–19] according to their representation. Parametric active contours (i.e., classical snakes) are represented explicitly as parametrized curves in a Lagrangian formulation, geometric active contours are implicitly represented as level sets [20] and evolve according to an Eulerian formulation [21]. The geometric model of active contours provides advantages such as eliminating the need to reparameterize the curve and automatic handling of topology changes via level set implementation [20].

In this section we present two cell segmentation approaches both using level setbased geometric active contours: (1) region-based segmentation and our extensions [1, 22]; and (2) edge-based segmentation and our extensions [23, 46]. The first approach is used for images where cells or nuclei appear as semi-homogeneous blobs with vague borders. The second approach is used for images where the interior of the cells appears highly heterogeneous or where interior and exterior intensity distributions are similar. The type of the cells, environmental conditions, and imaging characteristics such as zoom factor affect the appearance of cells, and thus choice of the approach.

3.2.1 Region-Based Active Contour Cell Segmentation

Segmentation of multiple cells imaged with a phase contrast microscope is a challenging problem due to difficulties in segmenting dense clusters of cells. As cells being imaged have vague borders, close neighboring cells may appear to merge. Accurate segmentation of individual cells is a crucial step in cell behavior analysis since both over-segmentation (fragmentation) and under-segmentation (cell clumping) produce tracking errors (i.e., spurious or missing trajectories and incorrect split and merge events).

In this section, we describe three different region-based level set segmentation methods and their performance for separating closely adjacent and touching cells in a dense population of migrating cells. The three techniques described in this section are all based on the "active contour without edges" energy functional [24] with appropriate extensions. These include a two-phase Chan and Vese algorithm [24] (`CV-2LS`), an N–level set algorithm with energy-based coupling by Zhang et al. [25] (`ZZM-NLS`), and an improved version of of our four-color level set algorithm with energy-based and explicit topological coupling [1] (`NBP-4LS-ETC`). The latter two techniques use coupling constraints to prevent merging of adjacent cells when they approach or touch each other.

Chan and Vese Active Contours Formulation

In [24], Chan and Vese presented an algorithm to automatically segment an image $I(y)$ into *two* distinct regions (or phases). The segmentation is done by minimizing the Mumford-Shah functional in (3.14) with respect to the scalar variables c_1, c_2, and ϕ; with $\phi(\mathbf{y})$ being the 2D level set function and $u(\mathbf{y})$ being the intensity image.

$$E_{pc}(c_1, c_2, \phi) = \mu_1 \int (u(\mathbf{y}) - c_1)^2 H(\phi(\mathbf{y}))\, d\mathbf{y} + \mu_2 \int (u(\mathbf{y}) - c_2)^2 (1 - H(\phi(\mathbf{y}))\, d\mathbf{y} + \nu \int |\nabla H(\phi(\mathbf{y}))|\, d\mathbf{y} \tag{3.14}$$

In (3.14), c_1 and c_2 model the gray values of the two phases used to define the segmentation, with the boundary for the phases described by regularized Heaviside function $H(\phi)$ [24]. ϕ models the interface separating these two phases. The first two terms in the functional aim at maximizing the gray value homogeneity of the two phases, while the last term aims at minimizing the length of the boundary separating these two phases; μ_1, μ_2, and v are adjustable constants associated with this functional. The Euler-Lagrange equation $\frac{\partial \phi}{\partial t}$, to evolve the level set curve, is

$$\frac{\partial \phi}{\partial t} = \delta(\phi)\left[v \operatorname{div}\frac{\nabla \phi}{|\nabla \phi|} - \mu_1(u(\mathbf{x}) - c_1)^2 + \mu_2(u(\mathbf{x}) - c_2)^2\right] \tag{3.15}$$

The scalars c_1 and c_2 are updated at each level set iteration step using (3.16) for $\phi(\mathbf{y}) > 0$ and $\phi(\mathbf{y}) < 0$, respectively.

$$c_1 = \frac{\int u(\mathbf{y})H(\phi(\mathbf{y}))\, d\mathbf{y}}{\int H(\phi(\mathbf{y}))\, d\mathbf{y}}, \quad c_2 = \frac{\int u(\mathbf{y})(1 - H(\phi(\mathbf{y}))\, d\mathbf{y}}{\int (1 - H(\phi(\mathbf{y}))\, d\mathbf{y}} \tag{3.16}$$

A regularized Heaviside function $H_{2,\epsilon}(\phi(\mathbf{y}))$ is used to obtain a numerically stable Dirac delta function $\delta(\phi)$ in (3.15) as described in [24]. A suitable stopping criterion (such as a small change in energy, $|\phi_{i+1} - \phi_i| < k_{thresh}$ where ϕ_{i+1}, ϕ_i are the level set functions at successive iterations) is used to terminate the curve evolution. In [26], in order to segment multiple (i.e., N distinct) objects, Chan and Vese extended their previous two-phase algorithm [24] by using $\lceil \log_2 N \rceil$ level sets.

Coupled N–Level Set Formulation

As observed by Zhang et al. [25], Dufour et al. [27] and Zimmer and Olivo-Marin [28], the two variants of the Chan and Vese algorithm are unsuitable for reliable cell segmentation, since they do not prevent apparent merges in cells. The $\lceil \log_2 N \rceil$ level set formulation improves on the performance of a single level set function, as more number of objects with varying intensities can be efficiently classified. However, it does not prevent under-segmentation (i.e., incorrect merges), when the objects have similar intensities (i.e., cells).

To overcome the drawbacks of classical Chan and Vese level set formulations, while at the same time solving the problem of apparent merging of cells during tracking, Zhang et al. [25] proposed a new model for segmenting cells, using N–level sets. Here, N is the number of cells at a given time instance. An a priori knowledge that cells do not merge during the evolution process was used to guide the segmentation process. This was achieved by a pair-wise energy-based coupling constraint on the level sets evolution process. A similar formulation was used by Dufor et al. in 3D cell segmentation [27]. The energy functional, $E_{nls}(c_{in}, c_{out}, \Phi)$, used to evolve N– coupled level sets is shown below [25]:

$$
\begin{aligned}
E_{nls}\ (c_{in}, c_{out}, \Phi) &= \mu_{in} \sum_{i=1}^{N} \int_{\Omega} (I - c_{in}^{i})^2 H(\phi_i)\ dy \\
&+ \mu_{out} \int_{\Omega} (1 - c_{out})^2 \prod_{\substack{i=1 \\ \forall\ i: H(\phi_i)<0}}^{N} (1 - H(\phi_i))\ dy + \nu \sum_{i=1}^{N} \int_{\Omega} |\nabla H(\phi_i)| dy \\
&+ \gamma \sum_{i=1}^{N} \sum_{j=i+1}^{N} \int_{\Omega} H(\phi_i) H(\phi_j)\ dy
\end{aligned} \tag{3.17}
$$

Here, $\Phi = \{\phi_1, \phi_2, \ldots, \phi_N\}$ represents $N-$level sets associated with N cells in the image; c_{in} represents the average intensities of cells for $\phi_i \geq 0$, while c_{out} is the average intensity of the background.[1] The first and second terms are homogeneity measures of the foregrounds and background of all level sets. The third term controls the lengths of interfaces $\phi_i = 0$, and minimizes the length of all level sets, while the fourth term of the functional penalizes pair-wise couplings or overlaps between level sets. The constants μ_{in}, μ_{out}, ν, and γ are weights associated with each of the terms.

Coupled 4-Level Set Formulation

The coupled N-level set formulation of Zhang et al. is successful in preventing apparent/false merges. But the $O(N^2)$ complexity of this approach where N is number of cells, makes it impractical for high-throughput screening studies, where each frame in the image sequence may contain hundreds to thousands of cells. Evolving thousands of level sets by solving thousands of separate partial differential equations with millions of coupling interaction terms is currently impractical or too expensive even using specialized hardware or parallel processing computing clusters. In [1], we presented an optimized version of the $N-$level set algorithm that scales to high number of cells/objects and demonstrated its performance in presence of an external tracking module in [22]. An overview of this optimized version is given below. And major implementation steps are presented in Algorithm 2.

Let N be the number of objects (i.e., cells) in an image. For any given object A_i, let $\mathcal{N}(A_i) = \{A_j, A_{j+1}, \ldots, A_L\}$, $L << N$ represent its L neighbors. Our optimization is based on the fact that: *during cell evolution, false merging in A_i can only occur within this local neighborhood of L objects*, and not over the entire image. Thus, to prevent apparent/false merges, it is sufficient to assign each object A_i to a level set different than the level sets of its neighbors. When the spatial layout of the cells in a frame is described using a planar neighborhood graph **P** where the vertices A_i^V represent the cells and the edges A_i^E represent the neighborhood relationship. The problem of assigning level sets to cells becomes equivalent to *graph vertex coloring.*

- *Graph vertex coloring:* assignment of labels or colors to each vertex of a graph such that no edge connects two identically colored vertices.

1. The region *exterior to all level sets* indicates the background.

Now the question becomes, *What is the minimum number of colors, thus the minimum number of level sets needed?* The *Four-Color Theorem* (FCT), states that every planar graph is four-colorable while insuring that neighboring vertices do not share the same color [29–31]. Since our neighborhood graph is a planar graph, the N-level set formulation of Zhang et al. [25] can be effectively reduced to a 4-level set formulation. Assignment of N objects to $k << N$ level sets (k = 4 for our case) is justified by the assumption that objects (in our case cells) have similar, or nearly similar characteristic features. In this case objects/cells can be randomly assigned to any of the four level sets. For a wide variety of applications (including our application of cell segmentation), the upper limit of the chromatic number for planar graphs is four. However, we do acknowledge that there may exist other planar graphs having even smaller chromatic numbers.

In order to evolve the four level sets, we propose minimizing the $N-$coupled level sets energy functional shown in (3.17), with $N = 4$. In addition, we replace the length minimizer term [third term in (3.17)] with its geodesic equivalent, and incorporate additional constraints in our proposed energy functional $E_{fls}(\mathbf{c_{in}}, c_{out}, \Phi)$ as shown below:

$$\begin{aligned}
E_{fls}(\mathbf{c_{in}}, c_{out}, \Phi) &= \underbrace{\mu_{in} \sum_{i=1}^{4} \int_{\Omega} (I - c_{in}^{i})^2 H(\phi_i)\, dy}_{\text{Homogeneity of Foregrounds}} \\
&+ \underbrace{\mu_{out} \int_{\Omega} (I - c_{out})^2 \prod_{\substack{i=1 \\ \forall\, i: H(\phi_i) < 0}}^{4} (1 - H(\phi_i))\, dy}_{\text{Homogeneity of Background}} + \underbrace{\nu \sum_{i=1}^{4} \int_{\Omega} |g(I) \nabla H(\phi_i)|\, dy}_{\text{Geodesic Length Minimizers}} \\
&+ \underbrace{\gamma \sum_{i=1}^{4} \sum_{j=i+1}^{4} \int_{\Omega} H(\phi_i) H(\phi_j)\, dy}_{\text{Energy-based Coupling}} + \underbrace{\eta \left\{ \sum_{i=1}^{4} \int_{\Omega} \frac{1}{2} (|\nabla \phi_i| - 1)^2\, dy \right\}}_{\text{Implicit Redistancing}}
\end{aligned} \tag{3.18}$$

where $g(I)$ is an edge indicator function given by [32]

$$g(I) = \frac{\alpha}{1 + \beta |\nabla G_\sigma \circledast I|}$$

with constants α, β. $|\nabla G_\sigma \circledast I|$ can also be replaced with a regularized version of our adaptive robust structure tensor (ARST) [33]. The last term in (3.18) enforces the constraint of $|\nabla \phi_i| = 1$, thus helping us avoid explicit redistancing of level sets during the evolution process [34], with η being a constant. As noted by Kim and Lim, addition of a geodesic length term improves the performance of an energy-only Chan and Vese level set formulation by preventing detection of regions with weak edge strengths in [35]. This property was also exploited by Dufour et al. in 3D cell segmentation [27].

Gradient-descent minimization can be used to iteratively solve the level set functions ϕ_i as

$$\phi_i(k+1) = \phi_i(k) + \Delta\tau \frac{\partial \phi_i(k)}{\partial k} \tag{3.19}$$

where k is an artificial evolution time, $\Delta\tau$ is the spacing between successive iterations, and $\frac{\partial \phi_i(k)}{\partial k}$ is the Euler-Lagrange equation associated with the ith level set ϕ_i, given by

$$\frac{\partial \phi_i(k)}{\partial k} = \delta(\phi_i) \Bigg\{ -\mu_{in}(I - c_{in}^i)^2 + \mu_{out}(I - c_{out})^2 \prod_{\substack{j=1 \\ \forall\, j: H(\phi_j)<0, j \neq i}}^{4} (1 - H(\phi_j))$$
$$+ \nu \left(g(I) \cdot \text{div}\left(\frac{\nabla \phi_i}{|\nabla \phi_i|} \right) + \nabla g(I) \cdot \frac{\nabla \phi_i}{|\nabla \phi_i|} \right) - \gamma \sum_{j=1; j \neq i}^{4} H(\phi_j) \Bigg\}$$
$$+ \eta \left\{ \nabla^2 \phi_i - \text{div}\left(\frac{\nabla \phi_i}{|\nabla \phi_i|} \right) \right\} \tag{3.20}$$

For numerical stability of Euler-Lagrange equations, the Heaviside functions (and in turn, the Dirac-delta functions) are regularized by a parameter $\epsilon > 0$ [24].

Accurate Neighborhoods Using Generalized Voronoi Diagrams

The success of apparent/false merge prevention in the 4-level sets approach depends on the accuracy of the neighborhood adjacency graph $\mathcal{N}(\mathbf{P})$. In our original formulation [1], we reported Delaunay triangulations as a means to obtain $\mathcal{N}(\mathbf{P})$. The Delaunay triangulation approach has been previously used in many biological image analysis applications, such as in quantitative spatial analysis of complex cellular systems [36], membrane growth analysis [37], histological grading of cervical intraepithelial neoplasia [38], segmentation of tissue architecture [39], diagnosis of malignant mesothelioma [40], bladder carcinoma grading [41], and characterization of muscle tissue [42].

Delaunay triangulation graphs obtained using distances between centroids of objects is adequate to represent the neighborhood relationships of circular objects of similar sizes. But since centroid-based distance measures are insensitive to shape, size, and orientation of objects, graphs derived from these measures *may* lead to incorrect neighborhood relationships for objects of irregular shapes and sizes, and ultimately to the failure of the 4-level sets segmentation (see Figure 3.4). Hence, rather than using centroid-based distance measures [i.e., Delaunay triangulation, and its geometric dual, the ordinary Voronoi diagram (OVD)], we propose using the *generalized Voronoi diagram* (GVD) that takes into account shape, size, and orientation of objects to compute accurate neighborhood relationships. The incorrect neighborhood relationships and subsequent failure of the 4-level sets segmentation illustrated in Figure 3.4 can then be avoided by extracting correct neighborhood relationships using generalized Voronoi boundaries as shown in Figure 3.5.

The GVD in any dimension can be precisely defined using point-to-object distance measures [43, p. 280]. Let $\mathbf{A} = \{A_1, A_2, \ldots, A_N\}$ be a set of arbitrarily shaped objects in a $d-$dimensional space $\mathbb{R}^d$. Now, for any point $\mathbf{p} \in \mathbb{R}^d$, let $\mathbf{D}(\mathbf{p}, A_i)$ denote a distance measure representing how far the point $\mathbf{p}$ is from the object A_i,

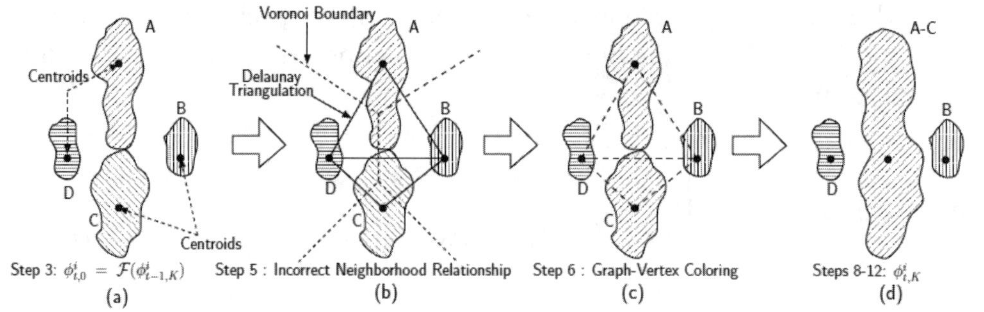

Figure 3.4 (a–d) Incorrect neighborhood relationships, resulting from Delaunay triangulation and its effects on Algorithm 2. Since the neighborhood of the cells A and C is not detected, both cells are assigned the same color, and hence to the same level set. This results in a *false-merge* taking place between cells A and C, at the end of the iteration process.

which is typically the minimum distance from $\mathbf{p}$ to any point in object A_i. The dominance region (also known as influence-zone) of A_i, is then defined as

$$\mathrm{Dom}(A_i, A_j) = \{\mathbf{p}|\mathrm{D}(\mathbf{p}, A_i) \leq \mathrm{D}(\mathbf{p}, A_j), \forall j, j \neq i\} \tag{3.21}$$

A generalized Voronoi boundary, between A_i and Aj, can then be defined as the loci of equidistant points between both objects, $\mathcal{L}(A_i, A_j)$, as

$$\mathcal{L}(A_i, A_j) = \{\mathbf{p}|\mathrm{D}(\mathbf{p}, A_i) = \mathrm{D}(\mathbf{p}, A_j)\} \tag{3.22}$$

with a corresponding influence zone, or Voronoi region, for A_i, $V(A_i)$, as

$$V(A_i) = \bigcap_{i \neq j} \mathrm{Dom}(A_i, A_j) \tag{3.23}$$

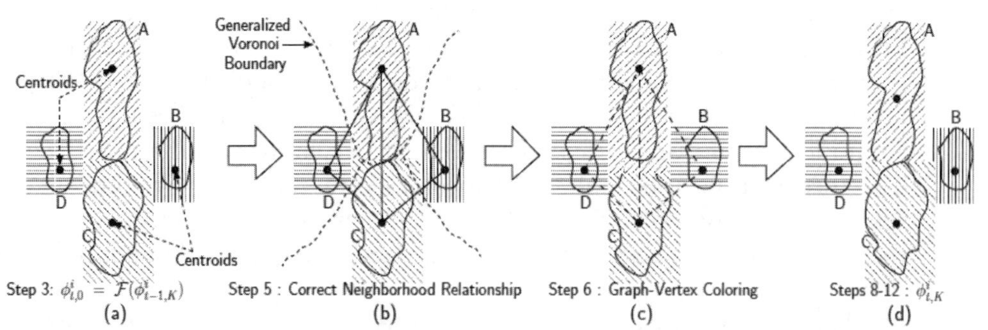

Figure 3.5 (a–d) The *false-merge* problem, shown in Figure 3.4, can be avoided in Algorithm 2 by extracting correct neighborhood relationships using generalized Voronoi boundaries. Cells A and C are correctly identified as neighbors, and hence assigned to different level sets.

Hence, the generalized Voronoi diagram of **A**, GVD(**A**) is given by the collection of such Voronoi regions, as

$$\begin{aligned} \mathrm{GVD}(\mathbf{A}) &= \bigcup_i V(A_i) \\ &= \bigcup_i \bigcap_{i \neq j} \{\mathbf{p} | \mathrm{D}(\mathbf{p}, A_i) \leq \mathrm{D}(\mathbf{p}, A_j), \forall j, j \neq i\} \end{aligned} \tag{3.24}$$

When **A** is a collection of points rather than sized objects, the GVD(**A**) reduces to an ordinary Voronoi diagram, OVD(**A**) (the Delaunay triangulation being the geometric dual of the OVD).

The GVD representation of a set of objects has a number of useful properties: (1) it is a thin set that partitions a space into connected regions; (2) it is homotopic to the number of objects; (3) it is invariant to global transformations applied to all objects in the image; and (4) each region is guaranteed to contain the entire object. For those interested in properties of the OVD for point objects, we direct them to the book by Okabe et al. [43] and the classic survey paper by Aurenhammer [44].

In [45] we present a new algorithm for efficiently and accurately extracting accurate neighborhood graphs in linear time by computing the Hamilton-Jacobi GVD using the exact Euclidean-distance transform with Laplacian-of-Gaussian, and morphological operators. We refer reader to [45] for further information and validation results using both synthetic and real biological imagery. In Section 3.3.3, we present a different GVD construction algorithm designed for implementation on graphical processing units (GPUs).

Spatiotemporal interactions between two neighboring cells can be grouped into four, depending on their spatial configurations (merged into a single connected component or separate components) and temporal configurations (one-to-one overlap, where each cell in the current frame overlaps only with a single cell in the previous frame; or temporally merged, where a single cell in the current frame overlaps with multiple cells in the previous frame). These cases are summarized in Table 3.1 and illustrated in Figure 3.6.

Figure 3.7 shows a sample synthetic case-2 sequence, its 4-level sets segmentation, and corresponding tracking results. Even though the two circles touch/merge for a while (Figure 3.6), the coupled 4-level sets segmentation succeeds in segmenting them as two objects.

Table 3.1 Spatiotemporal Interactions Between Two Neighboring Cells

Case	*Spatial*	*Temporal*	*N-LS*	*N-LS +coupling*	*Comments*
1	Separate	1-to-1 overlap	✓	✓	
2	Merged	1-to-1 overlap	×	✓	
3	Separate	Merged	×	✓×	Depending on temporal sampling rate
4	Merged	Merged	×	✓×	Depending on temporal sampling rate

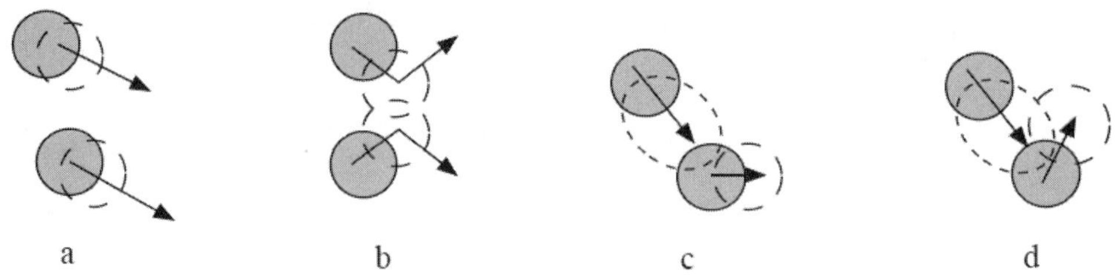

Figure 3.6 Cell-to-cell interaction cases: (a) case 1: separate cells, one-to-one temporal overlap; (b) case 2: merged cells one-to-one temporal overlap; (c) case 3: spatially separate and temporally merged; and (d) case 4: spatially merged and temporally merged.

3.2.2 Edge-Based Active Contour Cell Segmentation

While region-based active contours have been succesfully used in many cell and nucleus segmentation applications [1–3, 24], they require a bimodal image model to discern background and foreground which is not applicable in some high-resolution cell images. Figure 3.8 shows typical high-resolution phase contrast images of cells, where white regions surrounding the cells are the phase halos. A straightforward implementation of [24] converges to the boundaries between phase halo and the rest of the image. For images where intensity inside the cell boundaries is highly heterogeneous, and there is no clear difference between inside and outside intensity profiles, edge-based approaches are used since region-based active contour approaches are not applicable.

In this section, we describe two edge-based level set segmentation methods and their performance for handling phase halos. The first technique is the classical geodesic active contour approach [19]; the second technique is our extension directed edge profile-guided level set active contours. Compared to region-based

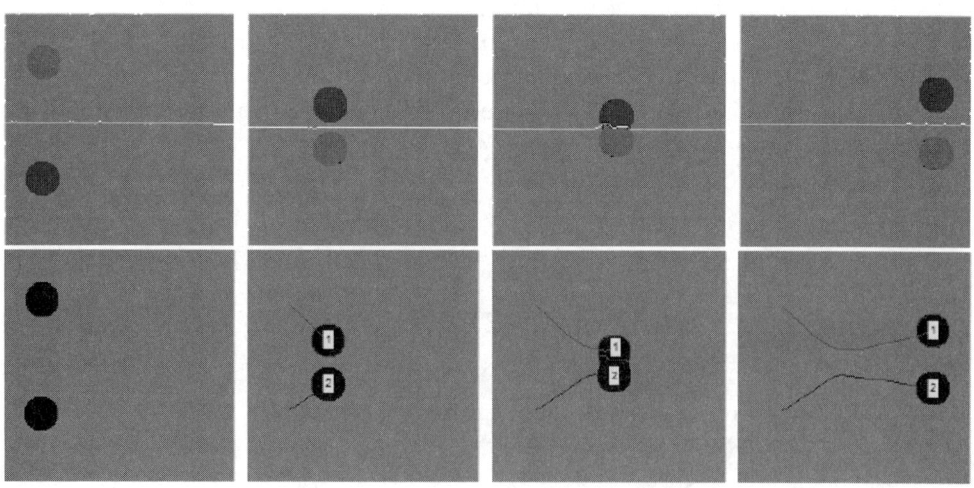

Figure 3.7 Cell-to-cell interaction case 2 test sequence frames #1, 8, 13, 20. Row 1: four color level set segmentation; Row 2: tracking results.

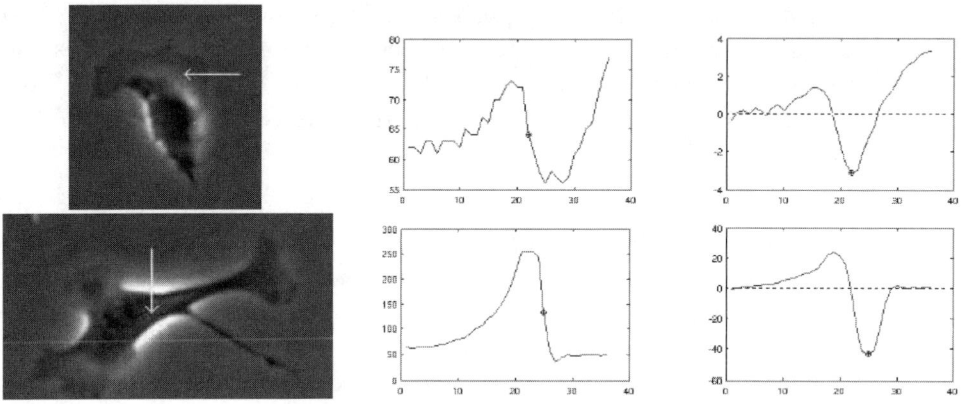

Figure 3.8 Phase contrast cell images. Left to right: sample cell, intensity profile at the marked direction, and derivative of the directed profile. The actual cell boundary point is marked on the graphs.

active contours, edge-based active contours are more sensitive to initialization. Since they are designed to stop at edges, they suffer from: (1) early stopping on background or foreground edges, (2) contour leaking across weak boundaries, and (3) for phase contrast microscopy imagery, early stopping at outer edges of phase halos. The first problem can be avoided by starting the contour evolution close to the object/cell boundaries. Initial contour can be obtained from a detection step using spatial properties as in ridge-based approach in [23], or spatiotemporal properties as in flux tensor-based approach in [46–48] described in Section 3.2. The latter two problems are addressed in our directed edge profile-guided level set active contours below.

Geodesic Level Set Active Contours

In level set based active contour methods, a curve $\mathcal{C}$ is represented implicitly via a Lipschitz function ϕ by $\mathcal{C} = \{(x,y)|\phi(x,y) = 0\}$, and the evolution of the curve is given by the zero-level curve of the function $\phi(t,x,y)$ [24]. In regular geodesic active contours [19] the level set function ϕ is evolved using the speed function,

$$\frac{\partial \phi}{\partial t} = g(\nabla \mathbf{I})(F_c + \mathcal{K}(\phi))|\nabla \phi| + \nabla \phi \cdot \nabla g(\nabla \mathbf{I}) \tag{3.25}$$

where F_c is a constant, $\mathcal{K}$ is the curvature term (3.26), and $g(\nabla \mathbf{I})$ is the edge stopping function, a decreasing function of the image gradient which can be defined as in (3.27).

$$\mathcal{K} = div\left(\frac{\nabla \phi}{|\nabla \phi|}\right) = \frac{\phi_{xx}\phi_y^2 - 2\phi_x\phi_y\phi_{xy} + \phi_{yy}\phi_x^2}{(\phi_x^2 + \phi_y^2)^{\frac{3}{2}}} \tag{3.26}$$

$$g(\nabla \mathbf{I}) = exp(-|\nabla G_\sigma(x,y) * \mathbf{I}(x,y)|) \tag{3.27}$$

The constant balloon force F_c pushes the curve inwards or outwards depending on its sign. The regularization term $\mathcal{K}$ ensures boundary smoothness, and $g(\mathbf{I})$ is used to stop the curve evolution at cell boundaries. The term $\nabla g \cdot \nabla \phi$ is used to increase the basin of attraction for evolving the curve to the boundaries of the objects.

Directed Edge Profile-Guided Level Set Active Contours

The major drawbacks of the classical edge-based active contours are the leakage due to weak edges and stopping at local maximums in noisy images. For the special case of cell segmentation, edge stopping functions used in regular geodesic active contours, if initialized outside of the cells, respond to the outer edges of phase halos and cause early stopping which produces an inaccurate segmentation. When the curve is initialized inside the cells, the texture inside the cell also causes premature stopping. Often phase halo is compensated by normalizing the image, but this weakens the cell boundaries further, especially in the areas where cells are flattened, and leads to leakage of the curve. One remedy is to enforce a shape model, but it fails for cells with highly irregular shapes. Nucleus-based initialization and segmentation [49] is also not applicable to images such as in Figure 3.8. To obtain an accurate segmentation, we propose an approach that exploits the phase halo effect instead of compensating it. The intensity profile perpendicular to the local cell boundary is similar along the whole boundary of a cell; it passes from the brighter phase halo to the darker cell boundary. We propose to initialize the curve outside the cell and phase halo, and guide the active contour evolution based on the desired edge profiles which effectively lets the curve evolve through the halos and stop at the actual boundaries. The existence of phase halo increases the edge strength at cell boundaries. We propose the edge stopping function g_d (3.30) guided by the directed edge profile, that lets the curve evolve through the outer halo edge and stop at the actual boundary edge. As shown in Figure 3.8, if initialized close to the actual boundary, the first light-to-dark edge encountered in the local perpendicular direction corresponds to the actual boundary. By choosing this profile as the stopping criterion, we avoid the outer edge of phase halo. This stopping function is obtained as follows. Normal vector $\vec{N}$ to the evolving contour/surface can be determined directly from the level set function:

$$\vec{N} = -\frac{\nabla \phi}{|\nabla \phi|} \tag{3.28}$$

Edge profile is obtained as the intensity derivative in opposite direction to the normal:

$$\mathbf{I}_{-\vec{N}} = \frac{\nabla \phi}{|\nabla \phi|} \cdot \nabla \mathbf{I} \tag{3.29}$$

Dark-to-light transitions produce positive response in $\mathbf{I}_{-\vec{\mathbf{N}}}$. We define the edge profile-guided edge stopping function g_d as

$$g_d(\nabla \mathbf{I}) = 1 - H(-I_{-\vec{N}})(1 - g(\nabla \mathbf{I})) \tag{3.30}$$

where H is the Heaviside step function:

$$H(x) = \begin{cases} 1 \text{ if } x > 0 \\ 0 \text{ elsewhere} \end{cases}$$

This sets g_d to 1 at regions where there is a dark-to-light (background-to-halo) transition perpendicular to boundary, and keeps regular edge stopping function everywhere else. Thus it lets the active contour evolve through the edges with a dark-to-light profile and stop at edges with light-to-dark profile. Equation (3.31) shows the speed function of the proposed curve evolution.

$$\begin{aligned} \frac{\partial \phi}{\partial t} &= (1 - \mathrm{H}(-\mathbf{I}_{-\vec{N}})(1 - g(\nabla \mathbf{I})))(c\mathbf{I} + \mathcal{K}(\phi))|\nabla \phi| \\ &+ \nabla \phi \cdot \nabla (1 - \mathrm{H}(-\mathbf{I}_{-\vec{N}})(1 - g(\nabla \mathbf{I}))) \end{aligned} \tag{3.31}$$

To prevent contour leakage on weak edges, we use a spatially adaptive force that slows down the curve evolution at boundaries. We replace the constant balloon force F_c in (3.25) with an intensity adaptive force F_A:

$$F_A(x, y) = c\mathbf{I}(x, y) \tag{3.32}$$

This intensity adaptive force increases the speed of contraction on bright phase halos and reduces it on thin dark extensions (pseudo-pods) and prevents leaking across weak edges.

3.2.3 GPU Implementation of Level Sets

Scientific Computing on Graphical Processing Units

The power of GPUs that can currently sustain hundreds of gigaflops is increasingly being leveraged for scientific computing. The exponential increase in GPU computing power is being driven by the commercially competitive and hardware intensive gaming market. The data parallel GPU architecture is emerging as a powerful environment for running many closely coupled but independent parallel threads. Both NVIDIA and AMD/ATI are expanding their market beyond entertainment and targeting dual use applications for their single chip multiple core processors with more than a billion transistors from the initial chip-design phase itself; this is often referred to as general purpose GPU (GPGPU) [50]. Early programs developed for GPUs had to use graphical APIs such as OpenGL or NVIDIA's Cg to control the hardware, but current general purpose APIs such as NVIDIA's CUDA, ATI's CTM, and Stanford's Brook for scientific computing [51], which avoid explicit graphics interfaces, are now commonly used for compute intensive tasks.

The GPU architecture is a SIMD pipeline for ingesting vertex and primitive shape information to construct filled, textured, and lit polygons for viewing on the screen. Initially, sets of processors called shader units were specialized for each stage, but with the advent of the Shader Model 4.0 specification the shader processor architecture has been generalized. This change maximizes the utilization of every available processor and reduces the economic cost of developing dedicated units. Three stages of the GPU pipeline are accessible via programming: vertex processing, geometry construction, and pixel shading. During a rendering pass, a complete set of geometry primitives passes through all these stages. Scientific applications are mapped to the architecture by loading data onto the GPU and performing multiple rendering passes in which different programs can be loaded to each shader stage for each pass.

The pixel (fragment) shader provides the most straightforward parallel processing mechanism for image processing and computer vision problems involving dense arrays of data. Take for example the task of applying a convolution kernel to an image. The image is loaded into texture memory of the GPU and a program is loaded which computes the convolution sum for a single pixel. A rendering target the same $n \times m$ size as the source image is set up to hold the results. On the GPU the rendering target is broken into $n \times m$ pixels which are each processed in parallel. The program takes the source image and the pixel's x/y location as parameters, computes the convolution, and writes the value to the pixel. The result is a filtered image. Additional filter passes are additional rendering passes using the previously generated output as the new input for a different shader program. The limitation of the pixel shader is the support of gather but not scatter, which means the shader can read values of its neighbors from previous time steps or source images but is limited to writing only to the value of the pixel it is currently processing. This makes certain tasks which require pixel level coherence such as connected component labeling less suitable.

Building Blocks for GPU Programs

Our implementation makes use of the pixel shader through the NVIDIA Cg API and a discussion of some of some useful techniques for formatting and handling data is merited. The GPU has rapid and convenient access to a large pool of (texture) memory, but the transfer of data to and from the CPU need to be explicitly coded. During computation a thread has limited access to memory, including (fragment) registers, some shared memory between threads, and a global memory without collision rules that allows general scatter/gather (read/write) memory operations. The data parallel algorithmic building blocks for GPU programming include map, scatter and gather, parallel reduce, parallel prefix scan, sort and search. Map applies a specified function, for example, local average or convolution operator, to all the elements in an array or stream. In CUDA the kernels specify the map or function which are applied in parallel by all executing threads. Another common GPU program building block is the split operation when many threads need to partition data as in sorting or building trees and the compact operator which can be applied to remove null elements from an array. Each thread can allocate a variable amount of storage per thread which is useful in marching cubes isosurface extraction algorithms and geometry creation. Since the scan operator can be used as the basis for implementing many of the other building blocks, we describe it in more detail.

The scan or parallel prefix sum operator is an efficient multithreaded way to compute partial results to build up the desired output based on all the values in the input stream. Given the array $A = [a_0, a_1, , a_{(n-1)}]$ and a binary associative operator $\oplus$ with identity element I, then, the *exclusive* scan produces the output stream [52],

$$scan(A) = [I, a_0, (a_0 \oplus a_1), , (a_0 \oplus a_1 \oplus \ \oplus a_{n-2})] \tag{3.33}$$

For example, if $\oplus$ is addition, then applying scan to the set $A = [9, 0, 2, 1, 8, 3, 3, 4]$, returns the set $scan(A) = [0, 9, 9, 11, 12, 20, 23, 26]$. Other common binary functions include min, max, logical AND, and logical OR. The scan operator turns out to be a surprisingly fundamental building block that is useful for a variety of parallel algorithms including histograms, run-length encoding, box-filtering (summed area tables), cross-correlation, distance transform, matrix operations, string comparison, polynomial evaluation, solving recurrences, tree operations, radix sort, quicksort, and lexical analysis. An extremely efficient GPU-based unsegmented scan algorithm implementation that uses $O(n)$ space to accomplish $O(N)$ work is given by Harris et al. [53] that was further extended to segmented scans in [52]. Lectures and resource materials on GPU programming can be found at [54, 55].

Data Structures. The GPU memory model is grid or array based for textures rather than linear, and data can be mapped accordingly for maximum efficiency. One-dimensional arrays are limited by the maximum size of a texture (8,192 elements for the NVIDIA GeForce G80 processor) and also suffer from slower lookup times in GPU memory due to the optimization for two-dimensional textures via caching. Large one-dimensional arrays can be packed into two-dimensional textures and the shader program can perform the address translation to access the data correctly. Two-dimensional floating point or integer data arrays map exactly onto textures with the exception of those that exceed the maximum allowable size limit. At that point the array can be split amongst multiple textures and correct indexing is managed by setting multiple rendering targets. Three-dimensional (or higher) arrays pose a difficulty since currently GPUs can read the data as a texture but the GPU cannot directly generate *x/y/z* pixels for the pixel shader to operate upon. Lefohn et al. [56] suggested three solutions: (1) use a 3D texture but only operate on a single texture at a time; (2) store each slice as a separate texture in graphics memory with the CPU activating the appropriate slice; or (3) place all data slices into a single 2D texture allowing for operation on the entire 3D array in a single pass.

Framebuffer Objects. A typical rendering pass sends its results to the framebuffer for display on the screen. Framebuffer Object (FBO) extensions were added to the OpenGL specification to enable more efficient rendering directly to texture memory rather than copying information from the framebuffer. The FBO acts like a pointer to allocated areas of texture memory. A single FBO can be used to manage multiple textures called attachments, and the number of attachments is dependent on the GPU. Prior to executing a shader pass, an FBO attachment is bound, which switches rendering from the screen framebuffer to texture memory. After a rendering pass is completed, the result can be used as input for another shader for further processing.

Parallel Reduction. The pixel shader does not have a mechanism for accumulating a result like a sum or product of elements, or other binary associative operations like minimum or maximum value over an array (texture). A solution is to write the texture to CPU memory and iterate over the array of values, but this

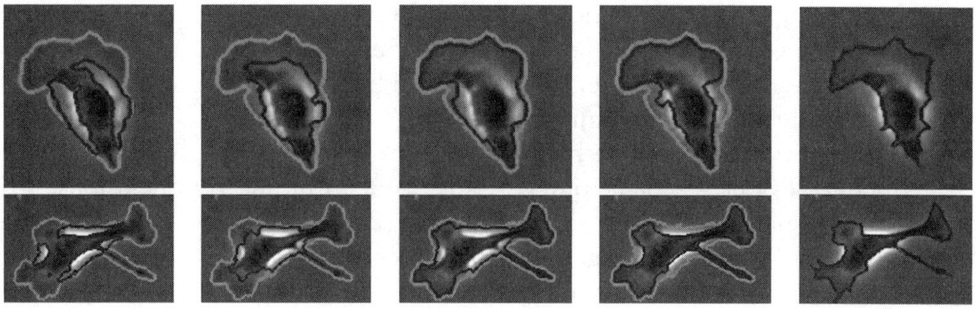

Figure 3.9 Magnified results on sample cells from BAEC dataset. From left to right: Chan and Vese, GAC, GAC with adaptive force, proposed method, and manual ground truth.

is an expensive operation. Using parallel reduction this kind of operation can be performed efficiently on the GPU [57]. The original texture is passed to a shader program parameter and the target rendering texture area is half its size. Every neighborhood of four texels is considered and evaluated for operation of interest. As shown in Figure 3.10 the maximum value of every four texels is written to the target texture. The result is passed back to the shader and the target render area reduced by half again. This operation is performed log N times where N is the width of the original texture and the result is the single maximum value among the texels. A tree-based parallel reduction using the ping-pong method of texture memory access is bandwidth bound. In CUDA, shared memory is used to reduce the bandwidth requirement and with sequential (instead of interleaved) addressing is conflict free. Parallel reduction of N elements requires $\log(N)$ steps that perform the same number of $O(N)$ operations as a sequential algorithm. With P processing cores (P physically parallel threads) the total time complexity is reduced to $O(N/P + \log N)$ compared to $O(N)$ for sequential reduction.

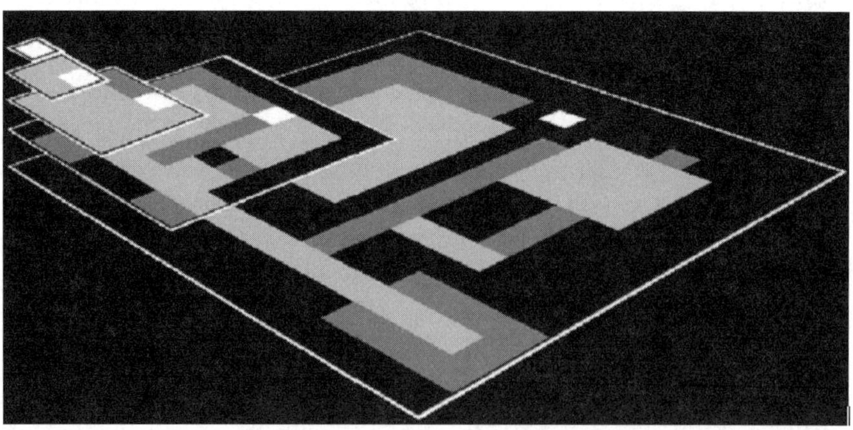

Figure 3.10 An example of using reduction to find the maximum value *n* the texture. The lightest shaded texel from each set of four neighboring texels is kept until the single white texel is isolated.

Level Sets on GPU

Overview. GPU-based implementation of level set algorithms is an active area of research due to the potential for scalable performance with large datasets or real-time requirements [50, 58–60]. Our GPU level set implementation is based on the Chan and Vese 3.34 formulation with geodesic terms as a series of pixel shader programs based on update equation (3.34).

$$\phi^{t+1} = \phi^t + \partial t \delta_\epsilon(\phi) \left[\mu \nabla \cdot g \left(\frac{\nabla \phi}{|\nabla \phi|} \right) - g\nu - \lambda_1 (u_o - c_1)^2 + \lambda_2 (u_o - c_2)^2 \right] \quad (3.34)$$

Each term is computed as a series of pixel shader programs using FBOs to store intermediate results. For the initialization step the source images are loaded into texture memory using *glTexImage2D* and the signed distance grid ϕ^0 is initialized by computing the distance of each pixel from an arbitrary circle contained within the bounds of the grid. Rather than discuss every term, I will focus on a few examples with accompanying shader code.

Computing the Energy Term. The energy term $-\lambda_1(u_o - c_1)^2 + \lambda_2(u_o - c_2)^2$ is computed in $4 + 2\log N$ rendering passes where N is the width of the source image. This term essentially computes the variance of the pixels in the source image that lie inside and outside of the zero level sets embedded in ϕ^t. The first shader computes

Algorithm 1 Moving Object Detection Using Flux Tensor

Input : Infrared image sequence $\mathbf{I}_{IR}(\boldsymbol{x}, t)$, Filters, trace threshold T_{flux}
Output : Foreground mask sequence $\mathrm{FG_M}(\boldsymbol{x}, t)$

// Filters.Dxy : Spatial derivative filter
// Filters.Sxy : Spatial smoothing filter
// Filters.D1t : Temporal derivative filter (first order)
// Filters.D2t : Temporal derivative filter (second order)
// Filters.Axy : Spatial averaging filter
// Filters.At : Temporal averaging filter
$nhDt = \lfloor size(Filters.Dt)/2 \rfloor$
$nhAt = \lfloor size(Filters.At)/2 \rfloor$
for each time t **do**
  $[\mathbf{I}_{xs}(t), \mathbf{I}_{sy}(t), \mathbf{I}_{ss}(t)] \leftarrow$ ComputeIxsIsyIss($\mathbf{I}(t)$, Filters.Dxy, Filters.Sxy)
  FIFO $\leftarrow$ Insert(FIFO, t, $\mathbf{I}_{xs}(t), \mathbf{I}_{sy}(t), \mathbf{I}_{ss}(t)$)
  $t_d = t - nhDt$
  $[\mathbf{I}_{xt}(t_d), \mathbf{I}_{yt}(t_d), \mathbf{I}_{tt}(t_d)] \leftarrow$ ComputeIxtIytItt(FIFO,td,Filters.D1t, Filters.D2t)
  $Trace(t_d) = \mathbf{I}_{xt}(t_d)^2 + \mathbf{I}_{yt}(t_d)^2 + \mathbf{I}_{tt}(t_d)^2$
  $Trace(t_d) =$ SpatialAveraging($Trace(t_d)$, Filters.Axy)
  FIFO $\leftarrow$ Insert(FIFO, t_d, $Trace(t_d)$)
  $t_a = t - nhDt - nhAt$
  $Trace(t_a) =$ TemporalAveraging(FIFO, t_a, *Filters.At*)
  Initialize motion mask, $FG_M(t_a) \leftarrow 0$
  $\mathrm{FG_M}(Trace(t_a) > T_{flux}) \leftarrow 1$
  $\mathrm{FG_M}(t_a) \leftarrow$ PostProcess($\mathrm{FG_M}(t_a)$)
end for

Algorithm 2 Four-Color Level Set Segmentation

Input : A time-varying sequence of N images and constants in (3.18)
Output : A time-varying sequence of N, "four-colored" masks

1: Initialize the segmentation process on the starting image, using level lines [66]. Isolate cells that may be touching or very close to each other to produce the segmentation mask (used in Step 4).
2: **for** $f = 1 \ldots N$ **do**
3: Project the segmentation mask (i.e., converged level sets) from the previous frame as an initial estimate, in order to speed up convergence (see [66, 94]).
4: Extract connected components from the *binary* segmentation mask.
5: Compute an adjacency graph for these connected components.
6: Apply a suitable graph-vertex coloring algorithm (see [95]) to partition the cells into *at most* four independent sets, and produce a *colored* segmentation mask so that neighboring cells belong to different colors
7: Associate one level set function with each mask color, and calculate the signed distance functions for each of the four initial zero level sets (i.e., ϕ_i^0 at evolution time t=0).
8: **for** $i = 1 \ldots K$ iterations **do**
9: Update c_{in} and c_{out}.
10: Evolve the level set within the narrow band of a cell using Euler-Lagrange equations.
11: Enforce an explicit coupling rule that the narrow band of a cell, ϕ_i *cannot overlap* with the level set of any of its neighbors.
12: **end for**
13: Generate a binary mask from the four-color segmentation and apply morphological filtering to remove spurious fragments.
14: Apply a spatially-adaptive level line-based coarse segmentation to the background (i.e., complement of the dilated colored segmentation mask), in order to detect objects not present in the previous frame (i.e., new objects entering the current frame).
15: **end for**

the Chan and Vese formulation of the regularized Heaviside function shown in (3.35) where x is the cell pixel value in ϕ^t and ϵ is a small constant (Algorithm 1).

$$H_\epsilon(x) = \frac{1}{2} + \frac{1}{\pi} tan^{-1}\left(\frac{x}{\epsilon}\right) \tag{3.35}$$

Table 3.2 Computing the Regularized Heaviside Function

```
void computeHeaviside(float2 coords : WPOS,
                      uniform samplerRECT phi,
                      uniform float epsilon,
                      out float4 color : COLOR)
{
   color = float4(0.0, 0.0, 0.0, 0.0);
   const float invpi = 0.318309886;
   float phiVal = texRECT(phi, coords).r;
   float hSide = 0.5 * (1 + (2 * invpi) * atan(phiVal / epsilon));
   color.r = hSide;
   color.g = 1 - hSide;
}
```

Table 3.3 Summing a Data Array

```
void arraySum2Df(float2 coords :WPOS,
                 uniform samplerRECT inputTex,
                 out float4 color : COLOR)
{
  color = float4(0.0, 0.0, 0.0, 1.0);
  float2 topleft = ((coords-0.5)*2.0)+0.5;
  float2 val1 = texRECT(inputTex, topleft).rg;
  float2 val2 = texRECT(inputTex, topleft+float2(1,0)).rg;
  float2 val3 = texRECT(inputTex, topleft+float2(1,1)).rg;
  float2 val4 = texRECT(inputTex, topleft+float2(0,1)).rg;
  color.rg = (val1 + val2 + val3 + val4);
}
```

The next shader performs a reduction operation to compute the sum of H_ϵ and $1 - H_\epsilon$, which roughly equates to the labeling pixels as interior or exterior to the zero level sets which smoothing numeric discontinuities (Algorithm 2).

The following shader reads in the full size texture array but uses a rendering target texture a quarter of its size. For each pixel its up-sampled northwest neighborhood is read from inputTex and the four values are summed. When the result of the Heaviside shader is provided as initial input, the result is the sum of pixels inside and outside the zero level set after $\log N$ iterations.

The rest of the shaders find the average intensities of the source image u_0 inside and outside the zero level set, compute the variance, and finally apply the constants and sum the interior and exterior terms.

Distance Transform. To compute the distance transform over the zero level set we use the jump flooding method described by [61]. They proposed a variation on a flood fill algorithm which provides an inexact result in $\log n$ rendering passes with additional passes resulting in fewer errors until it converges to the result of a standard flood fill. In their follow-up paper they offer improvements and variants on the algorithm to reduce errors and demonstrate its applicability to the 3D image domain.

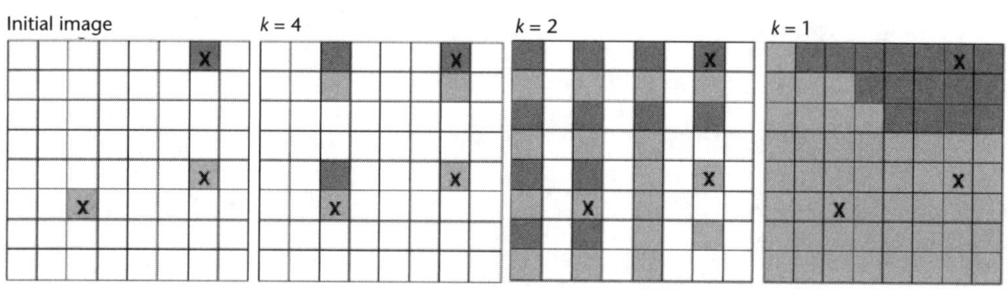

Figure 3.11 Example of performing multiple iterations of the jump flood algorithm. The initial image on the left is composed of three intial seed pixels and the subsequent images show nearest propagated seed pixel for each jump step size k.

Algorithm 3 Compute a 2D Neighborhood Adj. Graph

Input : **P**, a 2D mask with N labeled objects,
T_{LD}, threshold to detect ridges,
T_{HS}, threshold for max. hole size, and
σ, to control smoothing.
Output : $\mathcal{N}(\mathbf{P})$, the adjacency graph of **P**

1: Remove labeled 8-connected pixels in **P** that are adjacent to one or more different labels by applying a "*detach operator*".
2: Convert the processed mask into a binary image $\mathcal{B}$.
3: Compute the Euclidean distance transform (EDT), $\mathcal{D}$, of $\mathcal{B}$ using the FH-EDT algorithm [96].
4: Compute $\mathrm{E} = \nabla^2 G_\sigma \circledast \mathcal{D}$, the Laplacian of the smoothed EDT.
5: Obtain a binary image, E_{thr}, from E using a threshold value T_{LD}.
6: Fill holes using T_{HS}, the hole-size threshold.
7: Apply a suitable thinning algorithm (e.g., [97, 98]) on E_{thr} to obtain an image with 1-pixel thick GVD boundaries, E_{thr}^{thn}.
8: Apply any homotopy-preserving algorithm [99] to prune *branches* from the generalized Voronoi diagram, E_{thr}^{thn}.
9: Assign $\mathcal{Q} \leftarrow (\mathrm{E}_{thr}^{thn})^c$; the complementary image of E_{thr}^{thn}
10: Using 4-connectivity, label the connected components of $\mathcal{Q}$.
11: Update the neighborhood relationship map $\mathcal{N}(\mathbf{P})$ by checking a 3×3 neighborhood of each background pixel (i.e., boundary pixels of connected components) in $\mathcal{Q}$. $\mathcal{N}(\mathbf{P})$ is subsequently used in graph-vertex coloring.

The jump flood distance transform uses the filled pixels of a binary image mask as the seeds for the Euclidean distance calculations. For the initial rendering pass every pixel (x, y) in the image computes the minimum distance from its neighbors $(x \pm k,\ y \pm k)$ where k is initially the half width of the image. The minimum distance and the coordinates of the corresponding seed pixel are stored. For successive rendering passes the k is halved and each pixel repeats the operation but bases its distance calculation on the stored coordinates of the seed pixels propagated in earlier steps. Figure 3.11 demonstrates on an 8×8 image with an initial step size $k = 4$ coordinates of the initial seed pixels are propagated.

Table 3.4 Jump Flood Distance Transform Initialization

```
void initJumpFlood2Df(float2 coords : WPOS,
                      uniform samplerRECT source,
                      out float4 color : COLOR)
{
  float val = texRECT(source, coords).r;

  if(val > 0)
  {
    color = float4(0, coords.x, coords.y, 0);
  }
  else
  {
    color = float4(10000, -1, -1, 0);
  }
}
```

Algorithm 4 Tracking Algorithm

Input : A time-varying sequence of N, "colored" masks
Output : Trajectories and temporal cell statistics

1: For each frame $I(y, t)$ at time t, the tracking module receives four foreground mask layers, $\Omega_k(t)$, that correspond to level sets from the "four-color" segmentation algorithm.
2: Connected component analysis is performed on all refined foreground layers such that each $\Omega_k(t)$ is partitioned into n_k disjoint regions

$$\Omega_k(t) = \{\Omega_{k,1}(t), \Omega_{k,2}(t), \ldots, \Omega_{k}, n_k(t)\}$$

that ideally correspond to n_k individual cells.
3: For each disjoint region $\Omega_{k,i}$, features such as centroid, area, bounding box, and support map are extracted. Region information from all four layers are combined and relabeled as

$$\Omega_i(t), i \in [1 \ldots n], \text{ and } n = \sum_{k=1}^{4} n_k$$

without merging connected regions from different foreground layers. This preserves identities of previously disjoint cells and prevents false trajectory merges. This is similar to the technique described in Step 1 of Algorithm 3.
4: Relabeled region information is arranged in an "Object-Match" graph structure $\mathcal{O}_{\mathcal{GR}}$ that is used in tracking. Nodes in the graph represent objects $\Omega_i(t)$, while edges represent object correspondences.
5: Correspondence analysis searches for potential object matches in consecutive frames. $\mathcal{O}_{\mathcal{GR}}$ is updated by linking nodes that correspond to objects in frame $I(y, t)$ with nodes of potential corresponding objects in frame $I(y, t-1)$. $\mathcal{C}_{\mathcal{M}}(i, j)$, the confidence value for each match is also stored with each link.
6: The trajectory generation module analyzes $\mathcal{O}_{\mathcal{GR}}$ and generates cell trajectories.

Algorithm 5 Distributed Tracking Algorithm

Input : A time-varying sequence of $N-$frames
Output : Trajectories and temporal cell statistics

1: Divide $N-$frame input sequence into M stacks $S_{1 \ldots M}$ with $\{n_1, n_2, \ldots, n_M\}$ frames with k overlapping frames between each stack $(S_i(n_i - k + 1 : n_i) = S_{i+1}(1 : k))$.
2: **for** each stack S_i **do**
3:    Create a parallel job on a separate CPU.
4:    Run tracking algorithm described in Section 3.4 on frames of stack S_i.
5:    Produce trajectory segment lists for stack S_i.
6: **end for**
7: **for** i=2:M **do**
8:    **if** there is a coordinate system change between S_{i-1} and S_i **then**
9:       Register first k frames of S_i to the coordinate system of S_{i-1}.
10:    **end if**
11:    Find correspondences between objects in the last k frames of stack S_{i-1} to objects in the first k frames of stack S_i.
12:    Propagate trajectory labels of objects in stack S_{i-1} to the corresponding objects in S_i.
13:    Propagate the new labels in S_i to their children trajectories.
14: **end for**

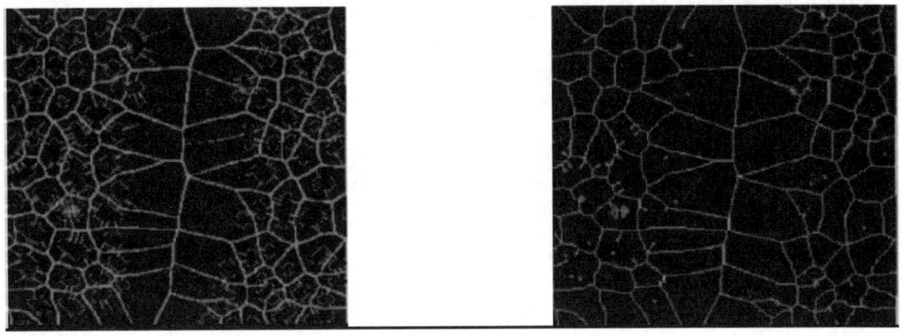

Figure 3.12 The image on the left shows the result of directly applying the Laplacian kernel to the distance transform. The image on the left is the result of thinning which eliminates most of the orphaned pieces.

For our implementation we first generate a binary mask from the zero level set to act as the seed image. For the jump flood distance transform the red, green, and blue texture color channels are used to store the computed distance from its nearest seed pixel, the seed x coordinate and the seed y coordinate. During an initialization step the distance for the seed pixels are set to 0 and their own texture coordinates are set. The nonseed pixels are given a large initial distance and their coordinates are set to (−1, −1) (Algorithm 3).

The second stage of the distance transform applies the description given above as the step size k is halved each rendering pass. Level sets are represented as signed distance fields and the jump flood can be used to produce this result by applying the binary mask used during initialization. After the distance transform is completed the mask is applied and any pixel beneath the mask has its value negated (Algorithm 4).

Generalized Voronoi Diagram. The generalized Voronoi diagram is constructed using the previously computed distance transform by first applying a 3×3 Laplacian kernel to find the local maxima with a threshold of 0.5 (Algorithm 5).

The result shown in Figure 3.12 has orphaned fragments with free ends and loops. To clean up the final result a thinning shader is applied to the image. For

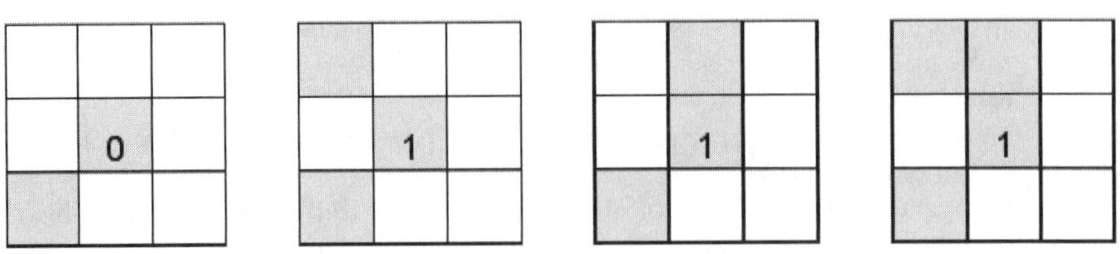

Figure 3.13 A subset of 3×3 structuring elements used to thin the Voronoi diagram.

each pixel, its eight neighbors are assigned a value with the upper left as 0 and the lower right as 128. A bit vector is formed with values from 0 to 255 by setting a bit if the corresponding pixel is filled and clearing it if empty. Figure 3.13 shows 4 out of the 256 possible pixel arrangements. The value of the vector is used to index a lookup table of all possible neighborhoods with the corresponding action. A 0 indicates the pixel should be cleared while a 1 leaves the pixel filled. The image after thinning is applied still contains isolated loops which could be eliminated by examining a larger neighborhood.

3.2.4 Results and Discussion

Figure 3.14 shows sample 4-level sets cell segmentation and GVD results. These sample result corresponds to a wound healing image sequence, obtained from phase-contrast microscopy, consisting of 136 frames of dimensions 300 × 300

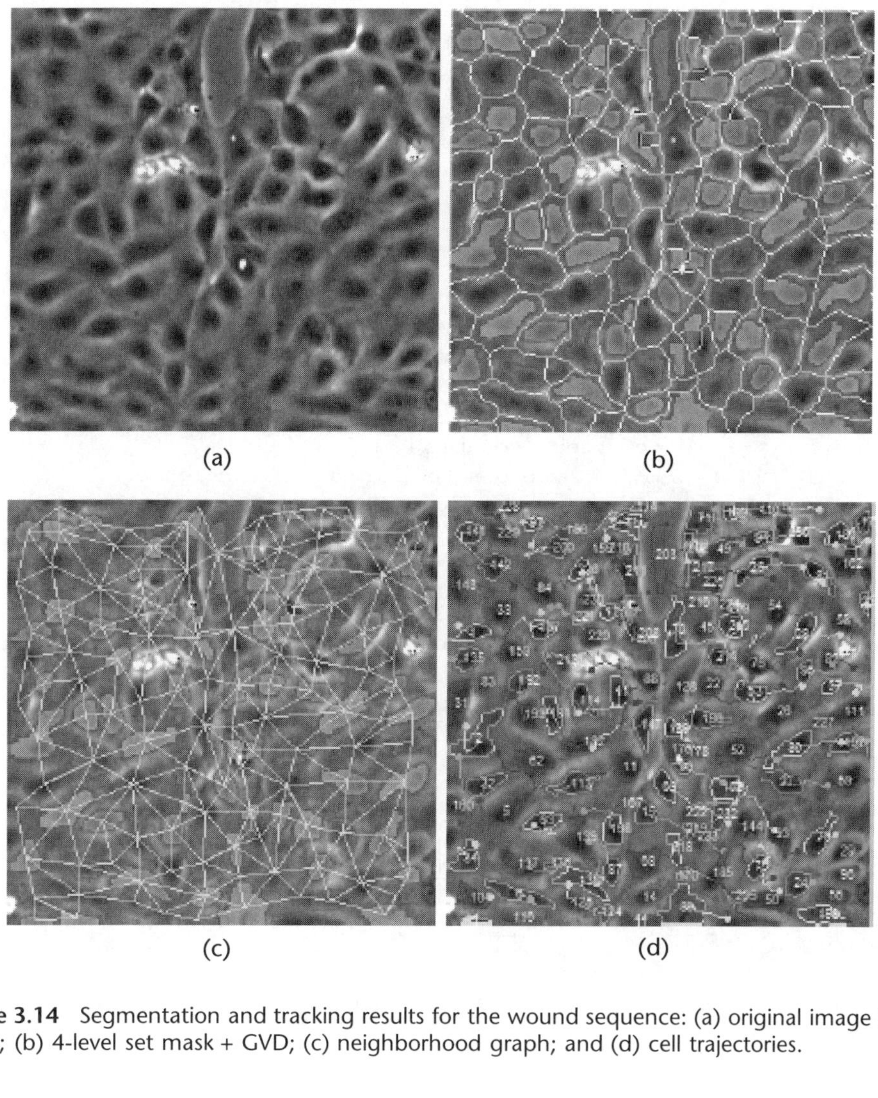

Figure 3.14 Segmentation and tracking results for the wound sequence: (a) original image (frame #135); (b) 4-level set mask + GVD; (c) neighborhood graph; and (d) cell trajectories.

Table 3.5 GPU Level Set Timing Results

Operation	*Timing(%)*
Initialization	0.01
Create masks	2.63
Compute distance transform	34.93
Create GVD	22.92
Redistance ϕ	2.08
Compute regularized dirac	2.51
Compute energy term	18.93
Compute length term	3.80
Compute area term	4.06
Update ϕ	4.98
Render image	3.15

Table 3.6 Jump Flood Distance Transform

```
void jumpFlood2Df(float2 coords : WPOS,
                  uniform float jumpStep,
                  uniform samplerRECT source,
                  out float4 color : COLOR)
{
  float4 currentPixel = texRECT(source, coords);

  //Initialize minimum distance and seed coordinates
  float minDist = currentPixel.r;
  float2 minCoords = currentPixel.gb;

  int stepSign[] = {1,0,-1};

  int counter = 0;

  if( currentPixel.r > 0){
    for(int i = 0; i < 3; i = i + 1){
      for(int j = 0; j < 3; j = j + 1){
        counter = counter + 1;
        // Get coordinates for each jump neighbor
        float2 jumpCoords = float2(coords.x +
          (stepSign[i] * jumpStep),
           coords.y + (stepSign[j] * jumpStep));

        // Get seed pixel coordinate for jump neighbor
        float2 seedCoords = texRECT(source, jumpCoords).gb;

        // Find nearest seed pixel from jump neighbors
        float seedDist = distance(seedCoords, coords);

        if(seedDist < minDist)
        {
         minDist = seedDist;
         minCoords = seedCoords;
        }
      }
    }
  }
  color = float4(minDist, minCoords.x, minCoords.y, 0);
}
```

(40 μm$\times$ 40 μm) with image intensities $I \in [0,255]$. The sequence has been obtained using a monolayer of cultured porcine epithelial cells, as described by Salaycik et al. in [61]. Images were sampled uniformly over a 9:00:48 hour period and acquired using a phase contrast microscope, with a 10$\times$ objective lens, and at a resolution of approximately 0.13 μm per pixel. Figure 3.14(a) shows frame #135 from the original sequence. Figures 3.14(b, c) show 4-level set masks, GVD boundaries, and resulting neighborhood graph.

Some comparative segmentation results for directed edge-profile guided active contours are shown in Figure 3.9. These magnified images correspond to sample Bovine aortic endothelial cells (BAECs; Cambrex East Rutherford, New Jersey) between passages 12 and 16. They were grown in DMEM (Invitrogen, Carlsbad, California) containing 10 streptomycin and kanamycin sulfate (Invitrogen, Carlsbad, California). Cell cultures were maintained in a humidified incubator at 37°C, with 5 every 3 to 4 days. Rat-tail collage I (BD Biosciences, Bedford, Massachusetts) at 0.3 mg/ml in acidic acid was absorbed onto the glass coverslip for 24 hours. Then the coverslip was rinsed in PBS to remove unbound collagen. A glass coverslip in a 35-mm petri dish with 1-ml cell culture medium is placed on an Olympus IX70 inverted microscope. 2 $\times$ 104 cells from cell suspension were added to the petri dish. Images were recorded in phase contrast mode with a CoolSNAPfx digital video camera (Photometrics, Tucson, Arizona), using a 10X objective.

A similar approach has also been used in segmentation of a low density culture of human melanoma cell line WM793 [23]. The proposed method was applied

Table 3.7 Laplacian Kernel

```
void laplacianFilter2Df(float2 coords : WPOS,
                        uniform samplerRECT img,
                        uniform float threshold,
                        out float4 color : COLOR)
{
  //3x3 Laplacian
  // -1 -1 -1
  // -1  8 -1
  // -1 -1 -1

  float sum = 0;
  sum += texRECT(img, float2(coords.x, coords.y + 1)).r;
  sum += texRECT(img, float2(coords.x - 1, coords.y + 1)).r;
  sum += texRECT(img, float2(coords.x + 1, coords.y + 1)).r;
  sum += texRECT(img, float2(coords.x + 1, coords.y)).r;
  sum += texRECT(img, float2(coords.x - 1, coords.y)).r;
  sum += texRECT(img, float2(coords.x, coords.y - 1)).r;
  sum += texRECT(img, float2(coords.x + 1, coords.y - 1)).r;
  sum += texRECT(img, float2(coords.x - 1, coords.y - 1)).r;

  float center = texRECT(img, coords).r * 8;
  if((center - sum) > threshold)
    color = float4(1,0,0,0);
  else
    color = float4(0,0,0,0);
```

to human melanoma cells. The local image variance and the Hamming texture measure are chosen as the robust features for this type of cells since the cells have visibly different variance and texture from the background whereas intensity does not constitute a discriminating feature. Figure 3.1 shows a sample frame from one field in the T25 plastic culture flask of a control experiment with a low density culture of human melanoma cell line WM793 to assess any toxicity to the highthroughput imaging system [3]. Pixel resolution is 0.67 micron 0.59 micron in X and Y using a 20 objective lens.

The GPU level set implementation ran approximately 3 to 15 times faster than a narrow band level set CPU implementation depending on the dimensions of the input images. However, it is quite difficult to make a valid comparison between different implementations on the basis of hardware platforms. Many novices to GPU programming have assumed they would immediately get large speed improvements simply by porting their code to a graphics processor with little regard for the type of problem or the quality of their equipment. As an analysis of the GPU level set implementation, Table 3.5 shows the percentage of total time taken by each operation within an iteration of the level set process. The three most expensive steps are computing the distance transform, creating the generalized Voronoi diagram, and computing the energy term. These operations represent steps that require extra processing to compute results not entirely suited for the GPU. The jump flood distance transform takes log N (where N is the width of a the side of a square image) rendering passes but still has to consider N^2 pixels each pass compared to CPU image based, such as by distance transforms by Dannielsson [62] and Felzenzwalb and Huttenlocher [63], which complete in linear time. The number of parallel stream processors and rapid memory access offset the cost, but this becomes less effective for increasingly larger images. The energy term requires computing the sum of all pixels within texture arrays twice using parallel reduction. While this method is efficient for the GPU it still requires reading every pixel in the initial step and then a quarter of the number of pixels for each succesive step. Obviously the CPU equivalent is a single for loop over an array of values once. The Voronoi diagram computation expense is accrued during the thinning step to remove spurs and orphaned fragments, but the standard implementation does not suffer in comparision to a typical CPU operation.

3.3 Cell Tracking

Tracking is a fundamental step in the analysis of long-term behavior of migrating cells. Various studies have been reported in the literature highlighting the importance of cell migration analysis such as: [64] in understanding drug treatments effects on cancer cells, [65] in assessing aggressiveness of cancer models, [66, 67] in study of inflammatory diseases, [68] in wound closure (which is important in embryonic development and tissue repair), [69] in quantification of protein activation, and other studies related to migrating cells [70, 71].

A majority of cell migration studies fall into one of the following classes:

- Cell motility through optic-flow analysis;

- Cell tracking through contour evolution;
- Cell tracking by detection and explicit correspondence analysis.

Optic flow methods provide dense motion flows, but not trajectories for individual objects, which is needed for the study of long-term behavior of individual cells. Active contour-based approaches provide trajectory information and are widely used in tracking biological objects with the underlying assumption of small displacements in objects (see [28, 66]). Generally, active contour tracking uses contour from the previous frame as an initial estimate for the current frame and evolves this estimate using velocity field, intensity boundaries, and some additional constraints such as shape priors (see [67]), or by processing 2D image slices taken over time as a spatiotemporal volume and applying 3D active contour segmentation [72].

In this section we present a detection-based cell tracking algorithm that extends our previous work in [2, 22, 73]. Detection-based tracking enables us to efficiently control each cell association and handle large amounts of displacements. Tracking is achieved by resolving frame-to-frame correspondences between multiple cells that are segmented as described in Section 3.3.

Recently, two similar tracking algorithms were presented by Al-Kofahi et al. [74] and by Li et al. [75]. In [74], tracking is implemented to analyze cell lineage. Cells are segmented using a combination of adaptive thresholding and watershed transformation, and tracked using explicit association analysis. Integer programming is employed to find optimal cell associations from a subset of possible associations. In Li et al. [75], active contours (with an auxiliary label function) are used as the first stage for tracking. Graph matching is subsequently implemented on cells and tracks that are unmatched in the first stage. In our approach, graph matching is used for all cells and tracks, without evolving any auxiliary label function. We can use graph matching for all the cells without suffering from matching ambiguities, thanks to the prevention of false merges during 4-color level set segmentation (Section 3.3.1), and the multistage overlap distance $\mathcal{D}_{\mathcal{MOD}}$ during tracking.

A detailed description is presented in Algorithm 4. Correspondence analysis and trajectory generation are key components of the tracking algorithm presented above and are further elaborated in the following subsections.

3.3.1 Cell-to-Cell Temporal Correspondence Analysis

Cell-to-cell temporal correspondence analysis consists of four major steps:

1. Object-to-object distance computation;
2. Match confidence matrix construction;
3. Absolute match pruning;
4. Bidirectional relative match pruning.

Appropriate and efficient object-to-object match distance computation is a key component of any tracking process. In the presented method, the distances between two cells Ω_i and Ω_j in consecutive frames are computed using a multistage overlap distance $\mathcal{D}_{\mathcal{MOD}}$ (3.36), which consists of three distinct distance functions $\mathcal{D}_{bbx}$,

$\mathcal{D}_{msk}$, and $\mathcal{D}_{olp}$ for three different ranges of cell motion. The distance selection is based on topological spatial relations between objects in consecutive frames, defined similar to the relation in Randell et al.'s region connection calculus (RCC) [76, 77] and Egenhofer et al.'s 9-intersection model [78, 79]. The correspondence analysis is largely based on proximity because intensity, texture, and shape are not distinguishing features for cells since they share the same appearance.

$$\mathcal{D}_{\mathcal{MOD}}(\Omega_i, \Omega_j) = \begin{cases} \mathcal{D}_{bbx} & \text{if} \quad \text{bbx}(\Omega_i) \cap \text{bbx}(\Omega_j) = \emptyset \\ \mathcal{D}_{msk} & \text{if} \quad \Omega_i \cap \Omega_j = \emptyset \\ \mathcal{D}_{olp} & \text{otherwise} \end{cases} \tag{3.36}$$

The *interbounding-box distance* $\mathcal{D}_{bbx}$ (3.37) quantifies long-range displacement between two regions (cells) in consecutive frames by the distance between their bounding boxes denoted as bbx().

$$\mathcal{D}_{bbx}(\Omega_i, \Omega_j) = K_{bbx} \times \sqrt{\mathcal{D}_{\mathcal{I}}(\text{bbx}_x(\Omega_i), \text{bbx}_x(\Omega_j))^2 + \mathcal{D}_{\mathcal{I}}(\text{bbx}_y(\Omega_i), \text{bbx}_y(\Omega_j))^2} \tag{3.37}$$

where $\text{bbx}(\Omega) \equiv < inf_x(\Omega), sup_x(\Omega), inf_y(\Omega), sup_y(\Omega) >$ is the minimum bounding box of a region; $\text{bbx}_x(\Omega) \equiv [inf_x(\Omega), sup_x(\Omega)]$ and $\text{bbx}_x(\Omega) \equiv [inf_y(\Omega), sup_y(\Omega)])$ are intervals obtained by projecting the bounding box $\text{bbx}(\Omega)$ onto x and y axes; $\mathcal{D}_{\mathcal{I}}$ is interval distance defined in (3.38); and K_{bbx} is a constant. In the cases of overlapping bbx_x or bbx_y intervals, $\mathcal{D}_{\mathcal{B}}$ corresponds to the minimum edge-to-edge distance; otherwise it corresponds to minimum corner-to-corner distance [Figure 3.15(a)].

$$\mathcal{D}_{\mathcal{I}}(\mathcal{I}_A, \mathcal{I}_B) = \begin{cases} 0 & \text{if } \mathcal{I}_A \cap \mathcal{I}_B \neq \emptyset \\ min\{|\mathcal{I}_A(1) - \mathcal{I}_B(2)|, |\mathcal{I}_B(1) - \mathcal{I}_A(2)|\} & \text{otherwise} \end{cases} \tag{3.38}$$

The *intermask distance* $\mathcal{D}_{msk}$ quantifies mid-range displacement between two regions (cells) in consecutive frames, by the minimum contour-to-contour dis-

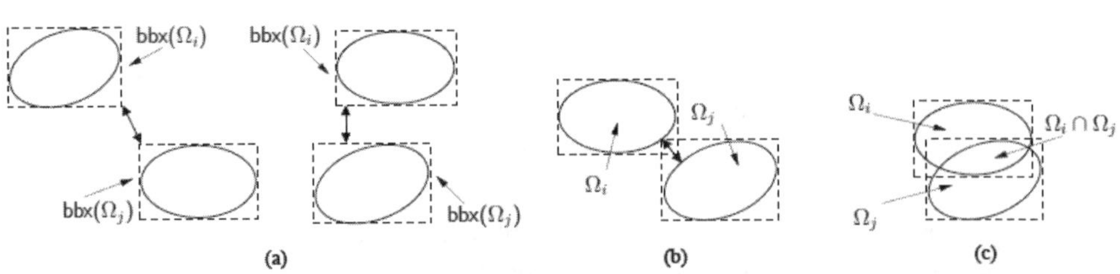

Figure 3.15 Object-to-object distance $\mathcal{D}_{\mathcal{MOD}}$: (a) $\mathcal{D}_{bbx}$: interbounding-box distance; (b) $\mathcal{D}_{msk}$: intermask distance; and (c) $\mathcal{D}_{olp}$: tonal-weighted overlap distance.

tance. This is computed in terms of the minimum number of morphological dilations needed to overlap the two regions [Figure 3.15(b)]. $\mathcal{D}_{msk}$ is defined as

$$\mathcal{D}_{msk}(\Omega_i, \Omega_j) = K_{msk} \times inf\{k : \delta_k(\Omega_i) \cap \Omega_j \neq 0 \ ; \ \Omega_i \cap \delta_k(\Omega_j) \neq 0\} \tag{3.39}$$

where δ_k denotes k-times dilation with a unit structuring element, and K_{msk} is a constant. $\mathcal{D}_{msk}$ is closely related to Hausdorff distance [80] (Figure 3.16), redefined in [81] using morphological operators as

$$H(A, B) = inf\{k : A \subseteq \delta_k(B) \ ; \ B \subseteq \delta_k(A)\} \tag{3.40}$$

The *tonal-weighted overlap distance* $\mathcal{D}_{olp}$ quantifies small-range displacement between two regions/cells [Figure 3.15(c)] by the degree of their overlap, in terms of shape *and* tonal dissimilarities. In order to emphasize overlap in nuclei (i.e., dark regions with low intensity values), and to de-emphasize cytoplasm overlap (i.e., light regions with high intensity values), overlapping and nonoverlapping regions are weighted by local tonal differences, as

$$\mathcal{D}_{olp}(\Omega_i, \Omega_j) = K_{olp} \times; \frac{\int_{\Omega_i \setminus \Omega_j} (1 - I_i(y))dy + \int_{\Omega_j \setminus \Omega_i} (1 - I_j(y))dy + \int_{\Omega_i \cap \Omega_j} |I_i(y) - I_j(y)|dy}{\int_{\Omega_i} I_i(y)dy + \int_{\Omega_j} I_J(y)dy} \tag{3.41}$$

where the intensity images, $I_i(y) = I_i(y,t)$ and $I_j(y) = I_j(y,t-1)$, are scaled such that $I \in [0,1]$, and K_{olp} is a constant. The first two terms in the numerator of (3.41) account for the distance due to uncovered regions in frames at time instants t and $t-1$, respectively. The complement of intensity images are used to obtain higher distances for uncovered low intensity regions (i.e., nuclei). The third term in the numerator accounts for the intensity dissimilarity within the overlapping region.

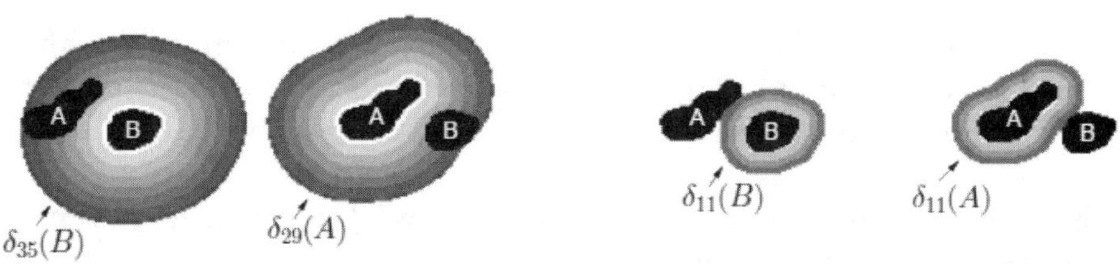

(a) $A \subseteq \delta_{35}(B)$ (b) $B \subseteq \delta_{29}(A)$ (c) $A \cap \delta_{11}(B) \neq 0$ (d) $\delta_{11}(A) \cap B \neq 0$

Figure 3.16 Hausdorff (a, b) and $\mathcal{D}_{\mathcal{M}}$ (c, d) distances for two cells A and B. $H(A, B) = min(35, 29)$, $\mathcal{D}_{\mathcal{M}} = min(11, 11)$.

The denominator is used to normalize the distance by the area of the two cells being compared.

This multistage distance measure depends on size and shape similarity of the compared regions, besides their proximity, and thus have several advantages over the widely used centroid distance measure, particularly for mitosis (cell split) cases. During mitosis, epithelial cells often become elongated and split across the minor axis. This produces a big increase in the centroid distance, and for dense cell populations the distances between a cell and its children become comparable to the distances between a cell and its neighboring cells. In such a scenario, a low gating threshold would result in parent-to-children matches being discarded, resulting in discontinuities in cell trajectories. However, if a high gating threshold is used, correspondence ambiguities may arise. Multistage overlap distance measure overcomes these problems. Depending on the appearance and motion characteristics of the cells being tracked, match distance can be modified to incorporate other features (i.e., shape, color, texture, and so forth).

Using the above information, for each frame $I(y,t)$ of the image sequence, a match matrix $\mathcal{M}$, and a confidence matrix $\mathcal{C}_{\mathcal{M}}$ are constructed. $\mathcal{M}(\Omega_i,\Omega_j)$ indicates whether the ith object in $I(y,t)$, corresponds to the jth object in $I(y,t-1)$. All the elements of $\mathcal{M}$ are initially set to one indicating a *match*. $\mathcal{C}_{\mathcal{M}}(\Omega_i,\Omega_j)$ indicates the confidence of this match and consists of weighted sum of two components:

- *Similarity confidence*, $\mathcal{C}_{\mathcal{SIM}}(\Omega_i,\Omega_j)$, a measure of the similarity between the matched objects, defined as

$$\mathcal{C}_{\mathcal{SIM}}(\Omega_i,\Omega_j) = 1 - \frac{\mathcal{D}_{\mathcal{MOD}}(\Omega_i,\Omega_j)}{\mathcal{D}^{\max}_{\mathcal{MOD}}} \tag{3.42}$$

  where the constant $\mathcal{D}^{\max}_{\mathcal{MOD}}$ is used to normalize $\mathcal{D}_{\mathcal{MOD}}$.
- *Separation confidence*, $\mathcal{C}_{\mathcal{SEP}}(\Omega_i,\Omega_j)$, a measure of the competition between possible matches for the current object, defined as

$$\mathcal{C}_{\mathcal{SEP}}(\Omega_i,\Omega_j) = \begin{cases} 1 & \text{no competitor,} \\ 0.5 - \dfrac{\left(\mathcal{D}_{\mathcal{MOD}}(\Omega_i,\Omega_j) - \mathcal{D}_{\mathcal{MOD}}(\Omega_i,\Omega_j^*)\right)}{2 \times \max\left(\mathcal{D}_{\mathcal{MOD}}(\Omega_i,\Omega_j), \mathcal{D}_{\mathcal{MOD}}(\Omega_i,\Omega_j^*)\right)} & \text{otherwise} \end{cases} \tag{3.43}$$

  where, Ω_j indicates the current candidate being compared with Ω_i, and Ω_{j^*} is its closest competitor in terms of distance. This measure favors matches without competitors, and matches with competitors having higher distances.

Final correspondences are determined after two types of match pruning, absolute and relative. During absolute pruning, matches whose match confidences are lower than a priori defined threshold T_A are pruned by setting corresponding elements in match matrix $\mathcal{M}$ to 0:

$$\mathcal{C}_{\mathcal{M}}(\Omega_i,\Omega_j) < T_A \Rightarrow \mathcal{M}(\Omega_i,\Omega_j) = 0 \quad \Omega_i \in I(y,t), \Omega_j \in I(y,t-1) \tag{3.44}$$

During backward relative pruning, for each cell $\Omega_i \in I(y,t)$, best matching cell $\Omega_j^* \in I(y,t-1)$ is determined. Matches for Ω_i, whose match confidences are lower than $T_R \times \mathcal{C}_{\mathcal{M}}(\Omega_i,\Omega_j^*)$, where T_R is a constant, are pruned:

$$\mathcal{C}_{\mathcal{M}}(\Omega_i,\Omega_j) < T_R \times \mathcal{C}_{\mathcal{M}}(\Omega_i,\Omega_j^*)) \Rightarrow \mathcal{M}(\Omega_i,\Omega_j) = 0 \quad \Omega_i \in I(y,t), \Omega_j \in I(y,t-1) \tag{3.45}$$

During forward relative pruning, the same relative pruning process is applied in the direction $I(y,t-1) \rightarrow I(y,t)$ by determining for each cell $\Omega_j \in I(y,t-1)$, best matching cell $\Omega_i^* \in I(y,t)$, and pruning accordingly.

3.3.2 Trajectory Segment Generation

Information on detected cells (centroid, area, shape, support map, and intensity distribution) and their temporal correspondences (obtained from match matrices $\mathcal{M}$ for each frame) are arranged in a graph structure ObjectMatchGraph $\mathcal{O}_{\mathcal{GR}}$ (Figure 3.17). Nodes in $\mathcal{O}_{\mathcal{GR}}$ represent cells and associated features, while edges represent frame-to-frame cell correspondences and associated match confidence values. The segment generation module traces links on $\mathcal{O}_{\mathcal{GR}}$ to generate trajectory segments. A multihypothesis testing approach with delayed decision is used. Rather than picking a single best match or some subset of the matches with limited information, many possible matches are kept and gradually pruned, as more information becomes available in the decision process. Besides one-to-one object matches, this scheme supports many-to-one, one-to-many, many-to-many, one-to-none, or none-to-one matches that may result from false detections, or false associations, segmentation errors, occlusion, entering, exiting, or dividing of cells. The segment generation module has three main steps:

1. *Node Classification:* The cells/objects in each frame (nodes in the ObjectMatchGraph $\mathcal{O}_{\mathcal{GR}}$) are classified into nine types based on the number of

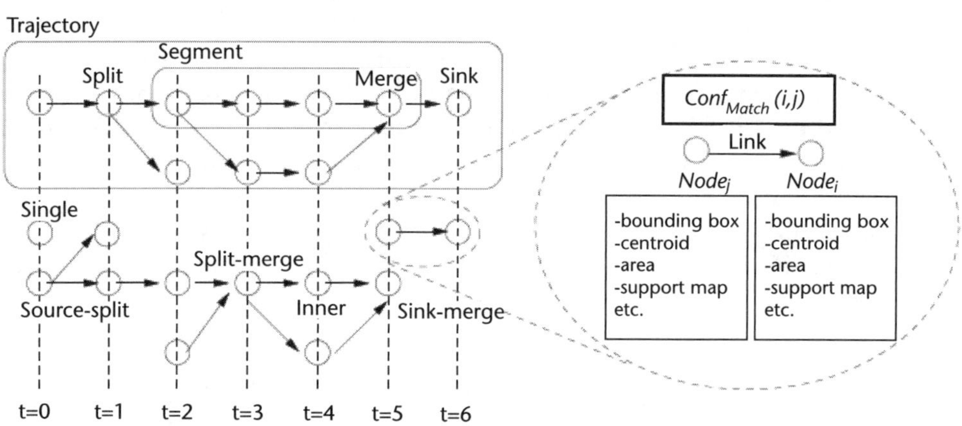

Figure 3.17 ObjectMatchGraph used in cell tracking. Nodes represent detected cells and associated features, while links represent frame-to-frame cell correspondences and associated confidence values.

parent and child objects: *single* (no parent-no child), *source* (no parent-one child), *source-split* (no parent-many children), *sink* (one parent-no child), *inner* (one parent-one child), *split* (one parent-many children), *sink-merge* (many parents-no child), *merge* (many parents-one child), *merge-split* (many parents-many children) (Figure 3.16).

2. *Segment Generation: Trajectory segments* are formed by tracing the nodes of the ObjectMatchGraph *ObjectMatch Graph*. A linked list of inner nodes (objects/cells), starting with a source or split type node and ending with a merge or sink type node, are identified and organized in a data structure called *SegmentList*.
3. *Segment Labeling:* Extracted trajectory segments are labeled using a method similar to connected component labeling. Each segment without a parent is given a new label. Segments that have parents inherit the parents' labels. In case of multiple parents, if the parents' labels are inconsistent, then the smaller label (older trajectory) is kept and a flag is set indicating the inconsistency. Trajectories are formed by joining segments sharing the same label.

Cell Trajectory Validation and Filtering

Factors such as vague cell borders, DNA unwinding during cell division, noise, fragmentation, illumination or focus change, clutter, and low contrast cause segmentation and association problems, which result in missing or spurious trajectory segments and false trajectory merges or splits. Filtering is performed at various levels of processing to reduce the effects of such factors. At the *image level*, small objects that may be due to noise, background clutter, or imaging and segmentation artifacts are removed by using morphological operators. At the *segmentation level*, use of coupling (Section 3.3.1) prevents false merges. At the *correspondence level*, unfeasible cell-to-cell matches with distances above a threshold are eliminated through gating. At the *confidence level*, matches with low confidence values are pruned with absolute and relative prunings. A final validation and filtering module analyzes trajectory segments using accumulated evidence such as temporal persistence, size consistency, and area ratios; and to take appropriate action (i.e., removal or merge) for spurious or incorrectly splitted trajectory segments.

Figure 3.14(d) shows individual cell trajectories for a sample wound healing image sequence. Segmentation of the cells [shown in Figure 3.14(b, c)] is accomplished by 4-level sets cell segmentation described in Section 3.3.1. Once trajectories are obtained various statistics can be obtained for individual or group of cells. Besides trajectories, the tracking module supplies information on trajectory events such as merges, splits, and disappearances. These events are useful for determining cell lineage and for computing mitosis (cell split) and apoptosis (cell death) rates.

3.3.3 Distributed Cell Tracking on Cluster of Workstations

With the rapid increase in the amounts of data generated by biomedical applications, high-performance computing systems are expected to play an increased role. Various high-performance computing strategies exist using systems ranging from accelerators (or many-core processors) such as graphics processing units (GPUs), cell broadband engines (Cell BEs), field-programmable gate arrays (FPGAs), to cluster computers. An intensive study of the parallel computing research landscape can be found in [82]. A large set of parallel computing case studies dealing with bioinformatics and computational biology related problems can be found in [83].

For high-throughput analysis of cell behaviors, our tracking algorithm (Algorithm 4) is parallelized for cluster computers using parallel and distributed job creation capabilities of MATLAB Distributed Computing Toolbox [84]. A cluster computer is a group of computers generally connected through fast local area networks that work together closely so that in many respects they can be viewed as though they are a single computer [85]. The approach is designed for the University of Missouri Bioinformatics Consortium (UMBC) cluster system [86], but can be adapted to other cluster computers. The UMBC cluster system consists of 128 compute nodes and 512 processors. Each node has 4 or 6 GB of memory and is attached to 50 TB of disk using an Infiniband high-speed interconnect infrastructure for both processor and storage access.

Parallel MATLAB

MATLAB is a widely used mathematical computing environments in technical computing. Very high-level languages (VHLLs) like MATLAB provide native support for complex data structures, comprehensive numerical libraries, and visualization along with an interactive environment for execution, editing, and debugging, resulting in codes short, simple, and readable [87]. Recently, there has been an increasing interest in developing a parallel MATLAB. In a 2005 article Ron Choy and Alan Edelman give an extensive survey of projects working on parallelization of MATLAB by providing communication routines (MPI/PVM), routines to split up work among multiple MATLAB sessions, parallel backend, or by compiling MATLAB scripts into native parallel code. They group and study 27 such projects (including pMatlab, MatlabMPI, MultiMATLAB, and Matlab Parallel Toolbox), some of which are no longer in use. Several recent applications use these parallel MATLAB tools. In [88], Krishnamurthy et al. concentrate on parallel MATLAB techniques with an emphasis on signal and image processing. In [89], Riviera et al. explore parallelization of 3D imaging applications using two approaches: parallelization on GPUs and parallelization on a cluster of multiple multicore processors using MATLAB and Star-P (a parallel implementation of the MATLAB programming language). In [87], Gilbert et al. describe their efforts to build a common infrastructure for numerical and combinatorial computing by using parallel sparse matrices to implement parallel graph algorithms using MATLAB and Star-P.

In our application, we use Mathworks' MATLAB Distributed Computing Toolbox [84]. MATLAB Distributed Computing Toolbox supports both data-parallel and task-parallel applications by use of a set of construct such as parallel for-loops, distributed arrays, message-passing functions, and distributed job creation commands. When used with MATLAB Distributed Computing Server, parallel MATLAB programs can be scaled to cluster computers.

Distributed Cell Tracking

Typically computation or memory extensive biomedical image processing applications involve iterative operations, images with high spatial resolution, or long video sequences. Particularly for the last two cases, parallelization by data partitioning can greatly improve performance. In the case of cell tracking, we chose to partition the data in time. Our decision is motivated by: (1) ease of partitioning, (2) degree of task independence (amount of interprocess communication), and (3) amount of code adaptation efforts needed. Distributed cell tracking module consists of two submodules:

1. Intrastack tracking module;
2. Trajectory stitching or consistent labeling module.

The original N-frame sequence is divided into M stacks with k overlapping frames. Objects/cells in each stack are tracked using our tracking algorithm (Algorithm 4) described in Section 3.4. Tracking of cells in each stack is performed independently in parallel using parallel and distributed job creation capabilities of MATLAB Distributed Computing Toolbox [84]. The major steps of the process are given below, where a task corresponds to the tracking of the cells in the assigned stack:

1. Find a job manager.
2. Create a job.
3. Create tasks.
4. Submit the job to the job queue.
5. Retrive job's results.

Each task produces trajectory segment lists. Beside object level information such as object IDs, centroids, sizes, and support maps, each trajectory segment contains IDs of their parents, IDs of their children, and a label indicating the trajectory to which they belong. Since cells in each stack are tracked independently, cell trajectory labels are not consistent in time. The *Trajectory Stitching/Consistent Labeling module* is a serial process that propagates trajectory labels from the previous stack to consistently relabel trajectory segments in the current stack. This process ensures that trajectory segments belonging to the same cell consistently have the same label in time, independent of how many stacks they cross. This problem is similar to the consistent labeling problem in multicamera tracking systems [90, 91]. Our trajectory stitching/consistent labeling process described in Algorithm 5 and illustrated in Figure 3.18 is originally designed for moving camera surveillance applications, and uses both geometry-based and appearance-based object matching and can handle changing field of views. For cell tracking, only single field of view geometry-based object matching is used. For each stack S_i trajectory segments that start within the first k frames T_i^S and trajectory segments that end within the last k frames T_i^E are identified. To consistently label (or stitch) trajectory segments across stacks, correspondences of the trajectory segments T_{i-1}^E to T_i^S are determined by matching the last instances of the objects in T_{i-1}^E to the first instances of the objects in T_i^S. For the general case both appearance and geometry are used in object matching, but for the case of static camera with temporally overlapping stacks, simple proximity

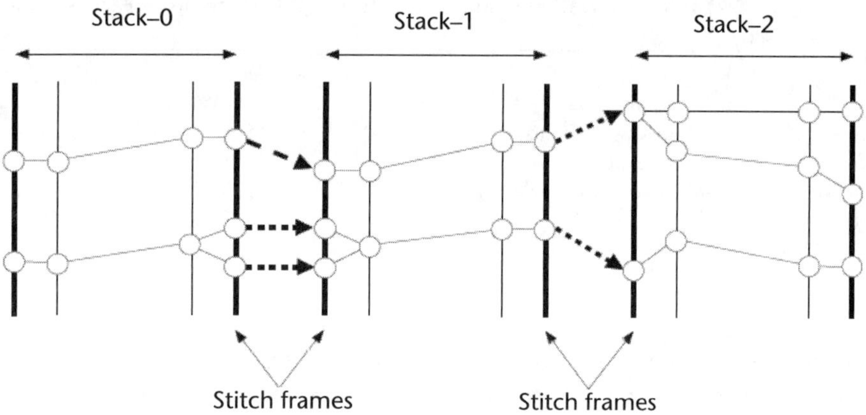

Figure 3.18 Multistack trajectory stitching.

based matching is adequate. Once correspondences are identified, T_{i-1}^E labels are propagated to the corresponding segments in T_i^S and to their children recursively. Temporal stack overlap amount k is typically set to 1 but may be set to $k \geq 1$ depending on trajectory discontinuity tolerance (i.e., temporal loss of the object due to incorrect segmentation, occlusion, and so forth).

3.3.4 Results and Discussion

Many factors—such as length of the image sequence, size of the frames, density and size of the tracked objects—affect the execution time of the tracking program. Here we show performance of our parallel cell tracking program on two sample cell sequences with different characteristics shown in Table 3.8.

Seq-1 is a wound healing image sequence with relatively small 300×300 (40 μm × 40 μm) images but with high density of cells. The sequence has been obtained from phase-contrast microscopy using a monolayer of cultured porcine epithelial cells, as described by Salaycik et al. in [61]. Images were sampled uniformly over a 9:00:48 hour period and acquired using a phase contrast microscope, with a 10× objective lens, and at a resolution of approximately 0.13 μm per pixel. The original sequence consists of 136 frames, but for testing purposes, it has been replicated into a four times longer 544-frame sequence using a forward-backward-forward-backward pattern. Segmentation of the cells is done using our coupled 4-level set cell segmentation program described in Section 3.3.1. Seq-2 is a sample

Table 3.8 Properties of the Image Sequences Used in Parallel Tracking Performance Evaluation

	Seq-1	*Seq-2*
Sequence type	Wound healing	Human Melanoma
Image size	300 × 300	672 × 512
# Frames	544	395
# Objects detected	59,564	6,021
Average density	109 object/frame	15 object/frame

Table 3.9 Tracking Timing Results for 100 Frames from Each Sample Sequence

	Seq-1	*Seq-2*
1-Initializations	1.08%	3.87%
2-Object feature extraction	15.40%	75.21%
3-Correspondence analysis	73.91%	4.29%
4-Trajectory generation & analysis	1.56%	1.83%
5-Saving results	8.05%	14.80%

human melanoma cell sequence with lower density of cells but with larger 672 × 512 images. The pixel resolution is 0.67 μm × 0.59 μm and the images are obtained with a 20× objective lens. The sequence consists of 395 frames from one field in the T25 plastic culture flask of a control experiment with a low density culture of human melanoma cell line WM793 [92]. Cells are detected using flux tensors as described in Section 3.2.1. Segmentation is refined using the multifeature geodesic active contour level set algorithm as described in [46].

The differences in the characteristics of the two sample test sequences are reflected in the timing of the tracking program (Table 3.9). For Seq-1, which contains a higher number of objects per frame, the majority of the tracking time (73.91%) is spent in Step 3, correspondence analysis. Since the processed images are smaller in size, image processing done to compute object features takes relatively less time (15.40%). For Seq-2, which contains a smaller number of objects in larger images, the majority of the tracking time is spent in image processing done in Step 2, object feature extraction. Trajectory generation and analysis takes a low percentage of the tracking time for both sequences (1.56% and 1.83%) because this step consists of graph operations and does not involve expensive image processing operations. For both sequences, saving results takes considerable amounts of time (8.05% and 14.80%) because for further analysis, besides trajectories we save a number of object features including support maps and intensity distributions, which require

Table 3.10 Timing Results for Parallel Tracking Program

		# Tasks				
		1	*4*	*8*	*16*	*32*
Seq-1	$Task_{min}$	618.70	158.63	67.88	34.16	20.22
	$Task_{avg}$	618.70	159.00	83.19	44.35	24.99
	$Task_{max}$	618.70	159.81	99.17	56.05	30.89
	Stitch	0	2.01	2.65	3.84	6.44
	Total	618.70	161.82	101.82	59.89	37.33
	Speedup	1	3.82	6.07	10.33	16.57
Seq-2	$Task_{min}$	146.65	28.17	16.88	11.43	9.14
	$Task_{avg}$	146.65	38.36	22.30	14.31	10.31
	$Task_{max}$	146.65	51.76	29.63	18.10	12.96
	Stitch	0	1.68	1.63	1.73	2.23
	Total	146.65	53.44	31.26	19.83	15.19
	Speedup	1	2.74	4.69	7.39	9.65

large amounts of IO. If no further analysis will be done, by limiting the output to trajectories and trajectory related features, time spent in Step 5 can be greatly reduced.

Timing results and parallelization performance of the tracking algorithm on the two sample test sequences Seq-1 and Seq-2 are given in Table 3.10. As described

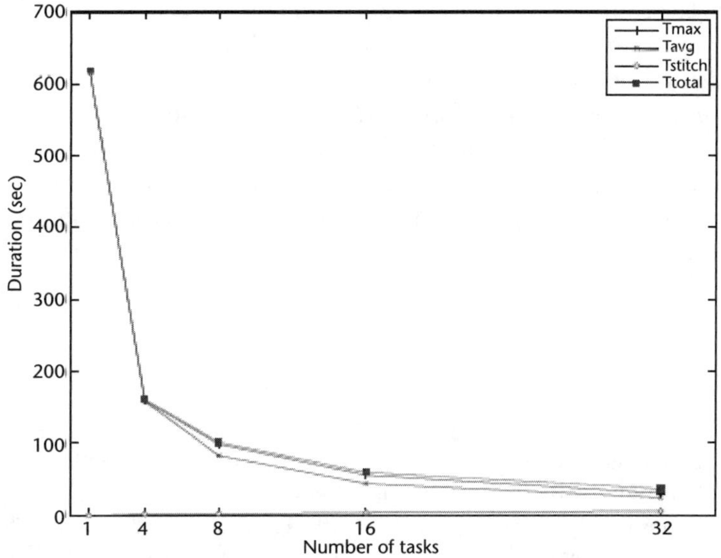

(a)

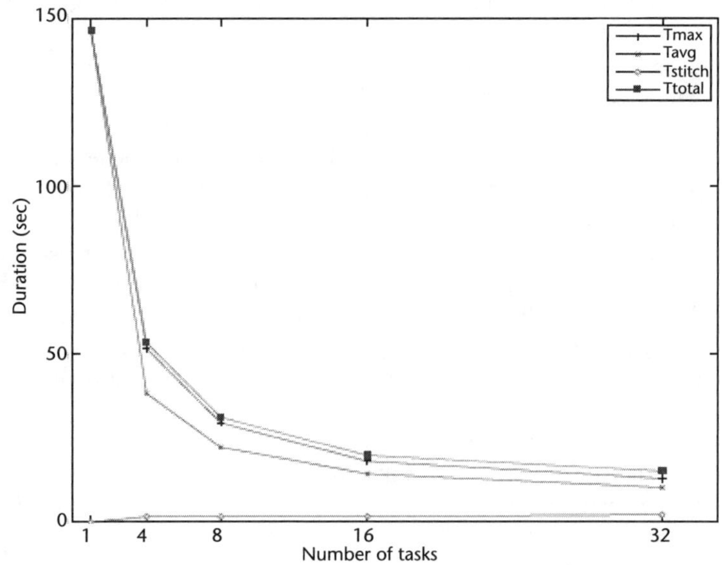

(b)

Figure 3.19 Timing of the parallel tracking algorithm for sample sequences: (a) Seq-1; and (b) Seq-2 (y-axes are scaled differently).

in Section 3.4.3, the original sequence is divided into M stacks, tracking of the objects in each stack is assigned to a task, and tasks are run independently in parallel on a cluster computer. Consistent labeling of the trajectories in time is ensured with a sequential stack-to-stack stitching step. Since stack-to-stack consistent trajectory labeling (or stitching) involves correspondence analysis on only $2 \times k \times (M - 1) << N$ frames and since label propagation does not involve computationally expensive image processing operations, as seen in Table 3.10, the sequential stitching part does not introduce a considerable overhead. As can be observed from the performance differences between Seq-1 and Seq-2, higher workloads result in higher efficiencies (Speedup/#Tasks). The increase in speedup slows down for both sequences for 32 processors; this result is expected since both of the sample jobs are quite small (619 and 147 seconds) to begin with; for larger jobs the slowdown is expected for a larger number of tasks. With a small amount of code adaptation efforts required (compared to accelerators such as GPUs, Cell BEs, FPGAs), coarse grain parallelization on cluster computers is a good candidate for data intensive jobs.

References

[1] S. Nath, K. Palaniappan, and F. Bunyak, "Cell segmentation using coupled level sets and graph-vertex coloring," in *LNCS - Proc. MICCAI 2006* (R. Larsen, M. Nielsen and J. Sporring, eds.), Springer-Verlag, 2006.

[2] F. Bunyak, K. Palaniappan, S.K. Nath, T.I. Baskin, and G. Dong, "Quantitive cell motility for *in vitro* wound healing using level set-based active contour tracking," *Proc. 3rd IEEE Int. Symp. Biomed. Imaging (ISBI)*, (Arlington, VA), April 2006.

[3] A. Hafiane, F. Bunyak, and K. Palaniappan, "Clustering initiated multiphase active contours and robust separation of nuclei groups for tissue segmentation," *IEEE ICPR'08*, December 2008.

[4] I. Ersoy, F. Bunyak, K. Palaniappan, M. Sun, and G. Forgacs, "Cell spreading analysis with directed edge profile-guided level set active contours," *LNCS - Proc. MICCAI 2008*, 2008.

[5] K. Palaniappan, H.S. Jiang, and T.I. Baskin, "Non-rigid motion estimation using the robust tensor method," in *IEEE Comp. Vision. Patt. Recog. Workshop on Articulated and Nonrigid Motion*, vol. 25, (Washington, DC), 2004.

[6] S. Nath and K. Palaniappan, "Adaptive robust structure tensors for orientation estimation and image segmentation," *LNCS-3804: Proc. ISVC'05*, (Lake Tahoe, Nevada), pp. 445–453, Dec. 2005.

[7] H. Nagel and A. Gehrke, "Spatiotemporally adaptive estimation and segmentation of OF-Fields," *LNCS-1407: ECCV98*, vol. 2, (Germany), pp. 86–102, June 1998.

[8] B. Horn and B. Schunck, "Determining optical flow," *Artificial Intell.*, vol. 17, pp. 185–203, Aug. 1981.

[9] K. Palaniappan, H. Jiang, and T. Baskin, "Non-rigid motion estimation using the robust tensor method," *IEEE Comp. Vision and Patt. Recog. Workshop on Articulated and Nonrigid Motion*, (Washington, DC), June 2004.

[10] J. Zhang, J. Gao, and W. Liu, "Image sequence segmentation using 3-D structure tensor and curve evolution," vol. 11, pp. 629–641, May 2001.

[11] H. Scharr, "Optimal filters for extended optical flow," *LNCS: First Int. Workshop on Complex Motion*, vol. 3417, (Berlin, Germany), pp. 66–74, Springer-Verlag, Oct. 2004.

[12] H. Scharr, I. Stuke, C. Mota, and E. Barth, "Estimation of transparent motions with physical models for additional brightness variation," *13th European Signal Processing Conference, EUSIPCO*, 2005.

[13] E. Dejnozkova and P. Dokladal, "Embedded real-time architecture for level set-based active contours," *EURASIP Journal on Applied Signal Processing*, no. 17, pp. 2788–2803, 2005.

[14] X. Wang, W. He, D. Metaxas, R. Matthew, and E. White, "Cell segmentation and tracking using texture-adaptive snakes," *Proc. 4th IEEE Int. Symp. Biomed. Imaging (ISBI)*, pp. 101–104, *IEEE Comp. Soc.*, April 2007.

[15] M. Kass, A. Witkin, and D. Terzopoulos, "Snakes: Active contour models," *Journal International Journal of Computer Vision*, vol. 1, pp. 321–331, January 1988.

[16] L. D. Cohen, "On active contour models and balloons," *Computer Vision, Graphics, and Image Processing. Image Understanding*, vol. 53, no. 2, pp. 211–218, 1991.

[17] V. Caselles, F. Catte, T. Coll, and F. Dibos, "A geometric model for active contours," *Numerische Mathematik*, vol. 66, pp. 1–31, 1993.

[18] R. Malladi, J. Sethian, and B. Vemuri, "Shape modelling with front propagation: A level set approach," *IEEE Trans. Patt. Anal. Mach. Intell.*, vol. 17, no. 2, pp. 158–174, 1995.

[19] V. Caselles, R. Kimmel, and G. Sapiro, "Geodesic active contours," *Int. Journal of Computer Vision*, vol. 22, no. 1, pp. 61–79, 1997.

[20] J.A. Sethian, *Level Set Methods and Fast Marching Methods: Evolving Interfaces in Computational Geometry, Fluid Mechanics, Computer Vision, and Materials Science*, Cambridge, UK: Cambridge University Press, 1999.

[21] C. Xu, A. Yezzi, and J. Prince, "A summary of geometric level-set analogues for a general class of parametric active contour and surface models," *Proc. IEEE Workshop on Variational and Level Set Methods in Computer Vision*, pp. 104–111, 2001.

[22] S. Nath, F. Bunyak, and K. Palaniappan, "Robust tracking of migrating cells using four-color level set segmentation," *LNCS - Proc. ACIVS 2006* (J. Blanc-Talon, D. Popescu, W. Philips and P. Scheunders, ed.), vol. 4179-0920, pp. 920–932, Springer-Verlag, 2006.

[23] I. Ersoy, F. Bunyak, M.A. Mackey, and K. Palaniappan, "Cell segmentation using Hessian-based detection and contour evolution with directional derivatives," 2008.

[24] T. Chan and L. Vese, "Active contours without edges," *IEEE Trans. Image Proc.*, vol. 10, pp. 266–277, Feb. 2001.

[25] B. Zhang, C. Zimmer, and J.C. Olivio-Marin, "Tracking fluorescent cells with coupled geometric active contours," *Proc. 2nd IEEE Int. Symp. Biomed. Imaging (ISBI)*, pp. 476–479, Arlington, VA: IEEE Comp. Soc., April 2004.

[26] L. Vese and T. Chan, "A multiphase level set framework for image segmentation using the Mumford and Shah model," vol. 50, no. 3, pp. 271–293, 2002.

[27] A. Dufour, V. Shinin, S. Tajbakhsh, N.G. Aghion, J.C. Olivio-Marin, and C. Zimmer, "Segmenting and tracking fluorescent cells in dynamic 3-D microscopy with coupled active surfaces," vol. 14, pp. 1396–1410, September 2005.

[28] C. Zimmer and J.C. Olivio-Marin, "Coupled parametric active contours," vol. 27, pp. 1838–1842, Nov. 2005.

[29] K. Appel and W. Haken, "Every planar map is four colorable. Part I. discharging," *Illinois. J. Math.*, vol. 21, pp. 429–490, 1977.

[30] K. Appel, W. Haken, and J. Koch, "Every planar map is four colorable. Part II. reducibility," *Illinois. J. Math.*, vol. 21, pp. 491–567, 1977.

[31] N. Robertson, D.P. Sanders, P.D. Seymour, and R. Thomas, "The four color theorem," *J. Combin. Theory, Ser. B*, vol. 70, pp. 2–44, 1997.

[32] V. Caselles, R. Kimmel, and G. Sapiro, "Geodesic active contours," vol. 22, no. 1, pp. 61–79, 1997.

[33] S.K. Nath and K. Palaniappan, "Adaptive robust structure tensors for orientation estimation and image segmentation," *Proc. 1st Int. Symp. Visual Computing (ISVC)* (G. Bebis, R. Boyle, D. Koracin and B. Parvin, ed.), vol. 3804, pp. 445–453, Lake Tahoe, NV: Springer, 2005.

[34] C. Li, C. Xu, C. Gui, and D. Fox, "Level set evolution without re-initialization: A new variational formulation," *IEEE Comp. Soc.*, vol. 1, pp. 430–436, 2005.

[35] S. Kim and H. Lim, "A hybrid level set segmentation for medical imagery," in *Proc. IEEE Nucl. Sci. Symp.*, vol. 3, pp. 1790–1794, IEEE Nucl. Plasma. Sci. Soc., 2005.

[36] R.F. Gonzalez, M.H. Barcellos-Hoff, and C. de Solórzano, "A tool for the quantitative spatial analysis of complex cellular systems," vol. 14, pp. 1300–1313, September 2005.

[37] C. Indermitte, T. Liebling, M. Troyanov, and H. Clemencon, "Voronoi diagrams on piecewise flat surfaces and an application to biological growth," *Theo. Comp. Sci.*, vol. 263, no. 1-2, pp. 263-274, 2001.

[38] S. Keenan, J. Diamond, W. McCluggage, H. Bharucha, D. Thompson, P. Bartels, and P. Hamilton, "An automated machine vision system for the histological grading of cervical intraepithelial neoplasia (CIN)," *J. Pathol.*, vol. 192, no. 3, pp. 351-362, 2000.

[39] J. Geusebroek, A.W.M. Smeulders, F. Cornelissen, and H. Geerts, "Segmentation of tissue architecture by distance graph matching," *Cytometry*, vol. 35, no. 1, pp. 11–22, 1999.

[40] B. Weyn, G. Wouwer, S.K. Singh, A. Daele, P. Scheunders, E. Marck, and W. Jacob, "Computer-assisted differential diagnosis of malignant mesothelioma based on syntactic structure analysis," *Cytometry*, vol. 35, no. 1, pp. 23–29, 1999.

[41] H. Choi, T. Jarkrans, E. Bengtsson, J. Vasko, K. Wester, P. Malmström, and C. Busch, "Image analysis based grading of bladder carcinoma. Comparison of object, texture and graph based methods and their reproducibility," *Anal. Cell. Path.*, vol. 15, no. 1, pp. 1–18, 1997.

[42] A. Klemenčič, F. Pernuš, and S. Kovačič, "Segmentation of muscle fibre images using voronoi diagrams and active contour models," vol. 3, pp. 538–542, *IEEE Comp. Soc.*, 1996.

[43] A. Okabe, B. Boots, K. Sugihara, and S.N. Chiu, *Spatial Tesselations: Concepts and Applications of Voronoi Diagrams*, 2nd Edition, West Sussex, UK: John Wiley & Sons Ltd., 2000.

[44] F. Aurenhammer, "Voronoi diagrams - A survey of a fundamental geometric data structure," *ACM Comp. Surveys*, vol. 23, no. 3, pp. 345–405, 1991.

[45] *Lecture Notes in Computer Science (SCIA 2007)*, vol. 4522, 2007.

[46] I. Ersoy and K. Palaniappan, "Multi-feature contour evolution for automatic live cell segmentation in time lapse imagery," *IEEE EMBC'08*, 2008.

[47] K. Palaniappan, I. Ersoy, and S. K. Nath, "Moving object segmentation using the flux tensor for biological video microscopy," *Lecture Notes in Computer Science (PCM 2007)*, vol. 4810, 2007, pp. 483–493.

[48] F. Bunyak, K. Palaniappan, S. K. Nath, and G. Seetharaman, "Flux tensor constrained geodesic active contours with sensor fusion for persistent object tracking," *Journal of Multimedia*, August 2007.

[49] P. Yan, X. Zhou, M. Shah, and S. Wong, "Automatic segmentation of high throughput RNAi fluorescent cellular images," vol. 12, no. 1, pp. 109–117, 2008.

[50] J. D. Owens, D. Luebke, N. Govindaraju, M. Harris, J. Krger, A. E. Lefohn, and T. Purcell, "A survey of general-purpose computation on graphics hardware," *Computer Graphics Forum*, vol. 26, no. 1, pp. 80–113, 2007.

[51] I. Buck, T. Foley, D. Horn, J. Sugerman, K. Fatahalian, M. Houston, and P. Hanrahan, "Brook for GPUs: stream computing on graphics hardware," *ACM Transactions on Graphics*, vol. 23, no. 3, pp. 777–786, 2004.

[52] S. Sengupta, M. Harris, Y. Zhang, and J. Owens, "Scan primitives for GPU computing," *Proceedings of the 22nd ACM SIGGRAPH EUROGRAPHICS Symposium on Graphics Hardware*, pp. 97-106, 2007.

[53] M. Harris, S. Sengupta, and J. Owens, "Parallel prefix sum (scan) with CUDA," in *GPU Gems 3* (H. Nguyen, ed.), Addison Wesley, 2007.

[54] http://www.seas.upenn.edu/~cis665/Schedule.htm.

[55] http://www.gpgpu.org.

[56] A. Lefohn, J. Kniss, and J. Owens, "Implementing efficient parallel data structures on GPUs," in *GPU Gems 2* (M. Pharr, ed.), pp. 521–545, Addison-Wesley, 2005.

[57] M. Harris, "Mapping computational concepts to GPUs," in *GPU Gems 2* (M. Pharr, ed.), pp. 493–508, Addison-Wesley, 2005.

[58] A. Lefohn, J. Kniss, C. Hansen, and R. Whitaker, "A streaming narrow-band algorithm: Interactive deformation and visualization of level sets," *IEEE Transactions on Visualization and Computer Graphics*, vol. 10, pp. 422–433, July/August 2004.

[59] P. Labatut, R. Keriven, and J.-P. Pons, "A GPU implementation of level set multiview stereo"

[60] Y. Shi and W. Karl, "A real-time algorithm for the approximation of levelset-based curve evolution," *IEEE Transactions on Image Processing*, vol. 17, pp. 645–656, May 2008.

[61] G. Rong and T. Tan, "Jump flooding in GPU with applications to voronoi diagram and distance transform," *ACM Symposium on Interactive 3D Graphics and Games*, pp. 109–116, 2006.

[62] P. Danielsson, "Euclidean distance mapping," *Computer Graphics and Image Processing*, vol. 14, pp. 227–248, 1980.

[63] P. Felzenszwalb and D. Huttenlocher, "Distance transforms of sampled functions,"

[64] X. Chen, X. Zhou, and S.T.C. Wong, "Automated segmentation, classification, and tracking of cancer cell nuclei in time-lapse microscopy," vol. 53, pp. 762–766, April 2006.

[65] C. Hauwer, F. Darro, I. Camby, R. Kiss, P. Ham, and C. Decaestecker, "In vitro motility evaluation of aggregated cancer cells by means of automatic image processing," *Cytometry*, vol. 36, no. 1, pp. 1–10, 1999.

[66] D.P. Mukherjee, N. Ray, and S.T. Acton, "Level set analysis for leukocyte detection and tracking," vol. 13, pp. 562–672, April 2001.

[67] N. Ray and S.T. Acton, "Motion gradient vector flow: An external force for tracking rolling leukocytes with shape and size constrained active contours," vol. 23, pp. 1466–1478, Dec. 2004.

[68] R. Farooqui and G. Fenteany, "Multiple rows of cells behind an epithelial wound edge extend cryptic lamellipodia to collectively drive cell-sheet movement," *J. Cell Science*, vol. 118, pp. 51–63, Jan. 2005.

[69] F. Shen, L. Hodgson, A. Rabinovich, O. Pertz, K. Hahn, and J.H. Price, "Functional proteometrics for cell migration," *Cytometry Part A*, vol. 69A, pp. 563–572, June 2006.

[70] O. Debeir, P.V. Ham, R. Kiss, and C. Decaestecker, "Tracking of migrating cells under phase-contrast video microscopy with combined mean-shift processes," vol. 24, pp. 697–711, June 2005.

[71] X. Ronot, A. Doisy, and P. Tracqui, "Quantitative study of dynamic behavior of cell monolayers during *in vitro* wound healing by optical flow analysis," *Cytometry*, vol. 41, no. 1, pp. 19–30, 2000.

[72] D. Padfield, J. Rittscher, N. Thomas, and B. Roysam, "Spatio-temporal cell cycle phase analysis using level sets and fast marching methods," *Proc. 1st Work. Micro. Imag. Anal. App. Bio. (MIAAB)* (D. Metaxas, R. Whitaker, J. Rittscher and T. Sebastian, eds.).

[73] F. Bunyak and S.R. Subramanya, "Maintaining trajectories of salient objects for robust visual tracking," *Proc. 2nd Int. Conf. Image Anal. Recog. (ICIAR)* (M.S. Kamel and A.C. Campilho, eds.), vol. 3212, pp. 820–827, Toronto, ON: Springer, 2005.

[74] O. Kofahi, R. Radke, S. Goderie, Q. Shen, S. Temple, and B. Roysam, "Automated cell lineage construction: A rapid method to analyze clonal development established with murine neural progenitor cells," *Cell Cycle*, vol. 5, no. 3, pp. 327–335, 2006.

[75] K. Li, E.D. Miller, L.E. Weiss, P.G. Campbell, and T. Kanade, "Online tracking of migrating and proliferating cells imaged with phase-contrast microscopy," *Proc. IEEE Comp. Vis. Patt. Recog. Workshop (CVPRW-06)*, pp. 65–72, IEEE Comp. Soc., 2006.

[76] D. A. Randell, Z. Cui, and A. Cohn, "A spatial logic based on regions and connection," *KR'92. Principles of Knowledge Representation and Reasoning: Proceedings of the Third International Conference* (B. Nebel, C. Rich, and W. Swartout, eds.), pp. 165–176, San Mateo, California: Morgan Kaufmann, 1992.

[77] A. G. Cohn and N. M. Gotts, "A theory of spatial regions with indeterminate boundaries," in *Topological Foundations of Cognitive Science* (C. Eschenbach, C. Habel, and B. Smith, eds.), 1994.

[78] M. Egenhofer and R. Franzosa, "Point-set topological spatial relations," *International Journal for Geographical Information Systems*, vol. 5, no. 2, pp. 161–174, 1991.

[79] E. Clementini, J. Sharma, and M. J. Egenhofer, "Modeling topological spatial relations: strategies for query processing," *Computers and Graphics*, vol. 6, pp. 815–822, 1994.

[80] D. Huttenlocher, D. Klanderman, and A. Rucklige, "Comparing images using the Hausdorff distance," *IEEE Transactions on Pattern Analysis and Machine Intelligence*, vol. 15, pp. 850–863, September 1993.

[81] J. Serra, "Hausdorff distances and interpolations," in *Mathematical morphology and its Applications to Image and Signal Processing (Proc. ISMM'98)*.

[82] K. Asanovic, R. Bodik, B. C. Catanzaro, J. J. Gebis, P. Husbands, K. Keutzer, D. A. Patterson, W. L. Plishker, J. Shalf, S. W. Williams, and K. A. Yelick, "The landscape of parallel computing research: A view from Berkeley," Tech. Rep. UCB/EECS-2006-183, EECS Department, University of California, Berkeley, Dec. 2006.

[83] A. Y. Zomaya, *Parallel Computing for Bioinformatics and Computational Biology, (Wiley Series on Parallel and Distributed Computing)*, Wiley-Interscience, 2005.

[84] "The mathworks distributed computing toolbox," http://www.mathworks.com/products/distribtb/index.html.

[85] "Computer cluster," http://en.wikipedia.org/wiki/Computer_cluster.

[86] "University of Missouri Bioinformatics Consortium (UMBC)," http://umbc.rnet.missouri.edu.

[87] J. Gilbert, V. Shah, and S. Reinhardt, "A unified framework for numerical and combinatorial computing," *Computing in Science & Engineering*, vol. 10, pp. 20–25, March-April 2008.

[88] A. Krishnamurthy, J. Nehrbass, and S. S. J.C. Chaves, "Survey of parallel matlab techniques and applications to signal and image processing," *IEEE International Conference on Acoustics, Speech and Signal Processing (ICASSP 2007)*, vol. 4, April 2007.

[89] D. Rivera, D. Schaa, M. Moffie, and D. R. Kaeli, "Exploring novel parallelization technologies for 3-d imaging applications," *International Symposium on Computer Architecture and High Performance Computing*, pp. 26–33, 2007.

[90] A. Gilbert and R. Bowden, "Incremental, scalable tracking of objects inter camera," *Computer Vision and Image Understanding*, vol. 111, p. 4358, 2008.

[91] S. Calderara, A. Prati, and R. Cucchiara, "Hecol: Homography and epipolarbased consistent labeling for outdoor park surveillance," *Computer Vision and Image Understanding*, vol. 111, pp. 21–42, 2008.

[92] P. Davis, E. Kosmacek, Y. Sun, F. Ianzini, and M. Mackey, "The large scale digital cell analysis system: An open system for non-perturbing live cell imaging," *J. Microscopy*, vol. 228, no. 3, p. 296308, 2007.

[93] N. Paragiois and R. Deriche, "Geodesic active contours and level sets for the detection and tracking of moving objects," vol. 22, pp. 266–280, March 2000.

[94] E.J. Cockayne and A.J. Thomason, "Ordered colourings of graphs," *J. Comb. Theory - Ser. B*, vol. 32, pp. 286–292, 1982.

[95] P. Felzenswalb and D. Huttenlocher, "Distance transforms of sampled functions," Tech. Rep. TR2004-1963, Dept. of Comp. Sci., Cornell University, Ithaca, NY, Sept. 2004.

[96] A. Rosenfeld, "A characterization of parallel thinning algorithms," *Inform. Control*, vol. 29, pp. 286–291, 1975.

[97] J.M. Cychosz, "Efficient binary image thinning using neighborhood maps," in *Graphics Gems IV*, pp. 465–473, San Diego, CA: Academic Press Professional, Inc., 1994.

[98] L. Lam, S. Lee, and C.Y. Suen, "Thinning methodologies A comprehensive survey," vol. 14, pp. 869–885, Sept. 1992.

CHAPTER 4

Utilizing Parallel Processing in Computational Biology Applications

John Wagner and Kirk Jordan

4.1 Introduction

In this chapter, we illustrate the use of parallel computing in computational biology [1] using an example from computational oncology, specifically, the modeling of solid tumor invasion. Tumor formation and growth are complex dynamical processes that depend on many diverse processes, including accumulated genetic alterations of tumor cell chromosomes and the interactions between tumor cells and their tissue environments [2]. Genetic mutations alter the cell's ability to respond to normal homeostatic controls, leading to increased proliferation, migration, and lifespan [3]. As these cells migrate and proliferate, a small mass, or tumor, develops, fed by extracellular nutrients diffusing into the tumor. This growth continues until the tumor size exceeds the ability of the tissue to supply the cells with sufficient nutrients (approximately 10^6 tumor cells); at this point, the tumor cells begin recruiting blood vessels to increase the supply of nutrients, a process that is required for continued tumor growth.

Large scale mathematical and computational models offer one avenue for quantifying and understanding the effects of gene expression and the microenvironment on tumor growth and morphology [4–6]. As a result, numerous mathematical models have been developed to study quantitatively the mechanisms underlying this complex interplay between tumor cells and their environment. One class of these models utilizes a hybrid approach [7–9], whereby the dynamics of the tissue enviroment are described by reaction-diffusion equations, and tumor cells are represented as discrete entities with either simple rules [9, 10] or continuous equations governing their state and interaction with the tissue, including tumor cell migration within the tissue. While these models differ in their descriptions of the tissue and tumor cells, as well as their dynamics and interactions, all can be fit into a single computational framework.

The utility of such models in a clinical setting is limited, however, by the large computational requirements. Given that tumors can grow as large as $O(1)$ to $O(10)$ cm, with tumor cell diameters in the range of $O(10)$ to $O(100)$ μm, the number of cells comprising a tumor can grow to between $O(10^9)$ and $O(10^{12})$. Moreover, assuming a tumor grows over the course of many months, if not years, with a typical cell division time of 8 to 24 hours, such simulations must compute as many as $O(100)$ to $O(1{,}000)$ generations. As a result, high-performance, distributed computing techniques are necessary for simulating clinically relevant solutions of hybrid models of tumor formation and growth.

Here we discuss an MPI [11] and OpenMP [12] implementation of a published model of solid tumor invasion [10] on both the Blue Gene/L and Blue Gene/P supercomputers [13] using nearest neighbor communication tuned to the physical network topology. On Blue Gene/L our timing studies show weak scaling out to a clinically relevant problem size [$O(10^9)$ tumor cells] on 8,192 processors. Moreover, for the same per node workload on Blue Gene/P, our implementation shows strong scaling across all four cores. Our results demonstrate that clinically relevant simulations using hybrid models of tumor invasion are possible with a real-time to simulation-time factor of approximately 100.

4.2 Algorithms

In this Chapter, we consider hybrid mathematical models of tumor growth comprised of a tissue environement and indivudual tumor cells, each described by an internal state governed by either discrete rules or continuous equations. As a concrete example, we consider a published model that incorporates simple rules for tumor state to explore how the selective pressure of the tissue environment determines tumor morphology and phenoytypic characteristics [10]. This model for a vascularized solid tumor includes the following components and mechanisms:

- *Tumor cells*, described formally by cell density $C(t, x, y, z)$ but tracked individually, can migrate, divide and die, and contain state that determines how they interact with their tissue environment and other tumor cells.
- Diffusible *oxygen*, denoted by $O(t, x, y, z)$, required for tumor cell survival and migration, is delivered by the vasculature, modeled simply as proportional to the level of extracellular matrix proteins, and consumed at both a basal rate and by tumor cells.
- Diffusible *matrix degradative enzymes*, denoted by $M(t, x, y, z)$, include the plasminogen activator system and the matrix metalloproteinases, and are either produced or activated by tumor cells, and degraded at a basal rate.
- Nondiffusible *extracellular matrix proteins*, denoted by $E(t, x, y, z)$, that are degraded by matrix degradative enzymes, include the collagens, laminin, fibronectin and vitronectin, and anchor cells in the tissue and regulate cell-cell adhesion.

This model decomposes the tissue into cubes of size Δs^3 which are just large enough to contain a single tumor cell. Consistent with this, we assume the model holds on a domain $[0, T] \times [0, X] \times [0, Y] \times [0, Z]$ which can discretized into $N \times I \times J \times K$ volumes such that $T = N\Delta t$, $X = I\Delta s$, $Y = J\Delta s$, and $Z = K\Delta s$. Tumor cell density and tissue concentrations are then approximated, using cell density as an example, $C^n_{i,j,k} \approx C(t_n, x_i, y_j, z_k)$, where $t_n = n\Delta t$, $x_i = (i - 0.5)\Delta s$, $y_j = (j - 0.5)\Delta s$, and $z_k = (k - 0.5)\Delta s$; and $n = 0[1]\ N$, $i = 1[1]\ I$, $j = 1[1]\ J$, and $k = 1[1]K$.

4.2.1 Tumor Cell Migration

Tumor cell migration is assumed to result from two processes, haptotaxis—directed migration of cells due to chemical gradients, in this case, the extracellular matrix proteins—and diffision,

$$\frac{\partial C}{\partial t} = -\nabla \cdot (\mathbf{J}_\mathrm{h} + \mathbf{J}_\mathrm{d}) = -H_\mathrm{c}\nabla \cdot (C\nabla E) + D_\mathrm{c}\nabla^2 C \tag{4.1}$$

where H_c and D_c are the haptotaxic and diffusion coefficients, respectively. Note that the haptotaxis coefficient, H_c, varies from point to point depending upon the state of the tumor cell at that location. Defining $\lambda = \Delta t/\Delta s^2$, this equation can be written in a discreet, stochastic representation,

$$\begin{aligned} C^{n+1}_{i,j,k} = {} & P_{0,0,0}C^n_{i,j,k} + P_{-1,0,0}C^n_{i-1,j,k} + P_{+1,0,0}C^n_{i+1,j,k} + P_{0,-1,0}C^n_{i,j-1,k} \\ & + P_{0+1,0}C^n_{i,j+1,k} + P_{0,0,-1}C^n_{i,j,k-1} + P_{0,0,+1}C^n_{i,j,k+1} \end{aligned} \tag{4.2}$$

where the coefficients which are proportional to the movement probabilities are given by

$$P_{-1,0,0} = \lambda\left[D_\mathrm{c} + H_\mathrm{c}\left(E^n_{i+1,j,k} - E^n_{i-1,j,k}\right)/6\right] \tag{4.3}$$

$$P_{+1,0,0} = \lambda\left[D_\mathrm{c} - H_\mathrm{c}\left(E^n_{i+1,j,k} - E^n_{i-1,j,k}\right)/6\right] \tag{4.4}$$

$$P_{0,-1,0} = \lambda\left[D_\mathrm{c} + H_\mathrm{c}\left(E^n_{i,j+1,k} - E^n_{i,j-1,k}\right)/6\right] \tag{4.5}$$

$$P_{0,+1,0} = \lambda\left[D_\mathrm{c} - H_\mathrm{c}\left(E^n_{i,j+1,k} - E^n_{i,j-1,k}\right)/6\right] \tag{4.6}$$

$$P_{0,0,-1} = \lambda\left[D_\mathrm{c} + H_\mathrm{c}\left(E^n_{i,j,k+1} - E^n_{i,j,k-1}\right)/6\right] \tag{4.7}$$

$$P_{0,0,+1} = \lambda\left[D_\mathrm{c} - H_\mathrm{c}\left(E^n_{i,j,k+1} - E^n_{i,j,k-1}\right)/6\right] \tag{4.8}$$

and the coefficient which is proportional to the probability of no movement is given by

$$\begin{aligned} P_{0,0,0} = 1 - \lambda\Big[6D_\mathrm{c} + H_\mathrm{c}\Big(&\mathrm{E}^n_{i-1,j,k} + E^n_{i,j-1,k} + E^n_{i,j,k-1} \\ & - 6E^n_{i,j,k} + E^n_{i+1,j,k} + E^n_{i,j+1,k} + E^n_{i,j,k+1}\Big)\Big] \end{aligned} \tag{4.9}$$

These seven coefficients are used to compute the probabilities of the stochastic migration of individual tumor cells. Defining

$$P_\mathrm{T} = P_{-1,0,0} + P_{0,-1,0} + P_{0,0,-1} + P_{0,0,0} + P_{+1,0,0} + P_{0,+1,0} + P_{0,0,+1} \tag{4.10}$$

the probability of a tumor cell at (x_i, y_j, z_k) remaining at that location is given by $P_{0,0,0}/P_{\mathrm{T}}$, and the probability of that tumor cell moving, for example, one volume in the positive x directions is given by $P_{+1,0,0}/P_{\mathrm{T}}$.

4.2.2 Tissue Environment

The tissue, including the components and mechanisms listed above, is described by

$$\frac{\partial O}{\partial t} = D_{\mathrm{o}}\nabla^2 O - k_{\mathrm{oo}}O - k_{\mathrm{oc}}OC + k_{\mathrm{oe}}E \tag{4.11}$$

$$\frac{\partial M}{\partial t} = D_{\mathrm{m}}\nabla^2 M + k_{\mathrm{mc}}C - k_{\mathrm{mm}}M \tag{4.12}$$

$$\frac{\partial E}{\partial t} = -k_{\mathrm{em}}EM \tag{4.13}$$

where C is 1 if a given location (x, y, z) is occupied by a tumor cell, and 0 otherwise. It is important to note that two of the parameters in the tissue model—the oxygen consumption rate, k_{oc}, and the matrix degradative enzyme production rate, k_{mc}—vary from point to point depending upon the state of the tumor cell at that location.

The tissue equations are solved using an explicit Euler method:

$$\begin{aligned} O^{n+1}_{i,j,k} &= O^{n}_{i,j,k} + \lambda D_{\mathrm{o}} L(O^{n}_{i,j,k}) \\ &\quad - k_{\mathrm{oo}}O^{n}_{i,j,k}\Delta t - k_{\mathrm{oc}}O^{n}_{i,j,k}C^{n}_{i,j,k}\Delta t + k_{\mathrm{oe}}E^{n}_{i,j,k}\Delta t \end{aligned} \tag{4.14}$$

$$M^{n+1}_{i,j,k} = M^{n}_{i,j,k} + \lambda D_{\mathrm{m}} L(M^{n}_{i,j,k}) + k_{\mathrm{mc}}C^{n}_{i,j,k}\Delta t - k_{\mathrm{mm}}M^{n}_{i,j,k}\Delta t \tag{4.15}$$

$$E^{n+1}_{i,j,k} = E^{n}_{i,j,k} - k_{\mathrm{em}}E^{n}_{i,j,k}M^{n}_{i,j,k}\Delta t \tag{4.16}$$

where

$$\begin{aligned} L(V^{n}_{i,j,k}) &= V^{n}_{i-1,j,k} + V^{n}_{i,j-1,k} + V^{n}_{i,j,k-1} - 6V^{n}_{i,j,k} \\ &\quad + V^{n}_{i+1,j,k} + V^{n}_{i,j+1,k} + V^{n}_{i,j,k+1} \end{aligned} \tag{4.17}$$

For stability, this method requires $\lambda \max(D_{\mathrm{c}}, D_{\mathrm{m}}) < 1/6$. Since $D_{\mathrm{c}}/D_{\mathrm{m}} \approx O(10^3)$, we solve M and E on the same mesh as that used for tumor cells, but solve C on a coarser mesh with spacing 32 times larger (since $32^2 \approx 10^3$). This requires averaging and interpolating values between fine and coarse meshes, but allows the use of a much larger timestep.

4.2.3 Processes Controlling Individual Tumor Cells

The effects of tumor cell heterogeneity are included in the model by assigning to each cell an immutable phenotype, or state, chosen at birth from any of 100 predetermined phenotypes. The phenotype determines certain aspects of the cell's behavior and interaction with the tissue, and can therefore be considered a measure of the tumor cell's aggressiveness. The assignment of a phenotype confers individual characteristics and behaviors upon each cell, a fundamental property of the hybrid approach, and permits the incorporation of several biological processes important in tumor formation and growth. Here we discuss these processes in detail, and how they are included in the model.

- *Apoptosis*. Because cell survival requires sufficient oxygen, cells are assumed to die when their local tissue oxygen levels fall below a cutoff, O_{cut}. Dead cells are removed from the computation.
- *Proliferation*. Since tumor cells have the potential to undergo cellular division provided the cell has reached maturity, with each phenotype is associated an age threashold, m; divsion occurs when a tumor cell's age exceeds this threshold, and there is sufficient space surrounding the cell. When cells divide, one daughter cell is placed in the same location as the parent cell, and the other is placed randomly in one of the empty neighboring locations.
- *Mutation*. Each cellular division has a mutation probability p, resulting in a new phenotype, selected randomly from among the 100 predetermined phenotypes, being propagated to both daughter cells. If no mutation occurs, the parent cell's phenotype is propagated to both daughter cells.
- *Adhesion*. With each phenotype is associated a cell-to-cell adhesion value a, indicating the number of neighbors a cell prefers to adhere to. A cell is not permitted to migrate if it has fewer than a neighbors.
- *Oxygen Consumption, Enzyme Production, and Haptotaxis*. Each tumor cell phenotype is associated with different rates of oxygen consumption (k_{oc}), matrix degradative enzyme production (k_{mc}), and haptotaxis (H_{c}).

Thus each phenotype consists of five parameter values (a, m, k_{oc}, k_{mc}, and H_{c}) corresponding to the five processes (adhesion, proliferation, oxygen consumption, matrix degradative enzyme production, and haptotaxis, respectively) that are controlled in a cell-specific manner. In constucting each of the 100 predetermined phenotypes, the values of these five parameters are chosen randomly from a range. As a result, each predetermined phenotype has a different combination of values for these five parameters, and therefore, a different level of tumor cell aggressiveness.

4.2.4 Boundary Conditions

For both tumor cells and extracellular tissue, we assume Neumann boundary conditions. Our implementation uses an extra layer of "ghost" cells around the outside of the computational domain whose values are set to satisfy the boundary conditions. Assuming these cells are indexed with $i = 0$, $i = I + 1$, $j = 0$, $j = J + 1$, $k = 0$, and $k = K + 1$, the ghost cells are set, using C as an example,

$$C^n_{0,j,k} = C^n_{1,j,k}, \quad C^n_{I+1,j,k} = C^n_{I,j,k} \tag{4.18}$$

$$C^n_{i,0,k} = C^n_{i,1,k}, \quad C^n_{i,J+1,k} = C^n_{i,J,k} \tag{4.19}$$

Table 4.1 Model Parameter Values Used in All Simulations, Except Where Varied or Indicated Otherwise

Δt	0.001	Δs	0.0025	H_c	0.01–0.03
D_c	0.0005	D_o	0.5	D_m	0.0005
k_{oo}	0.025	k_{oc}	0.57–1.71	k_{oe}	0.5
k_{mc}	1–3	k_{mm}	0.0	k_{em}	50.0
a	0–5	m	0.5–1.0	p	0.1

$$C^n_{i,j,0} = C^n_{i,j,1}, \qquad C^n_{i,j,K+1} = C^n_{i,j,K} \tag{4.20}$$

for $i = 1[1]\mathrm{I}$, $j = 1[1]J$, and $k = 1[1]K$. Identical equations hold for O, M, and E.

4.2.5 Nondimensionalization and Parameters

All model quantities are assumed to be dimensionless, with time scaled by the average time between cellular divisions (approximately 24 hours), space scaled by the overall dimensions of the domain (X, Y, and Z), and concentrations scaled by their basal physiological levels [9]. As a result, the initial conditions are given by $O(0, x, y, z) = 1.0$, $M(0, x, y, z) = 0.0$, $E(0, x, y, z) = 1.0$, and tumor cells, with randomly assigned phenotypes, are assigned initially to some region in space, typically a small sphere in the center of the tissue.

Model parameters are shown in Table 4.1. Ranges of values are shown for the five parameters determined by the tumor cell phenotype; the value for each of these parameters is chosen randomly when forming each of the 100 predetermined phenotypes. For example, with each phenotype is associated a proliferation age in the range of 0.5 to 1.0, and a haptotaxis coefficient in the range of 0.01 to 0.03.

4.2.6 Model Simulation

For each time step, the model simulation is solved in six steps:

1. *Boundaries:* set the ghost cells using (4.18) to (4.20) applied to C, O, M, and E;
2. *Apoptosis:* cells in low oxygen undergo apoptosis;
3. *Tissue:* compute $O^{n+1}_{i,j,k}$, $M^{n+1}_{i,j,k}$, and $E^{n+1}_{i,j,k}$ using (4.14) to (4.16);
4. *Migration:* compute $C^{n+1}_{i,j,k}$, for $i = 1[1]I$, $j = 1[1]J$, and $k = 1[1]K$, using (4.2) in conjunction with (4.3) to (4.10);
5. *Proliferation:* update ages, mutate (randomly, with probability p) and divide mature cells with sufficient space, and place daughter cells;
6. *Update:* set $C^n_{i,j,k} = C^{n+1}_{i,j,k}$, $O^n_{i,j,k} = O^{n+1}_{i,j,k}$, $M^n_{i,j,k} = M^{n+1}_{i,j,k}$, and $E^n_{i,j,k} = E^{n+1}_{i,j,k}$, for $i = 1[1]I$, $j = 1[1]J$, and $k = 1[1]K$.

4.3 Decomposition

Domain decomposition is a common method for adapting the numerical solution of partial differential equations to parallel, high-performance computing platforms.

In this approach, the solution of the problem on the full spatial domain is computed by splitting the problem into smaller, independent problems on subdomains, each of which is solved on a different processor. At each step, the solutions of the subproblems are coordinated by communicating between processors values on the boundaries of the subdomains. Because processors must only communicate with those processors responsible for neighboring subdomains, this produces a localized, nearest neighbor communication pattern.

The Blue Gene/L and Blue Gene/P supercomputers [13] are scalable, high-performance systems designed to tackle some of the largest problems in science, including protein folding and molecular dynamics. Blue Gene systems are built up in units of racks, each containing 1,024 dual-core (Blue Gene/L) or quad-core (Blue Gene/P) processors. MPI jobs can run in either coprocessor mode, where each MPI process is mapped to a processor, or virtual node mode, where each MPI process is mapped to a core. Additionally, coprocessor mode on Blue Gene/P (but not Blue Gene/L) allows processes to take advantage of the additional cores using multiple threads, for example, by using OpenMP. Communication between processors is performed via a system of five interconnect architectures, including low latency, high bandwidth tree and torus topolgies that the Blue Gene MPI implementation chooses between to optimize communication. Finally, the Blue Gene processors are subdivided into partitions on which computations are executed.

Our tumor implementation on Blue Gene/L and Blue Gene/P employs a domain decomposition on a 3D cartesian grid of $P_x \times P_y \times P_z$ processors identified by single indices, or ranks, $p_x P_y P_z + p_y P_z + p_z$, for $p_x = 0[1]P_x - 1$, $p_y = 0[1]P_y - 1$, and $p_z = 0[1]P_z - 1$. We assume that the discretized domain can be decomposed evenly along each axis of the grid of processors—that is, that I/P_x, J/P_y and K/P_y are whole numbers—and that these dimensions match the dimensions of the torus topology of the Blue Gene partition used to run the calculation. To simplify averaging and interpolating between the coarse mesh used in computing C and the fine mesh, we also require that the dimensions of the subdomain meshes be divisble by 16; that is, that I/P_x, J/P_y and K/P_z be multiples of 16. Each subdomain utilizes a local indexing scheme, with, for example, $O_{p,i,j,k}$ denoting the oxygen concentration at the mesh location (i, j, k) of the subdomain assigned to the processor with rank p. Each subdomain also utilizes its own layer of ghost cells for storing the subdomain's neighboring values necessary for computing C, O, and M; these cells are indexed by $i = 0$, $i = I/P_x + 1$, $j = 0$, $j = J/P_y + 1$, $k = 0$, and $k = K/P_z + 1$.

Each processor is responsible for computing values in the interior of its associated subdomain. In the case of tumor cells, the processor is said to "own" the tumor cells in the interior of its subdomain; if a tumor cell diffuses outside of the processor's subdomain, ownership is transferred to the new subdomain's processor. Moving tumor cell ownership from one processor to another is therefore one communication operation involved in solving the tumor cell equation. The other operation involves communicating tumor cells lying along the edge of the subdomain, so that neighboring domains can determine the occupancy of their ghost cells. This copying of tumor cells is analogous to the communication of the diffusive variables O and M, whose values along the subdomain boundary are mapped to the ghost cells of neighboring domains.

These communication operations are described in detail next. To ease that discussion, we note that with this domain decomposition, communication only occurs for all ordered pairs of processor ranks

$$(q, r) = \left(q_x P_y P_z + q_y P_z + q_z, r_x P_y P_z + r_y P_z + r_z\right) \tag{4.21}$$

such that

$$|q_x - r_x| + \left|q_y - r_y\right| + |q_z - r_z| = 1 \tag{4.22}$$

This definition of q and r are used in the sections that follow.

Finally several implementation issues should be noted. The variables O, M, and E are stored as arrays of double-precision floating point values. Tumor cells are implemented as C++ objects encapsulating age, phenotype, and owner rank, and the variable C is implemented as an array of pointers to tumor cell objects. This choice incurs the expense of dereferencing pointers when accessing the object's data, but reduces the expense of moving tumor cells. However, in contrast to the other variables, which are sent and received "in place" via a user-defined datatype created with MPI_type_vector(), tumor cells must be sent and received using buffers, which incurs the additional cost of copying objects. While we have opted for code simplicity, the alternative choice of implementing C as an array of objects is also possible, and could lead to a slight speed increase. The tumor cell buffers are of fixed length ($I/P_x \times J/P_y$, $JP_y \times K/P_z$, and $K/P_z \times I/P_x$), with dummy values indicating the absence of a cell, so that the tumor cell locations need not be sent. Alternatively, we could use variable sized messages that pack both tumor cells and locations into the buffer, but this approach yields a speed increase only when there are few tumor cells to move or copy, and in fact leads to larger message sizes and speed decreases when tumor cell density is high. Again, we have opted for code simplicity, using fixed sized buffers which yield constant communication times throughout the calclation.

4.3.1 Moving of Tumor Cells

When a tumor cell moves from the interior of one subdomain to another, ownership must be transferred to the new subdomain's processor. This operation, as shown in Figure 4.1(a), can be described by

$$C^n_{q,I/P_x,j,k} \leftarrow C^n_{r,0,j,k} \quad C^n_{q,I/P_x+1,j,k} \rightarrow C^n_{r,1,j,k} \tag{4.23}$$

when $r_x = q_x + 1$;

$$C^n_{q,i,J/P_y,k} \leftarrow C^n_{r,i,0,k} \quad C^n_{q,i,J/P_y+1,k} \rightarrow C^n_{r,j,1,k} \tag{4.24}$$

when $r_y = q_y + 1$; and

$$C^n_{q,i,j,K/P_z} \leftarrow C^n_{r,i,j,0} \quad C^n_{q,i,j,K/P_z+1} \rightarrow C^n_{r,i,j,1} \tag{4.25}$$

when $r_z = q_z + 1$. Because the ghost cells on a processor actually correspond to locations in the interior of another processor's subdomain, only those tumor cells

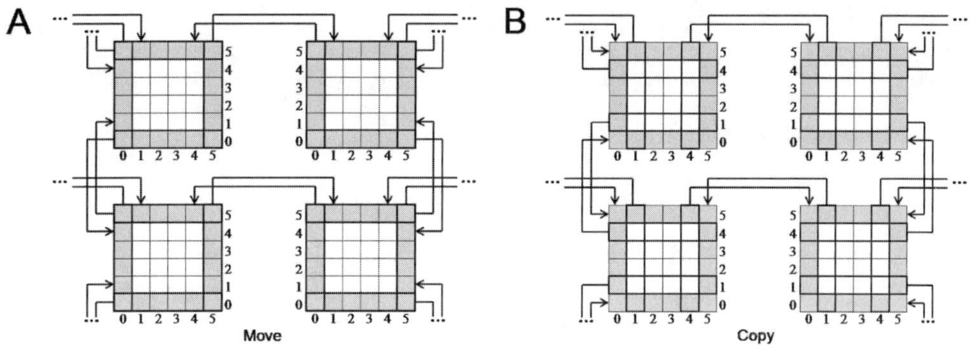

Figure 4.1 Two-dimensional analogue depicting inter-processor communication for (a) moving and (b) copying rows and columns of cells. In three dimensions these rows and columns correspond to sheet-like, two-dimensional slices through a cube.

owned by the sending process must be moved, and therefore communicated. Following the communication, the rank of the received tumor cells is set to that of the receiving process.

4.3.2 Copying of Tumor Cells

Computing the diffusion and lifecycle of tumor cells on the boundary of the subdomains requires knowledge of tumor cell occupancy in the subdomain's ghost cells. This requires copying tumor cells along the edge of the domain to the ghost cells of neighboring domains. This operation, as shown in Figure 4.1(b), can be described by

$$C^n_{q,I/P_x+1,j,k} \leftarrow C^n_{r,1,j,k} \quad C^n_{q,I/P_x,j,k} \rightarrow C^n_{r,0,j,k} \tag{4.26}$$

when $r_x = q_x + 1$;

$$C^n_{q,i,J/P_y+1,k} \leftarrow C^n_{r,i,1,k} \quad C^n_{q,i,J/P_y,k} \rightarrow C^n_{r,j,0,k} \tag{4.27}$$

when $r_y = q_y + 1$; and

$$C^n_{q,i,j,K/P_z+1} \leftarrow C^n_{r,i,j,1} \quad C^n_{q,i,j,K/P_z} \rightarrow C^n_{r,i,j,0} \tag{4.28}$$

when $r_z = q_z + 1$. In contrast to moving tumor cells, because the sending process owns all of the tumor cells inside its subdomain, all tumor cells must be copied, and therefore communicated, and the receiving process does not set the rank of the received tumor cells.

4.3.3 Copying of Continuous Variables

Values of the diffusive variables O and M along the edge of the domain must be copied to the ghost cells of neighboring domains, similar to the copying of tumor cells. This operation can therefore be described, using O as an example, by

$$O^n_{q,I/P_x+1,j,k} \leftarrow O^n_{r,1,j,k} \quad O^n_{q,I/P_x,j,k} \rightarrow O^n_{r,0,j,k} \tag{4.29}$$

when $r_x = q_x + 1$;

$$O^n_{q,i,J/P_y+1,k} \leftarrow O^n_{r,i,1,k} \quad O^n_{q,i,J/P_y,k} \rightarrow O^n_{r,j,0,k} \tag{4.30}$$

when $r_y = q_y + 1$; and

$$O^n_{q,i,j,K/P_z+1} \leftarrow O^n_{r,i,j,1} \quad O^n_{q,i,j,K/P_z} \rightarrow O^n_{r,i,j,0} \tag{4.31}$$

when $r_z = q_z + 1$. In contrast to the copying of tumor cells, however, this operation does not require buffers; instead, we use MPI_type_vector to create three user-defined datatypes, corresponding to the three (x, y, and z) directions values are communicated, and communicate the values directly between source and destination arrays.

4.3.4 Blue Gene Model Simulation

For each time step, the model simulation is solved in 11 steps:

1. Boundaries: set the ghost cells using (4.18) to (4.20) applied to C, O, M, and E.
2. Apoptosis: cells in low oxygen undergo apoptosis.
3. Tissue: compute $O^{n+1}_{i,j,k}$, $M^{n+1}_{i,j,k}$, and $E^{n+1}_{i,j,k}$ using (4.14) to (4.16).
4. Migration: compute $C^{n+1}_{i,j,k}$, for $i = 1[1]I$, $j = 1[1]J$, and $k = 1[1]K$, using (4.2) in conjunction with (4.3) to (4.10).
5. Move cells: move tumor cells between neighboring processors using (4.23) to (4.25).
6. Copy cells: copy tumor cells between neighboring processors using (4.26) to (4.28).
7. Proliferation: update ages, mutate (randomly, with probability p) and divide mature cells with sufficient space, and place daughter cells.
8. Move cells: move tumor cells between neighboring processors using (4.23) to (4.25).
9. Copy cells: copy tumor cells between neighboring processors using (4.26) to (4.28).
10. Update: set $C^n_{i,j,k} = C^{n+1}_{i,j,k}$, $O^n_{i,j,k} = O^{n+1}_{i,j,k}$, $M^n_{i,j,k} = M^{n+1}_{i,j,k}$, and $E^n_{i,j,k} = E^{n+1}_{i,j,k}$, for $i = 1[1]I$, $j = 1[1]J$, and $k = 1[1]K$.
11. Copy variables: copy the continuous variables between neighboring processors using (4.29) to (4.31) applied to O, M, and E.

4.3.5 Multithreaded Blue Gene Model Simulation

The multicore architecture of Blue Gene/P permits up to four threads per processor using OpenMP. Several of the steps involved in our Blue Gene model simulation can be performed in multiple threads without modification to the algorithm. In our implementation, however, we have elected to perform three steps in multiple threads: solving the tissue equations, migration, and proliferation. While other steps could be similarly performed in multiple threads, we have only focused on the most time-consuming steps where the speedup is most significant.

Because our implementation uses two arrays for each tissue equation, one for the current and one for the new time steps, multithreading the solution of the tissue equations is a straightforward use of an omp parallel for directive. The migration and proliferation steps, however, require more than the simple addition of an omp parallel directive. Both of these steps involve one thread examining and placing a tumor cell in tissue volumes that another thread is responsible for, and proliferation involves memory allocation when tumor cells divide. As a result, performing these steps in multiple threads requires the use of omp critical directives to coordinate the moves and allocation. Finally, we note that since both of these steps involve counting the number of neighboring tissue volumes that contain cells, an additional critical section is required to ensure proper counting; since we only need to test for null pointers when counting, however, we forego using a critical section and accept the additional stochasticity improper counting introduces, rather than incure the additional overhead.

4.4 Performance

To demonstrate the utility of our approach, we have investigated the performance and scaling of our implementation. Because memory on each processor of Blue Gene is limited, it is not possible to demonstrate strong scaling except on the smallest of problems. Instead, we examine the weak scaling characteristics of our implementation, using a test problem comprised of $64 \times 64 \times 64$ volumes on each processor, so the total problem size scales with the number of processors. Note that $64^3 = 2^{18}$ tissue volumes per processor, and $4{,}096 = 2^{12}$ processors, yielding a total problem size of $2^{30} = O(10^9)$, which corresponds to the lower end of clinicially relevant maximum tumor sizes. For simplicity and ease of comparison we restrict our calculations to square tissue on cube shaped partitions ($P_x = P_y = P_z$).

The total problem size also scales with the number of generations that are computed. Since different sized tumors require different numbers of generations be computed to generate the same number of tumor cells per processor, we focus instead on the per generation time. This requires that we assign each processor the same amount of work, independent of the number of processors. We have therefore chosen to measure the timing of the first generation ($T = 1$ with $\Delta t = 0.001$, so that $N = 1{,}000$ time steps), using a random initial condition where each tissue volume is assigned a tumor cell with probability one-half. As a result, our timings approximate a worst-case scenario in terms of per processor work, but a best-case scenario in terms of load balance.

The per generation timings on Blue Gene/L are shown for varying numbers of processors (32, 64, 128, 512, 1,024, 2,048, 4,096 and 8,192) in Figure 4.2. The parallel tumor invasion implementation demonstrates almost perfect weak scaling to one rack (1,024 processors) and reasonable weak scaling to eight racks (8,192 processors). The slightly reduced scaling on multiple racks is to be expected, due to the reduced communication bandwidth between racks, but the reduction in overall per processor, per generation performance is still quite good, yielding only a 24% reduction in scaling from 32 or 64 processors to 8,192 processors.

As described above, to keep the total problem size identical in our weak scaling runs, we used a random initial condition that yielded a worst-case scenario in

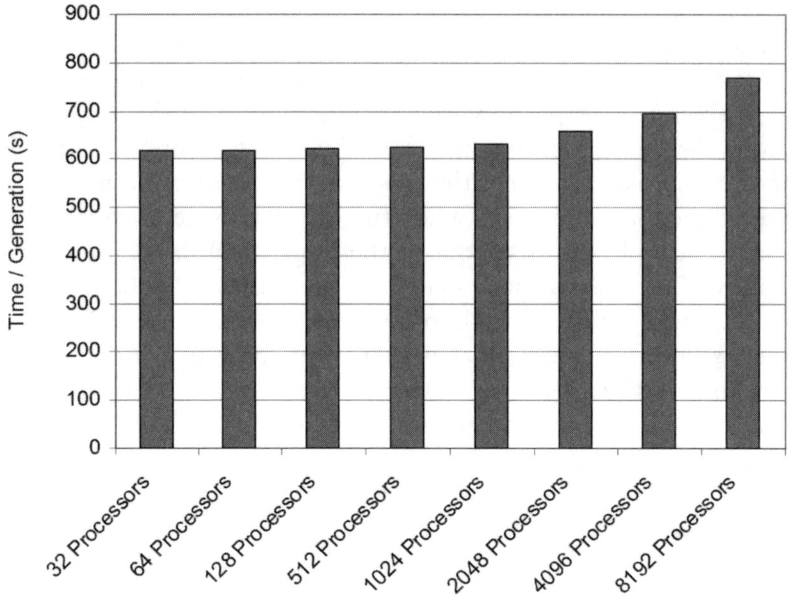

Figure 4.2 Per generation timings for 32, 64, 128, 512, 1,024, 2,048, 4,096, and 8,192 processors on Blue Gene/L.

terms of per processor work, but a best-case scenario in terms of load balance. It is important to note, however, that worst-case load balancing does not increase beyond a factor of two (due to assigning tumor cells to only one-half of the tissue volumes) the maximum load on any given processor, but only decreases the load on some processors. As a result, a poorly load balanced computation would make less efficient use of processors on average, but the per generation time of the calculation would not increase significantly. Indeed, we have tested this by comparing runs using the random initial condition with runs using an initial condition consisting of a spherically shaped tumor with a radius of 64 tumor cells. This choice of initial condition ensured that all tumor cells were initially restricted to the eight processors responsible for the central region of the tissue, producing the worst-case scenario where eight processors were maximally loaded and all other processors were minimally loaded. Our simulations demonstrated that poor load balancing did not affect significantly the computation time per generation relative to the random initial condition. For example, on 64 processors the computation time per generation changed by only 2 seconds, or about 0.3%, when the random initial condition was replaced by the spherical initial condition (676 seconds per generation for the random initial condition versus 678 seconds per generation for the spherical initial condition).

On Blue Gene/P, where one to four threads can be used to step each block of $64 \times 64 \times 64$ tissue volumes, our implementation shows reasonable weak scaling across processors, as on Blue Gene/L, but slightly reduced strong scaling with increasing numbers of threads. These results are shown in Figure 4.3, for 64, 512, 4,096, and 8,192 processors and one to four threads. As seen in the Blue

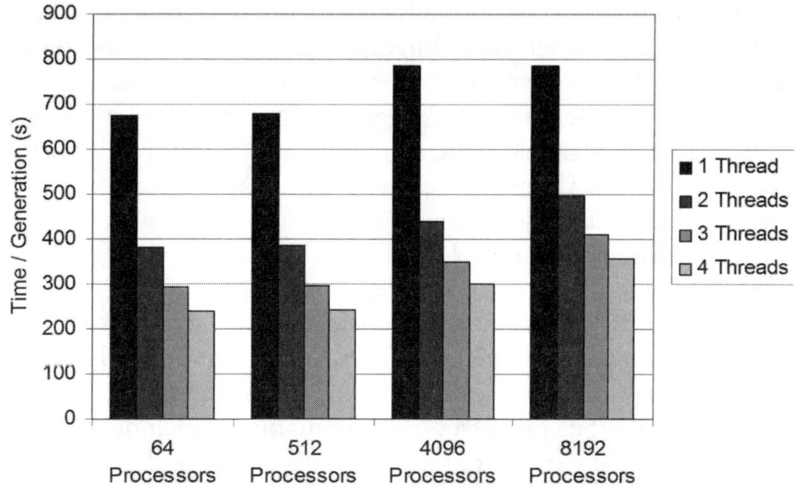

Figure 4.3 Per generation timings for 64, 512, 4,096 and 8,192 processors and 1 to 4 threads on Blue Gene/P.

Gene/L calculations, weak scaling is nearly perfect up to one rack, with slightly reduced weak scaling when the calculation spans multiple racks. On Blue Gene/P, however, we observe only a 16% reduction in scaling from 64 processors to 8,192 processors, far better than the 24% reduction observed on Blue Gene/L.

We also examined strong scaling across multiple threads on Blue Gene/P. On 64 and 512 processors, we observed a speedup of 1.8 from one to two threads, and a speedup of 2.8 from one to four threads. The speedup was slightly reduced on 4,096 processors (1.8 from one to two threads, and 2.6 from one to four threads), and even more reduced on 8,192 processors (1.6 from one to two threads, and 2.2 from one to four threads). While these results suggest it is possible to get increased performance with multiple threads, more work is required to maximize the performance gains. This is not surprising, given that we did not attempt to optimize the performance on multiple threads, opting instead to ensure portability between Blue Gene/L and Blue Gene/P, as well as other high-performance computing platforms. For example, we did not utilize critical sections when loading and unloading the buffers used to copy and move cells. Since moving and copying cells accounts for between 50 and 90 seconds of the total per generation time, this is a significant amount of work that does not scale at all with increasing numbers of threads, reducing the overall strong scaling performance on Blue Gene/P.

4.5 Conclusions

We have discussed an implementation of a published model of solid tumor invasion [10] on both Blue Gene/L and Blue Gene/P using nearest neighbor communication tuned to the physical torus communication network. This approach permits the numerical simulation of clinically relevant, large scale tumor models incorporating up to $O(10^9)$ discrete tumor cells in a reasonable time. On 8,192 processors on Blue Gene/L, each generation (corresponding to 8 to 24 hours) could be computed in just under 13 minutes. At this rate, a realistic simulation consisting of 1,000 generations would require 213 hours, producing a simulation time that is between 37 and 113 times faster than real time. On 8,192 processors on Blue Gene/P, using four threads, the per generation time was reduced to just under 6 minutes, corresponding to a realistic simulation time (1,000 generations) of 100 hours and a speedup between 80 and 242.

These timing results are encouraging for the use of large scale tumor calculatins in a clinical setting, but must be considered in the context of several limitations. The most glaring issue is the simplicity of this tumor model. While it is clear that tumor cell phenotype and tissue environment are important for tumor growth and invasion, many other processes are also involved, including the dynamical control of tumor cell behavior by the genetic network that this model represents with a simple phenotype. Indeed, many tumor models incorporate more complex genetic pathways represented by, for example, systems of ordinary differential equations. These more complex models require far more local computation, thus increasing the computational requirements beyond what is currently available.

Two additional and significant limitations, not addressed at all here, are the storage and visualization of the data generated by these types of calculations. Even for this simple model, where the tumor cell and tissue states are characterized by four values (tumor cell phenotype, oxygen, matrix degradative enzymes, and extracellular matrix proteins), storage of $O(10^9)$ to $O(10^{12})$ tumor cells every generation for 1,000 generations would require $O(10^{13})$ to $O(10^{16})$ bytes, or 10 terabytes to 10 petabytes. Such large data sizes are also a consideration for visualization and data analysis.

Despite these limitations, we have shown that simulation of clinically relevant models of tumor formation and growth are feasible. Larger, more complex models, however, may require the next generation of supercomputers. Our timing and scaling studies here demonstrate the utility of this approach for small but clinically relevant simulations, establishing a baseline for the computing requirements for moving forward in larger-scale tumor modeling.

Acknowledgments

We thank Alexander Anderson, Brian Skjerven, Maria Eleftheriou, and James Kozloski for helpful discussions. John Wagner dedicates this manuscript to the memory of Eric Gittings, a truly gifted teacher who proved Yeats' conjecture, "Education is not the filling of a pail, but the lighting of a fire."

References

[1] C. P. Fall, E. S. Marland, J. M. Wagner, and J. J. Tyson, *Computational Cell Biology*. Springer-Verlag, 2002.

[2] P. Nowell, "The clonal evolution of tumor cell populations," *Science*, vol. 194, p. 23–28, 1976.

[3] E. Fearon and B. Vogelstein, "A genetic model for colorectal tumorigenesis," *Cell*, vol. 61, p. 759–767, 1990.

[4] R. Araujo and D. McElwain, "A history of the study of solid tumour growth: the contribution of mathematical modeling," *Bull. Math. Biol.*, vol. 66, p. 1039–1091, 2004.

[5] M. Chaplain, "Avascular growth, angiogenesis and vascular growth in solid tumours: the mathematical modelling of the stages of tumour development," *Math. Comput. Model.*, vol. 23, p. 47–87, 1996.

[6] S. Sanga, H. B. Frieboes, X. Zheng, R. Gatenby, E. L. Bearer, and V. Cristini, "Predictive oncology: a review of multidisciplinary, multiscale in silico modeling linking phenotype, morphology and growth," *Neuroimage*, vol. 37 Suppl. 1, pp. S120–S134, 2007.

[7] A. R. Anderson and M. A. Chaplain, "Continuous and discrete mathematical models of tumor-induced angiogenesis," *Bull. Math. Biol.*, vol. 60, pp. 857–899, September 1998.

[8] A. Anderson, M. Chaplain, E. Newman, R. Steele, and A. Thompson, "Mathematical modelling of tumour invasion and metastasis," *J. Theoret. Med.*, vol. 2, p. 129–154, 2000.

[9] A. R. A. Anderson, "A hybrid mathematical model of solid tumour invasion: the importance of cell adhesion," *Math. Med. Biol.*, vol. 22, pp. 163–186, June 2005.

[10] A. R. A. Anderson, A. M. Weaver, P. T. Cummings, and V. Quaranta, "Tumor morphology and phenotypic evolution driven by selective pressure from the microenvironment," *Cell*, vol. 127, pp. 905–915, December 2006.

[11] W. Gropp, E. Lusk, and A. Skjellum, *Using MPI, 2nd Edition: Portable Parallel Programming with the Message Pssing Interface*. Scientific and Engineering Computation Series, Cambridge, MA: MIT Press, 1999.

[12] R. Chandra, R. Menon, L. Dagum, D. Kohr, and J. M. D. Maydan, *Parallel Programming in OpenMP*. Morgan Kaufmann, 2000.

[13] A. Gara, M. A. Blumrich, D. Chen, G. L.-T. Chiu, P. Coteus, M. E. Giampapa, R. A. Haring, P. Heidelberger, D. Hoenicke, G. V. Kopcsay, T. A. Liebsch, M. Ohmacht, B. D. Steinmacher-Burow, T. Takken, and P. Vranas, "Overview of the blue gene/l system architecture," *IBM Journal of Research and Development*, vol. 49, no. 2/3, p. 195, 2005.

PART II

Understanding and Utilizing Parallel Processing Techniques

CHAPTER 5

Introduction to High-Performance Computing Using MPI

Rahul Garg

5.1 Introduction

Computing technologies have undergone improvements at a remarkable pace over the last 50 years. The number of components in a chip and its frequency of operation has been increasing at an exponential rate. This trend is widely referred to as *Moore's Law*. In 1965, Gordon Moore published a famous paper in which he predicted that the transistor densities (i.e., the number of transistors in a chip) will roughly double in every 2 years due to technological improvements. Surprisingly, this trend has continued through 2008 and is expected to continue in the near future as well (see Figure 5.1).

Increases in transistor densities have historically led to increases in the operating frequencies of microprocessors, roughly at the same rate (i.e., doubling every 2 years). This trend, however, cannot continue forever. Smaller form-sizes have resulted in larger leakage currents leading to higher power consumption per unit area. This imposes limits on the frequency of operation of a microprocessor built using the current silicon-based technologies. The microprocessor frequency trends shown in Figure 5.2 suggest that a limit of around 3 GHz may have been reached in 2005. Between the 1980s and 2005, the desktop systems were mostly uniprocessor systems with steady improvements in operating frequencies. Since 2005, desktop systems are turning multicore with increasing number of processors, but little or no improvements in operating frequencies. While general purpose systems have shown this trend, some specialized systems such as the ones based on the IBM POWER6 processors have reached operating frequencies of more than 5 GHz. The price to be paid for this performance, however, is that specialized technologies (e.g., water cooling) may be needed to cool the chip. Unless there is a breakthrough in nanotechnology or optical computing technologies, future microprocessors are not expected to have significantly higher frequencies.

Consistent with transistor density and operating frequency trends, the performance of parallel systems has also maintained an exponential growth trend. According to the TOP500 project [34], the peak performance of supercomputers has been improving at the rate of thousand times every 11 years (see Figure 5.3). This corresponds to performance improvements of approximately 3.5 times every 2 years. The project also tracks the cumulative sum of the performance capabilities of the 500 fastest supercomputers in the world. The cumulative supercomputer performance also has the same trend as shown in Figure 5.3.

It is expected that most of the performance improvements in future supercomputers will be obtained using extreme scale parallelism. The NEC Earth simulator

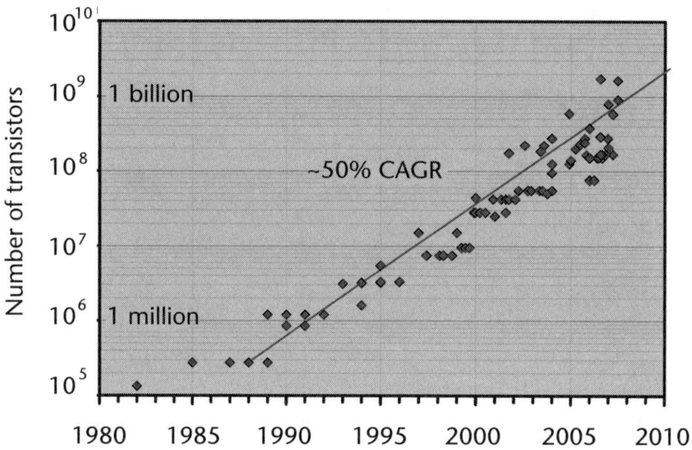

Figure 5.1 The number of transistors in a chip has continued to double every 2 years. According to estimates, this trend will continue in the future at least until 2010. (Courtesy of Timothy J. Dalton, IBM Corp.)

was the fastest supercomputer in the world from 2002 until 2004. It comprised 640 nodes, each with eight processors (i.e., a total of 5,120 processors) operating at 500 MHz. In comparison, the IBM Blue Gene/L supercomputer, which was the fastest supercomputer from 2004 until 2007, was made of 212,992 processors operating at 700 MHz. Future supercomputers are likely to have millions of processors, operating at similar clock frequencies.

These two trends of transistor density and clock frequency improvements have very important implications for the future of computing in general, and parallel computing in particular. Unlike in the past, performance improvements in the

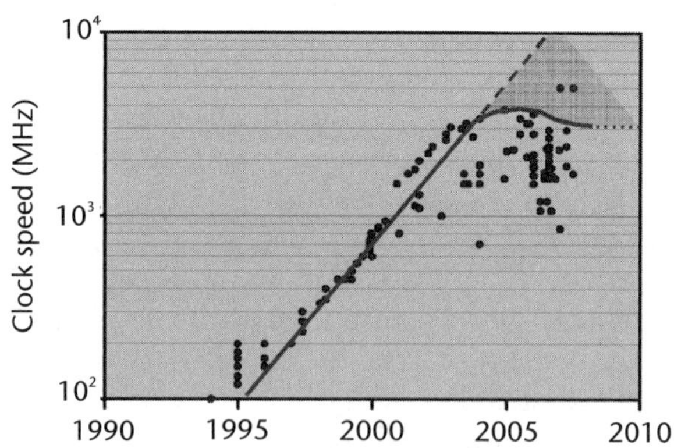

Figure 5.2 Due to increasing power consumption and cooling limitations, the chip frequencies cannot continue to increase at the rates witnessed historically. These are expected to stabilize in the range 1 to 10 GHz. In massively parallel multicore systems, the frequencies may actually decrease in the future. (Courtesy of Timothy J. Dalton, IBM Corp.)

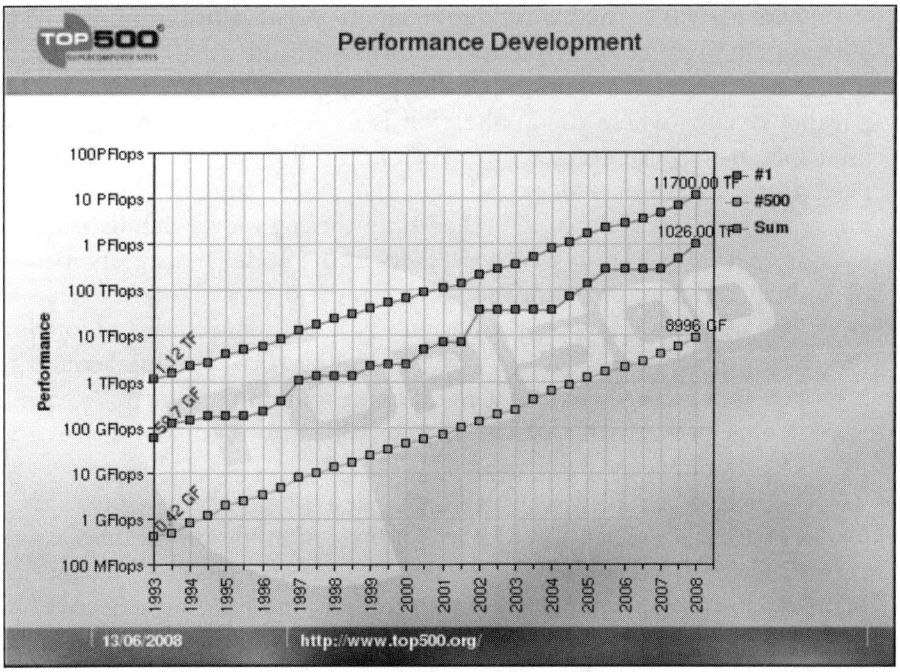

Figure 5.3 Trends in the performance of supercomputers. (Courtesy of TOP500 project [34].)

future will not be achieved due to increases in operating frequency of microprocessors. Instead, as the transistor density grows, chips are expected to contain multiple processors, possibly operating at smaller clock frequencies. Therefore, the future performance improvements at the level of desktops, servers, and supercomputers will be obtained primarily by increasing the level of parallelism in the applications.

The above trends amply illustrate the importance of parallel programming in the future. To deliver optimal performance on future systems, applications will need to run simultaneously on multiple processors while effectively utilizing the processing capabilities offered by the underlying hardware. Therefore, understanding the fundamentals of parallel computing will be extremely important for future programmers and users.

Improvements in imaging and microscopy technologies will continue to generate increasing amounts of data that will require more and more computational resources. Since the clock frequencies of microprocessors are not expected to improve significantly in the future, scientists and engineers will increasingly have to rely on parallel programming to satisfy the needs of future imaging applications.

It is vital therefore for scientists and engineers in the areas of microscopy and imaging to familiarize themselves with the basic concepts in parallel programming so that they can make the best use of the computing technologies that will be available in the future. This chapter is intended to serve the purpose of familiarizing the reader with basic concepts in parallel computing.

There are three basic building blocks of parallel programming. These are parallel system architectures, parallel algorithms, and parallel programming models. The first step towards writing parallel programs is the identification of the class of parallel systems where the application is to be run. If the parallel system is based on a special-purpose architecture (such as a cell-based system [22]), then its details need to be carefully understood. On the other hand, if the system is based on a general-purpose architecture, then only the high-level details such as amount of memory per node, number of processors in the node (sharing its memory), number of nodes, the interconnection network bandwidth, and latency need to be considered. The next step involves the design of an efficient parallel algorithm that solves the given problem on the target parallel architecture. The final step is the choice of a programming language and a parallel programming model. A parallel programming model provides high-level abstract operations (or primitives) with which a parallel program may be written.

The next section gives a bird's eye view of parallel architectures. This is followed by a brief description of parallel programming models. Section 5.4 contains an introduction to programming using the message passing interface (MPI), which is the most widely used parallel programming model. The objective of this section is to give the reader a flavor of parallel programming using MPI and highlight some common issues faced while writing parallel programs. This is neither intended to be a comprehensive reference nor a detailed tutorial on MPI. Instead, the objective is to enable the reader to write simple yet useful parallel applications that may arise in the microscopy or imaging domain. Nine basic MPI functions are illustrated and explained using simple programs. Using these functions, the reader should be able to write useful parallel MPI programs for processing the imaging data. These programs may be run on small to moderate sized high-performance parallel systems. For the sake of simplicity, some of the issues such as deadlocks are not discussed. More details can be found in [14, 19, 28, 32].

5.2 Parallel Architectures

Most of the computer systems of today are based on the *Von Neumann architecture*. In 1945, John von Neumann wrote a paper [36] describing a general-purpose stored-programmed system called EDVAC. The Von Neumann architecture as shown in Figure 5.4 is an abstraction of this design. It consists of a general-purpose memory, a general-purpose processing unit (also called central processing unit or

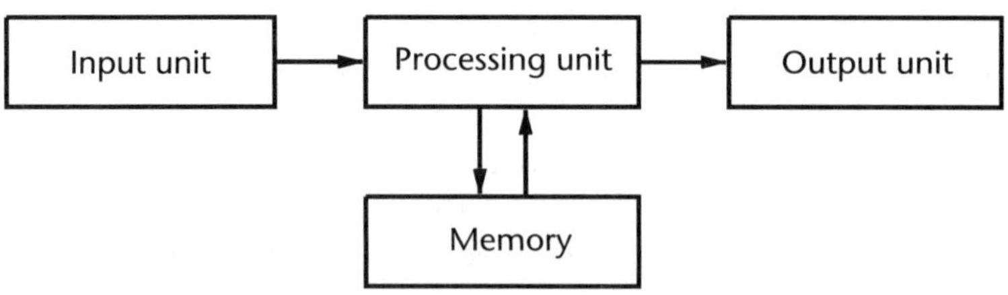

Figure 5.4 The classical Von Neumann architecture.

CPU), and I/O units. The memory stores the program as well as data needed for computations. The CPU manipulates the data stored in the memory according to the instructions given in the program. The I/O units carry out data transfer between the system and the external world such as display, keyboard, and permanent storage devices.

The modern computer systems are based on only minor modifications of the basic Von Neumann architecture. These modifications include the addition of a cache hierarchy, networks for communication between different systems, and parallel processing capabilities.

Figure 5.5 shows the modified Von Neumann architecture with memory hierarchy and communication network. While the processors have seen dramatic improvements in clock frequencies over the last four decades, improvements in the rate at which the main memory can be accessed have been relatively slow. The concept of a cache memory has been used to bridge the gap between processor clock frequency and memory access time. A cache is a small block of high-speed (and expensive) memory, set aside to store frequently accessed data. Most of the programs exhibit large temporal and spatial locality in their memory access patterns. This property is used to dynamically store blocks of memory in the cache that are likely to be accessed in the near future. As a result, the processor does not have to wait for the memory, if the required data is available in the cache. This leads to tremendous performance improvements in the programs. As the gap between processor and memory performance increased, the cache organization became hierarchical with multiple layers of caches of different sizes, performance levels, and costs. The L1 cache, which is the fastest, the smallest, and the most expensive, is closest to the processor (typically in the processor chip itself). The L2 cache is larger, slower, and less expensive. Similarly, the L3 cache is bigger and slower than the L2 cache.

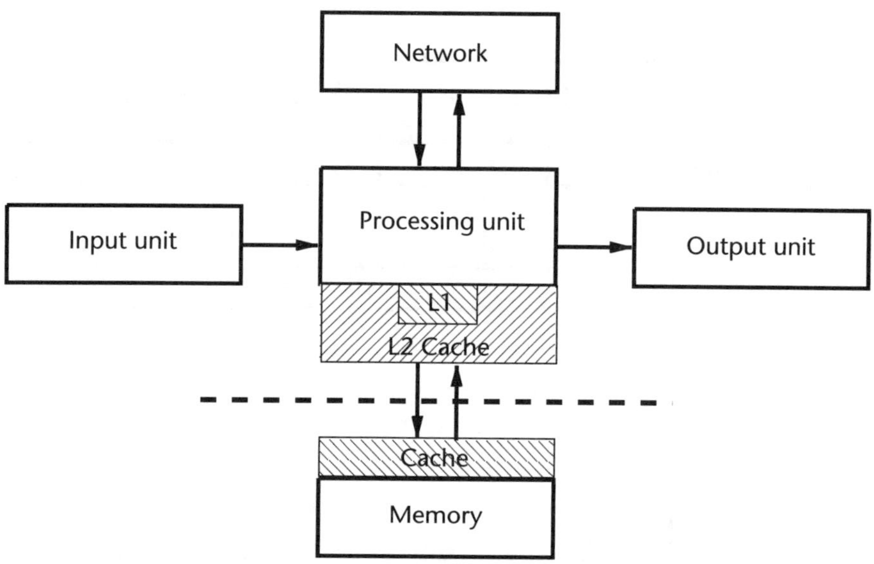

Figure 5.5 The Von Neumann architecture with memory hierarchy and communication network.

The communication networks give the system the ability to communicate with other systems over long distances. These networks are also utilized to carry out major parts of the I/O and storage.

A parallel system comprises many processing units interconnected with each other in different ways. A *shared memory system* has multiple processing units sharing the same memory as shown in Figure 5.6. These processors typically have a private L1 cache which is made consistent with the caches of other processors by the hardware. Higher levels of cache such as L3 may be shared among the processors. In addition, each processor may also have some private memory not shared with other processors. Most of the communication among the processors may be done via the shared memory, but the system may also employ specialized hardware for interprocessor communication.

Due to cache consistency overheads, the shared memory systems cannot have a very large number of processors. Current state-of-the-art systems, such as the IBM Power 595, support up to 64 shared memory processors in a node.

A *distributed memory system* consists of multiple compute nodes (which may themselves be shared memory systems), interconnected using a high-speed network as shown in Figure 5.7. Each of the nodes has its own memory which is local to the node and is not directly accessible to the other nodes. The nodes may exchange data among each other using the high-speed interconnection network.

General purpose high-performance computing systems are typically based on a hybrid shared/distributed memory architecture wherein several shared memory nodes are connected using a high-performance interconnection network. The configurations of such systems can differ widely. For example, the Blue Gene/L supercomputer, installed at the Lawrence Livermore National Labs, Livermore, California, USA, is based on 106,496 dual-processor nodes, whereas the EKA supercomputer installed at the Computational Research Laboratories, Pune, India is based on 1,800 nodes, where each node is an eight-processor shared-memory system.

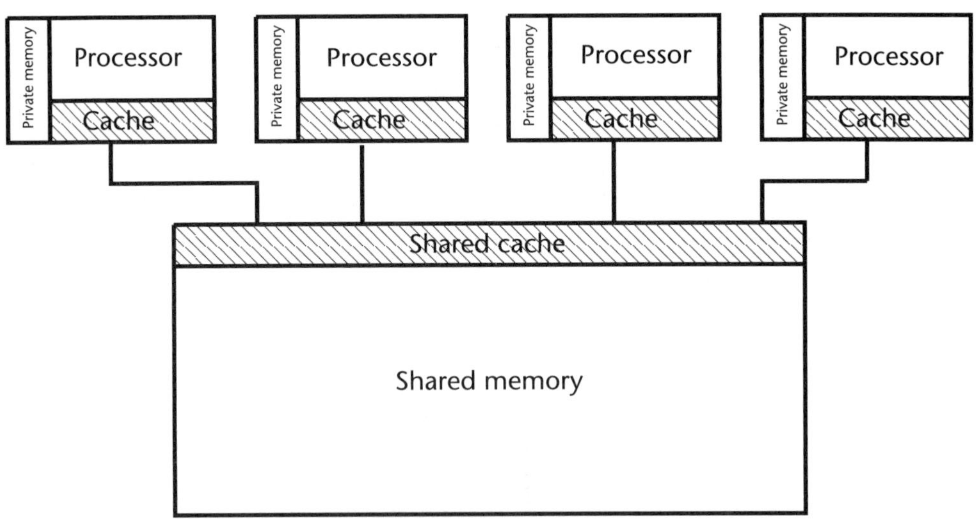

Figure 5.6 Architecture of a shared-memory machine.

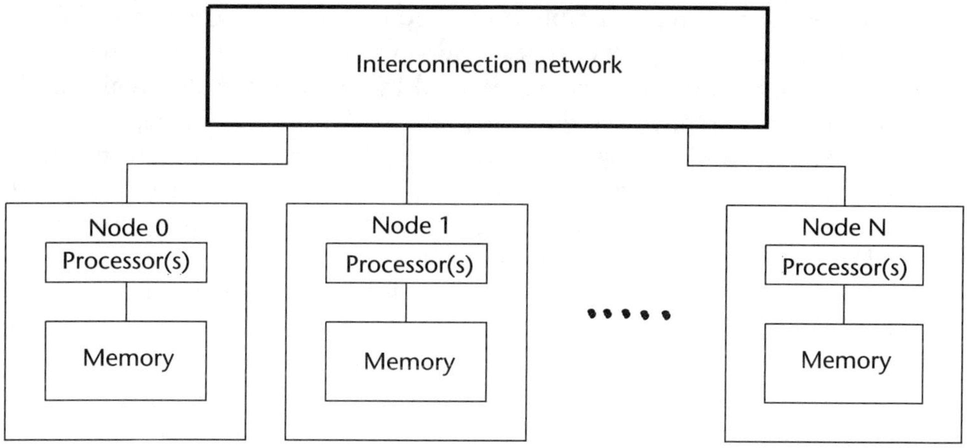

Figure 5.7 Architecture of a distributed-memory machine.

In addition to the general purpose, shared memory and distributed memory parallel systems, another class of special-purpose parallel systems is emerging. Such systems are usually designed for specific applications and are capable of providing exceptional computational performance. However, programming such a system often requires a detailed understanding of the underlying hardware and a careful tuning of the application to optimally utilize its capabilities.

One such system is the IBM Roadrunner supercomputer, the first system to achieve a Petaflop performance, installed at the Los Alamos National Laboratory in New Mexico. This system is based on an improved version of the special-purpose Cell processor [13, 22] which was originally used to power Sony's PlayStation 3 video game machine. In addition to a generic 64-bit PowerPC processor, the Cell chip contains eight special purpose units (SPUs) that may operate in a single instruction multiple data (SIMD) mode [22]. These units are key to the performance that the Cell-based systems can offer.

5.3 Parallel Programming Models

Parallel programming models are designed to ease the task of parallel programming. Without the support of a parallel programming model, programming a parallel system is not simple. Even after a problem has been decomposed into a parallel algorithm, its implementation requires a very precise coordination of activities across different parallel processors. It needs to read and distribute data across multiple nodes or processors, assign work to each processor and instruct it to carry out its assigned work on the assigned data, exchange intermediate results among processors in a coordinated fashion, synchronize processors at different stages in the algorithm, maintain a consistent view of data distributed across processors, and resolve race conditions, if any. Moreover, all this coordination has to be done efficiently, in a way that processors spend most of their time carrying out the "real" computation, and minimize the time spent in overheads of coordination, message exchanges, and waiting for intermediate results from other processors. This problem is exacerbated by the fact that different parallel systems have different architectures (vector versus scalar processors, shared memory versus distributed memory, single core versus multi-core) and architectural parameters, such as number of processors, amount of memory per processor, and network bandwidth. Ideally one should be able to write parallel programs that are independent of such system specific details.

This situation is analogous to the old days of programming a uniprocessor computer system in machine or assembly language, where the programmer was required to keep track of contents of the specific machine registers, perform machine-specific operations on them, and move the data across registers or main memory, while maintaining a consistent view of the data and the program. The evolution of high-level languages such as COBOL, FORTRAN, C, C++, and Java has given a tremendous boost to the programmability of computer systems by providing an abstract model of a computer system which exposes all its essential features that are common across a variety of computer architectures and generations. As a result, the programmer can focus on only encoding the algorithm using a generic high-level machine independent language, without worrying about the daunting architectural details of the specific computer system on which the program is to be run. The task of converting the high-level representation of an algorithm (in the form of a program) to a low-level machine and architecture-specific form has been relegated to the compilers for the machine. In addition to the ease of programming, the high-level languages also allow users to run a program on different computers of different architectures and generations with little effort. For running a program on a new system, one simply needs to recompile it using the system-specific compiler.

5.3.1 The Three P's of a Parallel Programming Model

The goal of a parallel programming model, in the same vein, is to expose the essential elements of a parallel system in an abstract manner, making it easier for a programmer to code a parallel algorithm into a high-level parallel program using the primitives and abstractions provided by the model. Such a program can then be translated into a low-level representation for execution on the parallel system. Such a model must be intuitive, easy to understand, and must make the task of coding the parallel algorithm into a high-level parallel program as simple as possible. Moreover, it should hide the unnecessary details of the underlying parallel system, so that a programmer can focus exclusively on the task of mapping a parallel algorithm to the abstract parallel programming model, without having to worry about the details of the specific system on which the program has to be run.

All such abstractions however, come at a cost. There is an inherent trade-off between the levels of abstraction and the program performance. Ideally, one should not even have to bother writing a parallel program. One should just be able to write a sequential program in a high-level language and a parallelizing compiler should take care of the task of parallelizing the program. While this may be possible for some specific applications running on specific systems, writing a general-purpose compiler that carries out this mapping for all possible input programs, and also give reasonable system performance, is next to impossible. Such a compiler will also have to design an efficient parallel algorithm for every sequential program that it may take as its input!

A parallel programming model must strike a balance between the level of abstraction provided to the programmers and the levels of performance it can achieve. In addition, it should provide a reasonably accurate model for parallel systems that exposes the issues involved in designing efficient parallel algorithms and forces the programmer to think about them. For example, providing a shared memory abstraction on a distributed memory system would hide the important issues of data distribution and interprocessor communication, which may need to be tackled through algorithmic improvements. While such an abstraction will make it very easy to program massively parallel systems, the programs themselves will be of little use as they will hardly be able to achieve any reasonable performance. A programmer can never write efficient code if a truly opaque shared-memory abstraction is provided on a distributed memory system. There will be no way to carry out efficient data distribution or to estimate the interprocess communications involved in accessing data stored at any given memory location.

It is also important that the model be flexible enough to allow a determined programmer to use certain aspects of the parallel system that would enable a performance close to the peak system performance. For example, on the 64 Racks Blue Gene/L system which has 128K processors, application programs were able to achieve a performance of 207.3 TFLOPS (10^{12} floating point operations per second) which is 56.5% of the theoretical peak system performance. While such performance levels are exceptional and require programmers and researchers to utilize system-specific elements, it is still possible for programs to achieve a good fraction of the peak system performance while using only the standard programming models. For example, the Linpack benchmark achieved 22% of peak performance in the the HPC Challenge benchmark runs [21] using MPI and BLAS alone.

In a nutshell, productivity, portability, and performance are the three P's of a parallel programming model.

- *Productivity:* The model should enable the programmer to write parallel programs with the least effort. While being intuitive and simple, it should also enable to programmer to focus on critical issues pertaining to algorithmic design while hiding unnecessary system-specific details.
- *Portability:* The model should be abstract enough to have applicability on a wide variety of parallel systems. It should be possible to make it available on multiple systems so that programs written using the model can be ported to multiple parallel systems with little or no effort.
- *Performance:* The model should incorporate the essential elements of parallel systems in a way that enables the programmer to write efficient codes. Moreover, it should provide enough flexibility to the programmer to exploit the capabilities of the underlying parallel system optimally. Finally, it must be possible to implement the model on multiple parallel systems with little performance overheads.

The next section introduces parallel programming using the MPI. The MPI is the de facto standard for parallel programming today. Most of the current parallel systems support MPI. The objective of this section is to introduce the reader to MPI to a degree such that she/he can begin writing parallel programs using MPI. This section also discusses some of the issues faced by a parallel MPI programmer. This section is not intended to be a reference to MPI. Nor is this section intended to be a detailed tutorial on MPI. The complete reference to the MPI standard (versions 1.1 and 2.0) can be found in [14, 28]. The reader is referred to [15, 19, 32] for an in-depth understanding of advanced parallel programs.

5.4 The Message Passing Interface

MPI has evolved as one of the most widely used standards for writing parallel programs. If comprises a single program multiple data (SPMD) programming model, where multiple instances of the same program run on different processors of a parallel system, as different MPI processes. Each MPI process has its own local memory which is not shared with other processes. These processes are identified using a *rank* that ranges from zero to one less than the total number of MPI processes (*size*). These processes may exchange messages with each other using a carefully defined application programming interface (API) called the message passing interface.

The use of *rank* and *size* simplifies the task of naming and addressing different processes in the application. It also hides the details about name and address of the host system, the identity of the processor running a specific MPI process, and the communication protocol, and gives a convenient method to exchange data among processors.

In addition to naming and addressing, MPI provides an abstraction for several basic data-types such as integers, real numbers (single or double precision), complex numbers, booleans, and characters. The MPI processes can communicate these data-types among each others using simple send/receive function calls. As a result, the programmer does not need to know the lower level details such as size and encoding of these data-types (which is typically needed while writing general communicating programs). The implementation of these functions is done in a system-specific MPI library, which contains an optimized code that runs efficiently on the host system.

MPI provides several intuitive primitives to exchange data among application processes. These include functions for point-to-point message exchanges, collective communication operations within a group of processors, synchronization operations, asynchronous communication (to facilitate computation-communication overlap), among others. These primitives have been designed keeping the needs of various scientific applications in mind. They provide a rich set of methods to exchange data with varying degrees of complexity.

MPI is available in the form of a library of functions in C/C++ or Fortran. Wrappers around MPI have been built to enable it on other programming environments such as MATLAB and Octave [29]. The MPI version 1.1 [14] has over 90 functions, which was followed by MPI version 2.0 [28] which has more than

200 functions. Two popular public-domain implementations of MPI are MPICH [18, 20] and LAM-MPI [17, 33]. In addition, grid implementations of MPI are also available [23]. Most of the parallel system vendors supply their own optimized implementations of MPI.

5.4.1 The Nine Basic Functions to Get Started with MPI Programming

One should not get intimidated by such a large number of MPI functions in the formal MPI specifications [14, 28]. In practice, only a small subset of these functions is used in real applications. In fact, by using only nine MPI functions one can begin to write fairly complex MPI programs. These nine MPI functions, namely, `MPI_Init`, `MPI_Finalize`, `MPI_Comm_size`, `MPI_Comm_rank`, `MPI_Send`, `MPI_Recv`, `MPI_Bcast`, `MPI_Barrier`, and `MPI_Allreduce`, are described in more detail in rest of this section. These functions will be introduced through a series of MPI programs written in C, starting with a simple MPI-based "hello world" program, and ending with a program that computes all-pair correlations for a dataset with multiple time-series.

An MPI Version of a "Hello World" Program

Figure 5.8 contains the listing of an MPI version of the famous "hello world" program. Every process in the program carries out the following operations: (1) find out the number of MPI processes running as a part of the job; (2) find out its *rank*—an integer identifying the process; and (3) print the "hello world" message, the rank of the process, size of the job, and a message spreading happiness. Four MPI functions used in this program, `MPI_Init`, `MPI_Finalize`, `MPI_Comm_size`, and `MPI_Comm_rank`, and these are explained in detail in the following discussion.

MPI_Init and MPI_Finalize

The program mpi-hello.c, as listed in Figure 5.8, includes the file `mpi.h` at line 2. This file contains the MPI-specific prototypes and definitions. At line 10, the program makes a call to the function `MPI_Init`. As the name suggests, `MPI_Init` initializes the MPI subsystem and its data-structures. All MPI programs need to make a call to the function `MPI_Init` before carrying any other operation. The argument to this function are the pointers to the variables `argc` and `argv` passed to the function `main` by the system. Note that `MPI_Init` should be called even before carrying out any command-line parsing. This ensures that the variables `argc` and `argv` refer to the correct command-line arguments.

All the MPI processes in a program should call `MPI_Finalize` before terminating. This function call performs a shutdown of the MPI-subsystem, releases the system resources allocated earlier, and reports errors, if any. Line 17 of the program mpi-hello.c contains call to the function `MPI_Finalize`.

MPI_Comm_size and MPI_Comm_rank

As discussed earlier, the MPI follows a single program multiple data programming model. MPI programs are typically run as MPI jobs (or simply, jobs). When an MPI job is launched on a parallel system, many identical copies of the program

```
 1: #include <stdio.h>
 2: #include <mpi.h>
 3:
 4: char *msg = "May everyone be happy. May everyone be h$$
 5:
 6: int main(int argc, char *argv[])
 7: {
 8:    int size, rank;
 9:
10:    MPI_Init(&argc, &argv);
11:    MPI_Comm_size(MPI_COMM_WORLD, &size);
12:    MPI_Comm_rank(MPI_COMM_WORLD, &rank);
13:
14:    printf("Hello World. There are %d MPI processes. \
15: My rank is %d. My message is: %s\n", size, rank, msg);
16:
17:    MPI_Finalize();
18:    return 0;
19: }
```

```
$  mpirun -partition wR0 -exe mpi-hello.rts  -np 4000
Hello World. There are 4000 MPI processes. My rank is 3794. My
message is: May everyone be  .... May everyone Hello World. Th
ere are 4000 MPI processes. My rank is 3538. My message is:...
```

Figure 5.8 An MPI version of a "hello world" program and excerpts of its output.

are created and executed as different MPI processes on different processors. The number of MPI processes in a job and the details of the parallel system where these processes will be executed are not specified in the MPI program. Instead, these are specified at the time the job is created. The number of processes in the job is usually referred to as the job size. When a job starts executing, all its MPI processes start executing simultaneously in parallel. In the example, all the MPI processes created for the job will make a call to `MPI_Init`.

The function `MPI_Comm_size` may be used in a program to determine the number of MPI processes running as a part of the current MPI job. Line 11 of mpi-hello.c contains a call to this function. `MPI_Comm_size` takes two arguments. The first argument is the communicator, which is set to a predefined default MPI communicator called `MPI_COMM_WORLD`. MPI commuincators are discussed later in Section 5.4.2. The second argument is a pointer to the integer where the job size is to be returned.

Each MPI process in an MPI job is identified by a unique number called its *rank*. The rank of an MPI process in a job varies from zero to one less than the number of processes in the job. The function `MPI_Comm_rank` is used by an MPI process to determine its rank. The first argument to this function is the communicator, which is set to the default MPI communicator called `MPI_COMM_WORLD`. The second argument is a pointer to the integer where the rank is to be returned.

Line 12 of the program contains a call to the `MPI_Comm_rank` function. Finally, lines 14 and 15 print the ranks and sizes along with the message.

Figure 5.8 also shows a session of running the program mpi-hello.c, on a 4,096-node Blue Gene/L system system and excerpts of its output. The first line shows the `mpirun` command, used to launch the executable binary file `mpi-hello.rts` on the Blue Gene/L system. The first argument `wR0` represents a 4,096-node partition (i.e., it specifies the specific set of processes on the Blue Gene/L System where this job is to be run). The argument `-np 4000` specifies the number of MPI processes that need to be created. The argument `-exe mpi-hello.rts` specifies the compiled executable `mpi-hello.rts` of the mpi-hello.c program to be run. The job launch command and parameters are not standardized and vary from system to system. Even for a single system, there may be many ways to launch a job. One must consult the system administrator to get details of how MPI jobs may be created.

Each MPI process of the program prints its message simultaneously to the standard output. MPI does not define any ordering in which the outputs of individual MPI processes are displayed. These outputs are generally printed in a first come first serve (FCFS) manner. As a result of this, the output of the MPI processes is garbled as shown in Figure 5.8.

How can this problem be fixed? Since the output is printed in FCFS manner, imposing an explicit ordering by appropriately timing the output of the MPI processes may solve this problem. One naive (though not advisable) way to impose order is by introducing different delays at different MPI processes before printing. The amount of delay at each MPI process could be made proportional to its rank (as in the program mpi-hello2.c listed in Figure 5.9), thereby ensuring that the MPI processes print their message in a staggered way. In mpi-hello2.c, lines 2 and 15 were added to stagger the `printf` by introducing delays proportional to the rank of the MPI processes. In the output of this program, messages from different MPI processes are printed in separate lines, as desired.

Instead of adding delays proportional to the ranks, one could add random delays as in mpi-hello3.c listed in Figure 5.10. Quite surprisingly, the output of this program was garbled and looked like that of mpi-hello.c. This happens due to the fact that when an MPI job is started, identical copies of the same program are created. The calls to the `rand` function return a sequence of *pseudo random numbers* determined by a *seed*. When the MPI job starts, all its processes are identical and therefore have the same seed. Thus, the pseudo random sequence generated at each process is the same. So each of the MPI processes, waits for exactly the same time, and then prints out the message at exactly the same time, resulting in the garbled output.

This problem can be fixed by initializing the seeds of the random number generators at each process with different numbers. One can use the rank of the process to carry out this initialization. Adding the call `srand(rank+1)` before line 16 solves this problem. With this change, output of mpi-hello3.c program is nicely staggered with no mixing of messages printed by different MPI processes.

Any solution that requires processes to wait, without synchronizing with each other, is not guaranteed to work. The delays required to stagger the output are highly dependent on the system and its state at the time the processes print their output. The MPI subsystem guarantees that identical copies of the program will

```
#include <stdio.h>
#include <unistd.h>
#include <mpi.h>

char *msg = "May everyone be happy. May everyone be h$$

int main(int argc, char *argv[])
{
    int size, rank;

    MPI_Init(&argc, &argv);
    MPI_Comm_size(MPI_COMM_WORLD, &size);
    MPI_Comm_rank(MPI_COMM_WORLD, &rank);

    usleep(1000*rank);
    printf("Hello World. There are %d MPI processes. \
My rank is %d. My message is: %s\n", size, rank, msg);

    MPI_Finalize();
    return 0;
}
```

Figure 5.9 In the program mpi-hello2.c, delays proportional to ranks are used to stagger the output.

be created and executed in parallel. However, it does not provide any guarantee about when these processes will start execution and the rates at which these will be executed. So when the program starts, some of the processes might be behind others while some may execute at a rate faster than that of others. Since, in general, the performance of a program is determined by the performance of its slowest process, the MPI subsystem does try to ensure that all the processes of a job start at the same time and execute at the same rate. However, it does not provide any such guarantee. Therefore, MPI programs can make no assumptions about the relative start times or the rates of execution of its processes.

MPI_Send and MPI_Recv

To find a reliable solution to the problem of mixing of outputs of different MPI processes, one may carry out explicit coordination among processes. For this, a *token passing* strategy may be used. Under this scheme, a process prints its message only if it has the token. Initially, the token belongs to the first process which prints its message and passes the token to the next process. The second process, after receiving the token, prints its message and forwards the token to the next process. This continues until the last process gets the token and prints its message. To implement this strategy, two new MPI functions are used. The first function is `MPI_Send`, which is used to send a message from one process to another. The second is `MPI_Recv`, which is used to receive messages at the processes. Figure 5.11

```
#include <stdio.h>
#include <stdlib.h>
#include <unistd.h>
#include <mpi.h>

char *msg = "May everyone be happy. May everyone be h$$

int main(int argc, char *argv[])
{
    int size, rank;

    MPI_Init(&argc, &argv);
    MPI_Comm_size(MPI_COMM_WORLD, &size);
    MPI_Comm_rank(MPI_COMM_WORLD, &rank);

    usleep(1.0 * rand() / RAND_MAX);
    printf("Hello World. There are %d MPI processes. \
My rank is %d. My message is: %s\n", size, rank, msg);

    MPI_Finalize();
    return 0;
}
```

Figure 5.10 In the program mpi-hello3.c, random delays are used to stagger the output.

shows the listing of the mpi-hello4.c program, which implements this token-passing strategy. Here, the token is implemented by an integer of value zero. The process of rank `0` prints its message and sends this integer to the rank `1` process. A process of rank between `1` and `size - 2` first waits for a message from the process of one less rank, then prints its message and then sends the integer to the process of one greater rank. The process with the largest rank, which is `size - 1`, waits for the message from the process of rank `size - 2`, and then prints its message without passing the token to any other process.

The `MPI_Send` function takes six arguments. Its first argument is a pointer to the data to be sent. In the example mpi-hello4.c (see Figure 5.11), it is set to the pointer to the integer `token`. The second argument is the number of data elements that are to be sent. In the above example, it is set to 1, since only one integer is to be sent. The third argument is the type of data being sent. MPI defines 15 primitive data-types for the C and FORTRAN languages. These are listed in Table 5.1.

In addition, MPI provides the data-type `MPI_PACKED`, for sending arbitrary noncontiguous data. This data-type is useful for developing additional communication libraries using MPI. Additional MPI functions such as `MPI_Type_vector`, `MPI_Type_indexed`, `MPI_Type_struct`, `MPI_Type_contiguous`, `MPI_Address`, `MPI_Type_extent`, and `MPI_Type_size` are provided for creation of more complex data-types such as vectors, arrays, indexed structures, contiguous data-types, and general data structures.

```
#include <stdio.h>
#include <mpi.h>

char *msg = "May everyone be happy. May everyone be h$$

#define TOKEN_TAG 1

int main(int argc, char *argv[])
{
   int size, rank;
   int token = 0;
   MPI_Status stat;

   MPI_Init(&argc, &argv);
   MPI_Comm_size(MPI_COMM_WORLD, &size);
   MPI_Comm_rank(MPI_COMM_WORLD, &rank);

   if (rank != 0)
     MPI_Recv(&token, 1, MPI_INTEGER, rank - 1, \
       TOKEN_TAG, MPI_COMM_WORLD, &stat);

   printf("Hello World. There are %d MPI processes. \
My rank is %d. My message is: %s\n", size, rank, msg);

   if (rank <= size - 2)
     MPI_Send(&token, 1, MPI_INTEGER, rank + 1,
       TOKEN_TAG, MPI_COMM_WORLD);

   MPI_Finalize();
   return 0;
}
```

Figure 5.11 In the program mpi-hello4.c, token passing is used to stagger the output.

The fourth argument to `MPI_Send` is the rank of the destination process, to which the message is to be sent. In the example above, every process passes the token to the process with one greater rank. Thus, this argument is set to `rank + 1`. The fifth argument is called the *tag* in MPI terminology. It is an integer which identifies the "channel" on which the data is being sent. Between any pair of communicating processes, the data may be sent on multiple channels. The receiving process can choose to receive message on a particular channel identified by the tag. This functionality is very useful if multiple types of messages are exchanged. In the example above, if the processes were also communicating data (say, in floating point) in addition to the token, then a separate tag could have been used for sending the additional data. This enhances modularity and simplyfies the task of bookkeeping of the messages. For example, libraries built on top of MPI may use a disjoint set of tags for exchanging messages. This will ensure that the messages sent by the libraries do not interfere with other messages of the programs.

Table 5.1 MPI Data-Types and the Corresponding C and FORTRAN Data-Types

MPI Name	*C Data Type*	*MPI Name*	*FORTRAN Data-Type*
MPI_CHAR	signed char	MPI_CHARACTER	character(1)
MPI_SHORT	signed short int		
MPI_INT	signed int	MPI_INTEGER	integer
MPI_LONG	signed long int		
MPI_UNSIGNED_CHAR	unsigned char		
MPI_UNSIGNED_SHORT	unsigned short int		
MPI_UNSIGNED	unsigned int		
MPI_UNSIGNED_LONG	unsigned long int		
MPI_FLOAT	float	MPI_REAL	real
MPI_DOUBLE	double	MPI_DOUBLE_PRECISION	double precision
MPI_LONG_DOUBLE	long double		
		MPI_COMPLEX	complex
		MPI_DOUBLE_COMPLEX	double complex
		MPI_LOGICAL	logical
MPI_BYTE	8 binary digits	MPI_BYTE	8 binary digits

The last argument to the `MPI_Send` function is the communicator to be used. In the program discussed above, it is set to the default MPI communicator `MPI_COMM_WORLD`.

The `MPI_Recv` function takes seven arguments. The first three arguments are, respectively, the buffer in which the data is to be received, the maximum number of data elements that can be received, and the data-type to be received. The fourth and fifth arguments are the rank of the sender and the tag (channel) on which the message is to be received. In the present example, each process (except process of rank 0) receives the message from the process of one less rank. The sixth argument is the communicator presently set to `MPI_COMM_WORLD`.

The last argument is a pointer to the status object that contains more information about the message transfer. In many cases, a process may not know the sender or tag of the message that it should receive. To enable an MPI process to receive messages from a sender and/or a tag, without explicitly specifying its identity, the constants `MPI_ANY_SOURCE` and `MPI_ANY_TAG` may be used. If `MPI_ANY_SOURCE` is used to identify the sender in the `MPI_Recv` function call, then a message from any sender that matches the tag may be returned. Likewise, if `MPI_ANY_TAG` is used to identify the tag, then any message that matches the sender, may be returned irrespective of the tag. If `MPI_ANY_SOURCE` and `MPI_ANY_TAG` are used in a call to the function `MPI_Recv`, then any available message may be returned. In these cases, the status object returns the identity of the sender and the tag of the message received.

`MPI_Recv` is called a *blocking receive* function. The function call does not return until a matching message is available at the process. The token passing strategy can be implemented using `MPI_Send` and `MPI_Recv` in program mpihello4.c of Figure 5.11, precisely because of this property. The rank 1 process gets *blocked* at `MPI_Recv`, until the message sent by the rank 0 process is available, which happens only after the rank 0 process prints its message and calls `MPI_Send`.

Thus, the rank 1 process prints its message only after returning from `MPI_Recv`. As a result, the `printf` function at lines 22–23 is called by each MPI process in a strictly sequential order.

Note that the data-type of message sent (integer, floating point, etc.), its size, or its actual value is not important for the correct functioning of the program. Only the synchronization induced by the `MPI_Send` and blocking `MPI_Recv` function calls is required for its correctness.

Unfortunately, even after enforcing explicit ordering by the token passing method, the program mpi-hello4.c is not guaranteed to work correctly. To understand why this is the case, one must carefully examine the semantics of the `printf` function (and of I/O functions, in general) in an MPI program. In the mpi-hello4.c program, all the MPI processes write to the same file (the standard output). According to the MPI semantics, all the data written sequentially to a file (such as by `printf`) is guaranteed to be actually written to the file. However, the MPI standard does not provide any guarantee about preserving the order in which different MPI processes write their data. Although, most systems will produce a well-separated output of the program, it is a perfectly legitimate behavior if a system chooses not to do so, as long as it does not lose any part of the output.

At this point, it is worth noting that the concept of ordering of events in a distributed system acquires some relativistic dimensions. It is conceptually similar to that of ordering events at two distant points in space, as in the special theory of relativity. A set of events is said to be *totally ordered* if a consistent "happened before" relationship between any two events can be established. Here consistency implies that, if an event A happened before another event B and event B happened before the event C, then the event A happened before the event C. The set is called *partially ordered* if the happened-before relationship is consistent but not defined for all pairs of events. In a general distributed system, the events form a partially ordered set (see [26] for a detailed discussion). Therefore, one may not be able to define which of the two events (such as a call to the function `printf`) occurring on two different MPI processes happened before.

Figure 5.12 shows one way to program the hello-world application which guarantees the correct operation. Here, each MPI process sends its output to the rank `0` process. The rank `0` process receives `size - 1` messages, and prints each of the messages to the standard output (`size` is the number of MPI processes in the job). In fact, many real applications use a similar strategy to output results. These applications designate a *master* process to perform the input and output functions. All the other MPI processes send their results to the master process which writes them to files. Although this technique is guaranteed to work correctly, applications requiring large I/O may not scale well on large parallel systems. Essentially, this method serializes the I/O operations using the master node. Thus, on large systems the master process may become the bottleneck, slowing down the application. In such cases, it may be better to use MPI_IO [28]. If the amount of I/O required by the application is small, then the above technique may be best suited because of its simplicity and correctness.

Note that the above example was chosen to illustrate the issues involved in writing parallel programs using MPI. Alternatively, each process can print its output to a separate file (say, filename indexed by the process rank). This approach

```
#include <stdio.h>
#include <mpi.h>

char *msg = "May everyone be happy. May everyone be h$$

#define MSG_TAG 2

int main(int argc, char *argv[])
{
   int size, rank;
   int i;
   MPI_Status stat;
   char buffer[2048];

   MPI_Init(&argc, &argv);
   MPI_Comm_size(MPI_COMM_WORLD, &size);
   MPI_Comm_rank(MPI_COMM_WORLD, &rank);

   sprintf(buffer, "Hello World. There are %d MPI \
processes. My rank is %d. My message is: %s\n", \
    size, rank, msg);

   if (rank == 0) {
     printf(buffer);
     for (i = 0; i < size - 1; i++) {
       MPI_Recv(buffer, 2048, MPI_CHAR, \
 MPI_ANY_SOURCE, MSG_TAG, MPI_COMM_WORLD, &stat);
       printf(buffer);
     }
   } else {
     MPI_Send(buffer, 2048, MPI_CHAR, 0,
       MSG_TAG, MPI_COMM_WORLD);
   }

   MPI_Finalize();
   return 0;
}
```

Figure 5.12 The program mpi-hello5.c uses rank 0 process to exclusively carry out input and output. It is guaranteed to produce correct results.

may also work well if the amount of I/O as well as the number of MPI processes is small.

Computing All Pair Correlations

Applications in many domains including biology, economics, and functional magnetic resonance imaging involve time-series analysis. The data in these domains often contains multiple time-series. Computing correlations between every pair of time-series is a fundamental operation needed by these applications. In this subsection, a parallel MPI code to compute all-pair correlations will be presented and explained. In the process three more MPI functions, namely `MPI_Bcast`, `MPI_Allreduce`, and `MPI_Barrier` will be introduced.

The interest in functional connectivity analysis of fMRI data has been growing in the recent years [35]. One way to identify functional connectivity between different brain regions is by computing correlations among every voxel pair and declaring pairs with high correlations as "functionally connected." In this section, a parallel MPI program to implement this will be described in detail. The program works on fMRI data of fixed size. It computes the temporal correlation coefficient between every pair of voxels in the data. If the correlation is larger than a given threshold, then the pair of voxel and the corresponding correlation is printed in a text file. The program also computes the sum of absolute value of correlations between all pairs of voxels and prints the result to its standard output. Such a program can be used to carry out network analysis on fMRI data (such as the one described in [35]).

Data and Work Distribution

Before writing the code, it helps to think a little bit about the data and the work distribution. A simplifying assumption made here is that all the MPI processes have sufficient memory to store the complete input data. This is a reasonable assumption in the context of fMRI data analysis, where a session data typically occupies 100 to 500 MB. Thus, the task of data distribution becomes very simple. The complete data can be replicated on all the MPI processes.

The work distribution also is not too difficult. If there are N voxels, then the objective is to compute correlations for each of the $N(N-1)/2$ voxels pairs. For distributing this computation, voxels are assigned to MPI processes such that the process which has been assigned voxel i, computes correlations of voxel i with all other voxels j, such that $j < i$. Now, the task reduces to assigning the voxels to processes.

Let there be P processes available to carry out computation. The N voxels may be divided among P processes in blocks of size `blk_sz = ⌈N/P⌉`. In this case, the process of rank r is assigned voxels from `blk_sz*r` to `min(blk_sz*(r+1)-1, N-1)`. This work distribution is pictorially depicted in Figure 5.13. The entry c_{ij}, shown in the correlation matrix, represents the correlation between voxel i and voxel j. Since the correlations are symmetric (i.e., $c_{ij} = c_{ji}$), only the upper triangular part of the matrix needs to be computed. Each of the trapezoids represents the entries of the correlation matrix computed by the corresponding process. It is evident that the above work distribution is very uneven. It assigns very little work to processes of small ranks and progressively more work to processes of larger ranks. Thus, low rank processes will finish their work earlier and waste their computational resources while waiting for higher rank processes to complete the work assigned to them.

An alternate work distribution, as shown in Figure 5.14, assigns the voxels r, $r+P$, $r+2P$, ... to process ranked r. With this distribution, every process is assigned short as well as long columns of the correlation matrix. This leads to a more even load balancing and efficient utilization of computing resources.

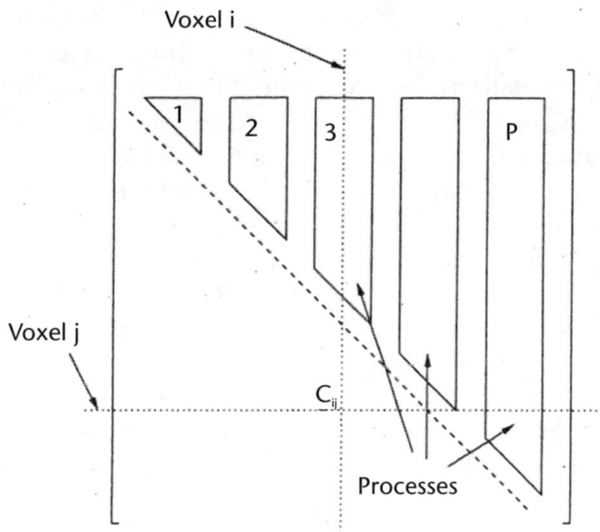

Figure 5.13 Work distribution with poor load balancing. Here, trapezoidal blocks of correlation matrix are distributed to the processors.

If the number of processes, P, is comparable to the number of voxels, N, or the amount of memory available to each of the process is not enough to store the entire data, alternate data and work distributions schemes can be designed. For example, instead of assigning the columns to the correlation matrix to the processes, blocks

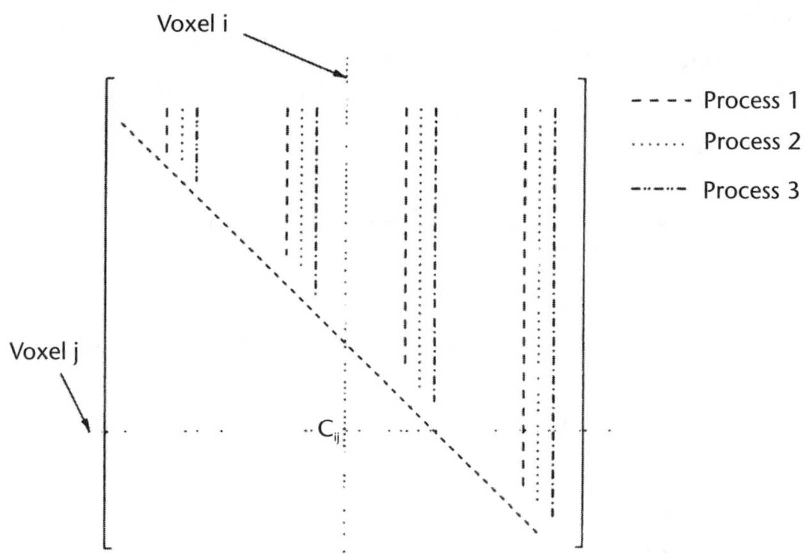

Figure 5.14 Work distribution with good load-balancing. Here, columns correlation matrix are distributed to the processors, cyclically.

of the matrix could be distributed, with suitable choice of block sizes and shapes. Such a scheme can be designed to work efficiently even if $P > N$, provided the number of correlations to be computed is significantly larger than the number of processes (i.e., $N * (N - 1)/2 >> P$). Moreover, since each process needs to store only the data for the voxels corresponding to its block, such a distribution would also allow for very large datasets where the entire data cannot be stored in the memory available at each of the nodes.

The program correlations.c listed in Figures 5.15, 5.16, and 5.17, carries out the work distribution shown in Figure 5.14. In this program, the rank `0` process has been designated as the *master process*, the rest of the processes from rank `1` to rank `mpi_size - 1` are available for computation. In the `main` program shown in Figure 5.15, the rank `0` process calls the function `master` (at line 90) that carries out all the input and output for the program. The rest of the processes call the function `compute` (line 91) to compute the all pair correlations.

MPI_Bcast

The function `master`, as shown in Figure 5.16, first reads the fMRI data from a file (lines 20 to 22) and then sends it to all the other *compute processes* using the function `MPI_Bcast` (lines 25 to 26). After receiving the data from the master, the compute processes implicitly partition the work and compute correlations for the voxel pairs assigned to them. The information about the voxel pairs, having correlations larger than the threshold, is sent back to the master process, which writes the voxel indices and the corresponding correlation value to a file (as in program mpi-hello5.c).

The function `master` (see Figure 5.16) makes a call to `MPI_Bcast` at lines 25-26 to distribute the fMRI data to all the other processes. The `MPI_Bcast` function takes five arguments. The first argument is a pointer to the buffer where the data is to be sent or received. The second and the third arguments are, respectively, the number of data elements and the data-type to be sent. The fourth argument is the rank of the node distributing the data (also called the "root" of the broadcast). In the example above, it is set to the rank of the master process (which is 0).

```
84: int main(int argc, char *argv[]) {
85:   int mpi_rank;
86:
87:   MPI_Init(&argc, &argv);
88:   MPI_Comm_rank(MPI_COMM_WORLD, &mpi_rank);
89:
90:   if (mpi_rank == 0) master(argv[1]);
91:   else compute();
92:
93:   MPI_Finalize();
94:   return 0;
95: }
```

Figure 5.15 Function `main` of the program correlations.c.

```
void master(char *filename) {

  FILE *fp;
  float Data[NVOLS][NVOXELS];
  int i, mpi_size, voxel_pair[2];
  float correlation, sum;
  MPI_Status stat;

  fp = fopen(filename, "rb");
  i = fread(Data, sizeof(float), NVOLS * NVOXELS, fp);
  fclose(fp);
  fp = fopen("Output.txt", "w");

  MPI_Bcast(Data, NVOLS*NVOXELS, MPI_FLOAT, 0,
            MPI_COMM_WORLD);
  MPI_Comm_size(MPI_COMM_WORLD, &mpi_size);
  int count = mpi_size - 1;

  while (count > 0) {
    MPI_Recv(voxel_pair, 2, MPI_INT, MPI_ANY_SOURCE,
             PAIR_TAG, MPI_COMM_WORLD, &stat);
    if (voxel_pair[0] >= 0) {
      MPI_Recv(&correlation, 1, MPI_REAL,
               stat.MPI_SOURCE, CORRELATION_TAG,
               MPI_COMM_WORLD, &stat);
      fprintf(fp, "%d %d %f\n", voxel_pair[0],
               voxel_pair[1], correlation);
    }
    else count--;
  }
  fclose(fp);
  sum=0;
  MPI_Allreduce(MPI_IN_PLACE, &sum, 1, MPI_REAL,
                MPI_SUM, MPI_COMM_WORLD);
  printf("Sum absolute correlation is %f\n", sum);
}
```

Figure 5.16 Function `master` of the program correlations.c.

The last argument of `MPI_Bcast` is the communicator, which is set to the default communicator (`MPI_COMM_WORLD`) in this program.

The semantics of `MPI_Bcast` require that all the processes (that belong to the communicator) call this function. Any program that does not ensure this is an incorrect MPI program and its behavior is undefined by the MPI standard. In reality, such a program may crash, give errors, deadlock, or give incorrect results.

Unlike `MPI_Send` or `MPI_Recv`, which are point-to-point communication functions, `MPI_Bcast` is called a collective communication function. It is defined on a group of MPI processes as defined by the communicator. The MPI semantics require that all the MPI processes in the group make call to this function. The

```
void compute() {
  float Data[NVOLS][NVOXELS], correlation, sum;
  int mpi_size, mpi_rank, i, j, k, voxel_pair[2];

  MPI_Bcast(Data, NVOLS*NVOXELS, MPI_FLOAT, 0,
            MPI_COMM_WORLD);
  MPI_Comm_rank(MPI_COMM_WORLD, &mpi_rank);
  MPI_Comm_size(MPI_COMM_WORLD, &mpi_size);

  Normalize(Data);

  sum = 0;
  for (i = (mpi_rank - 1); i < NVOXELS; i += mpi_size){
    for (j = 0; j < i; j++) {
      correlation = 0;
      for (k = 0; k < NVOLS; k++) {
        correlation += Data[k][i] * Data[k][j];
      }
      correlation = correlation / NVOLS;
      sum += fabs(correlation);
      if (fabs(correlation) > THRESHOLD) {
        voxel_pair[0] = i; voxel_pair[1] = j;
        MPI_Send(voxel_pair, 2, MPI_INT, 0, PAIR_TAG,
                 MPI_COMM_WORLD);
        MPI_Send(&correlation, 1, MPI_REAL, 0, \
                 CORRELATION_TAG, MPI_COMM_WORLD);
  } } }
  voxel_pair[0] = -1;
  MPI_Send(voxel_pair, 2, MPI_INT, 0, PAIR_TAG,
           MPI_COMM_WORLD);
  MPI_Allreduce(MPI_IN_PLACE, &sum, 1, MPI_REAL,
                MPI_SUM, MPI_COMM_WORLD);
}
```

Figure 5.17 Function `compute` of the program correlations.c.

programs should ensure that none of the processes in the group get blocked indefinitely waiting for an I/O or a blocking send/receive operation to complete. For example, before calling `MPI_Bcast`, a process should never get blocked waiting for a message that will be sent by another process after it comes out of `MPI_Bcast`. Such situations are likely to end up in deadlocks.

In general, for any collective operation, all the processes in the group are required to make call to the same collective function, with a consistent set of arguments. For example, in the program above, every MPI process in the group should call `MPI_Bcast`, with the same root node and the same communicator. For efficiency, all the processes in the group should call this function nearly at the same time, otherwise the last process to call this function may cause all the other processes to wait, thereby wasting computational resources.

Instead of using `MPI_Bcast`, the master could have sent data to the rest of the processes using `MPI_Send`. While such a program will give correct results, it will neither be efficient not scalable. To understand the reason for this, consider a seven node system connected in the form of a tree as shown in Figure 5.18. Assume that the node at the root needs to broadcast data to all the other nodes. The most efficient way of doing this is to send the data once along the solid lines. If the programmer does not have access to details of the underlying interconnection network, it is impossible to achieve such a communication pattern using point-to-point communication. A sequence of calls to `MPI_Send` for the six processes will lead to the communication pattern as shown by the dashed lines in Figure 5.18. In this case, the same data needs to be sent along some links three times! A system-specific implementation of `MPI_Bcast` could utilize the knowledge of interconnection network to give optimal performance (by using the communication pattern given by solid lines) without exposing the system-specific details to the programmer.

In an alternate approach, each MPI process could have opened the same file to read the data. This approach will also not be very efficient. Although all the processes read the same data, there is no easy way for the system to recognize this and optimize the communication. Thus, each read operation is likely to be treated independently by the system, leading to a communication pattern resembling that of an implementation done using point-to-point send calls (i.e., dashed lines in Figure 5.18).

The function `compute` of the program correlations.c is listed in Figure 5.17. The `for` loop at line 62 cycles through all the voxels i assigned to the compute process. The nested for loop at line 63 cycles through all the voxels $j < i$. The third

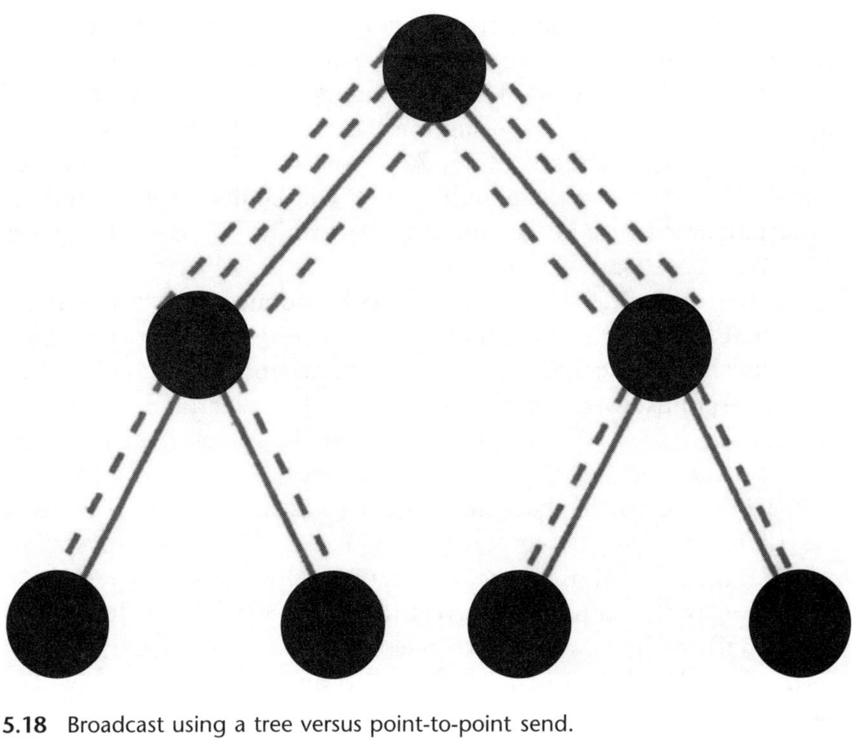

Figure 5.18 Broadcast using a tree versus point-to-point send.

loop at line 65 computes the correlations between voxels *i* and *j*. Note that the data was normalized at line 59 to have zero mean and unit variance. Therefore, the task of computing correlations reduces to computing the dot product of the two time-series.

After the correlation between a pair of voxels *i*, *j* has been computed, its absolute value is compared against a given threshold (line 70). If the correlation is greater than the threshold, the indices *i* and *j* and the corresponding correlation value is sent (lines 71 to 75) to the master process for printing to the output file. The call to `MPI_Send` at lines 72-73 sends the integer indices of the two voxels, and the call at lines 74-75 sends the (floating point) correlation value. Two separate MPI tags (`PAIR_TAG` and `CORRELATION_TAG`) are used for these two calls to ensure that the messages are matched appropriately at the master process.

The number of voxel pairs having a correlation above the threshold is dependent on the data. It is not possible to know in advance how many voxel pairs the master will receive from each of the compute processes. If the data has no significant correlations, it may receive none. Therefore, a mechanism is needed through which the master process can reliably infer that all the compute processes have finished their work and the master process has received all the data sent by the compute processes. To signal that a compute process has finished computation, a special "finish message" with negative index value for the first voxel is sent to the master process using the `PAIR_TAG` (lines 78-79 of Figure 5.17).

The master process maintains a `count` of the number of compute processes that are active and computing the correlations. This `count` is initialized to `mpi_size - 1` at line 28. The `while` loop at line 30 is executed as long as there are active compute processes. Inside this loop, at lines 31-32, the master waits for a message from any compute process using a call to the `MPI_Recv` function. At this stage, the master does not know the rank of the sender, from which the next message will arrive. So, instead of specifying the sender's rank explicitly, `MPI_ANY_SOURCE` is used. This instructs the `MPI_Recv` function to return a message sent from any sender, using the `PAIR_TAG`. Such a message could be sent by a compute process either at lines 72-73 or at lines 78-79. If the message was sent at lines 72-73, its first voxel index value will be nonnegative and the message will be accompanied by another message containing the correlation value on the tag `CORRELATION_TAG`.

If the first voxel value is found to be nonnegative, the master process posts another receive at lines 34-36 to get the corresponding correlation value. It needs to ensure that the correlation value obtained by the second receive (at lines 34-36) corresponds to the voxel pair received earlier (at lines 31-32). For this, in the second receive, the sender rank is explicitly specified to be the rank of the sender of the message received earlier at lines 31-32. This information is returned by the first receive in the field `stat.MPI_SOURCE`. The MPI semantics guarantees that the message sent on the same tag from one process to another, will not be reordered. Thus, if the sender of the two messages received at lines 31-32 and 34-36 is the same, the correlation value received at lines 34-36 will indeed correspond to the voxel pair received at lines 31-32. The master process, in this case, prints the indices of the two voxels and the correlation value to the output file.

A negative voxel index value indicates a "finish message" signaling that the sender process has finished its work. In this case, there is no companion message containing the correlation value. So, the master simply decrements the `count` and continues to the next iteration of the loop. When the master receives the "finish messages" from all the compute processes, the `count` becomes zero. Finally, the while loop of the master ends, and it closes the output file and proceeds to call `MPI_Barrier`.

MPI_Barrier and MPI_Allreduce

The compute processes also call `MPI_Barrier` after completing their work. `MPI_Barrier` is a collective operation that takes a MPI communicator as its only argument. This is set to the default communicator `MPI_COMM_WORLD`. All the processes calling `MPI_Barrier` wait in the function until all the processes in the group have made a call to `MPI_Barrier`. This ensures that a process can begin the post-barrier part of its work only after all the other processes have finished the pre-barrier part of their work. This functionality is very useful for synchronizing processes. Using this, different phases of computations can be cleanly separated from one another to ensure that MPI messages sent from a later phase do not interfere with messages sent during the earlier phases of computation.

Finally, all the processes call `MPI_Allreduce` to compute the sum of the absolute values of correlations between every voxel pair. While computing the correlations, each process maintains a running sum of the absolute value of correlations that it computes (line 69). In the end, these running sum values stored at all the compute processes, need to be added to compute the final sum. This is done efficiently using the collective function `MPI_Allreduce`. This function takes a number of data elements from each process as its input, applies an associative global *reduction* operator and distributes the results to all the processes. For example, if the input at each of the processes is an array of 10 integers and the global reduction operation is the *sum*, then the output will be an array of 10 integers containing, element-wise, the sum of the corresponding integers at all the processes.

The first argument to this function is a pointer to the buffer containing the input data. The second argument is a pointer to the buffer where the output is to be stored. To save memory and copying overhead, many applications desire the output at the same place as the input. To achieve this, `MPI_IN_PLACE` can be used as the first argument. In this case, the second argument is a pointer to the input and output buffer. The third and the fourth arguments to this function are the number of elements and the data-type. The fifth argument is the reduction operation to be performed. The sixth argument is the communicator.

The set of predefined reduction operators in MPI is quite rich. These include: `MPI_MAX` for maximum operation, `MPI_MIN` for minimum operation, `MPI_SUM` for sum operation, `MPI_PROD` for product operation, `MPI_LAND` for logical-and operation, `MPI_BAND` for bit-wise-and operation, `MPI_LOR` for logical-or operation, `MPI_BOR` for bit-wise-or operation, `MPI_LXOR` for logical xor operation, `MPI_BXOR` for bit-wise xor operation, `MPI_MAXLOC` for the location of the maximum value, and `MPI_MINLOC` for the location of the minimum value. A reduction operator outside this set is rarely needed. However, MPI allows users to create new reduction operators using MPI_Op_create. The details of this function may be found in any MPI reference such as [19, 32].

After the MPI_Allreduce returns, the variable sum at every process contains the sum of the absolute values of the correlations between every pair of voxels. The master process prints this value to the standard output and returns to the function main. All the compute processes also return to the function main, which calls MPI_Finalize before returning. This completes the description of the MPI-based parallel program that computes all-pair correlations.

This section has introduced nine basic MPI functions that are sufficient to write simple but useful parallel programs. These functions, also summarized in Table 5.2, are MPI_Init, MPI_Finalize, MPI_Comm_size, MPI_Comm_rank, MPI_Send, MPI_Recv, MPI_Bcast, MPI_Barrier, and MPI_Allreduce. These were introduced through a series of programs starting with a simple hello-world program. Through these programs, some of the issues faced by the parallel programmers were discussed. These issues were related to interprocess synchronization, consistency, importance of MPI semantics, data and work distribution, load balancing, and program efficiency. The programs presented in this section demonstrated how some of the issues may be addressed.

5.4.2 Other MPI Features

While it is possible to write fairly complex MPI programs using the nine MPI functions described earlier, the rest of the MPI functions provide several other useful functionalities for more advanced programmers. Some of the key features are briefly summarized next. More details can be found in a detailed MPI reference [19, 32] or formal MPI specifications [14, 28].

Communicators

All the MPI functions described earlier (with the exception of MPI_Init and MPI_Finalize) require a *communicator* argument to be specified. The communi-

Table 5.2 A Summary of Nine MPI Functions Introduced in This Chapter

Functionality	*MPI Function*
Initialization	MPI_Init(int *argc, char ***argv)
Cleanup	MPI_Finalize(void)
Get job size	MPI_Comm_size(MPI_Comm comm, int *size)
Get process rank	MPI_Comm_rank(MPI_Comm comm, int *rank)
Point-to-point send	MPI_Send(void* buffer, int count, MPI_Datatype datatype, int destination, int tag, MPI_Comm comm)
Point-to-point receive	MPI_Recv(void* buffer, int count, MPI_Datatype datatype, int sender, int tag, MPI_Comm comm, MPI_Status *status)
Synchronization	MPI_Barrier(MPI_Comm comm)
Broadcast	MPI_Bcast(void* buffer, int count, MPI_Datatype datatype, int root, MPI_Comm comm)
Global reduction	int MPI_Allreduce(void* sendbuffer, void* receivebuffer, int count, MPI_Datatype datatype, MPI_Op operation, MPI_Comm comm)

cator provides a context in which the communication is carried out. A communicator identifies a group of MPI processes along with association of a unique rank to each of the process in the group and a communication context. The default communicator `MPI_COMM_WORLD` contains all the MPI processes and assigns a unique rank from 0 to $P - 1$, where P is the number of processes in the job.

An MPI process can have different rank in the context of different communicators. Similarly, different communicators may have different sizes and contain different groups of processes. The rank of a process in the context of a communicator and the size of the communicator can be queried using the functions `MPI_Comm_rank` and `MPI_Comm_size` described earlier, with a suitable communicator argument.

Communicators can be dynamically created using `MPI_Comm_dup` which duplicates a given communicator into another communicator, `MPI_Comm_split` which partitions the group of processes in a given communicator into several disjoint communicators, `MPI_Comm_create` which creates a new communicator using the concept of *MPI groups*, and `MPI_Intercomm_create` which creates an *intercommunicator* to link two distinct communicators. The communicators can be destroyed dynamically using `MPI_Comm_free`.

Messages sent or received in the context of a communicator cannot get mixed with messages in the context of another communicator. Thus, communicators provide a modular way of matching point-to-point send and receive function calls, in addition to the message tags. For collective operations, communicators provide a modular way to hierarchically decompose MPI processes into smaller groups. Processes in two disjoint communicators can independently make calls to collective operations in the context of their own (disjoint) communicators, without interfering with each other. These functions are very useful while writing MPI-based reusable parallel libraries.

Process Topologies

MPI communicators may also be used to define virtual process topologies. These topologies act as hints to the MPI subsystem indicating the expected communication pattern of the application. The MPI subsystem may use this information to map the MPI processes to the hardware processors in a way that the specified virtual process topology matches well with the underlying communication network. This allows the application to optimally utilize the underlying network of the system without losing its portability.

The function `MPI_Cart_create` creates a virtual topology of a general d-dimensional torus with associated functions `MPI_Cart_get`, `MPI_Cart_rank`, and `MPI_Cart_coords` to get various attributes of the topology. The function `MPI_Cart_sub` is used to create a Cartesian subspace (similar to `MPI_Comm_split`) and the function `MPI_Cart_shift` is used to get information needed to communicate with neighbors. Such topologies are very useful in many applications. For example, 3D Cartesian topologies can be used for 3D image filtering operations as demonstrated in Chapter 6.

For working with arbitrary graph topologies, functions `MPI_Graph_create`, `MPI_Graph_neighbors_count`, and `MPI_Graph_neighbors` are available.

Asynchronous Communication

The functions MPI_Send and MPI_Recv provide the functionality of a *blocking* communication. The MPI_Send call blocks until the data has either been sent on the network or copied into the system buffers. The application is free to modify the buffer given as the input argument to the function MPI_Send, as soon as the call returns. Similarly, the MPI_Recv function call blocks until the required data is available and copied into the receive buffer specified in the call.

For performance reasons and to obtain better overlap of computation and communication, it is desirable to have functions using which the applications can initiate a communication asynchronously, monitor its progress, check the availability of the data, and get information about the state of the network. Several asynchronous communication functions are available to achieve this goal. These functions include MPI_Isend, MPI_Ibsend, MPI_Issend, and MPI_Irsend for different flavors of asynchronous send operations, MPI_Irecv for asynchronous receive operation, MPI_Test, MPI_Testany, MPI_Testall, and MPI_Testsome for monitoring the progress of a previously initiated communication operation, MPI_Wait, MPI_Waitany, MPI_Waitall, and MPI_Waitsome for blocking until the specified communication operation is complete, MPI_Probe for checking if there is a message ready to be delivered to the process, and MPI_Cancel for canceling a previously initiated asynchronous communication.

Collective Communication Operations

In addition to MPI_Bcast and MPI_Allreduce, several other collective communication operations are available in MPI. The function MPI_Gather is used to aggregate data from all the processes of a communicator into a single large buffer of a specified *root* process. The scatter operation, using the function MPI_Scatter, is the opposite. It slices an array stored at a root process and distributes it to the other processes in the communicator. The functions MPI_Gatherv and MPI_Scatterv may be used when the size of the data at different processes is different. The functions MPI_Allgather and MPI_Allgatherv are variants of MPI_Gather and MPI_Gatherv, where the data collected at the root process is simultaneously broadcast to all the processes. The functions MPI_Alltoall and MPI_Alltoallv are used for more general communication pattern, in which each node has data for every other node. For example, the function MPI_Alltoall may be directly used for the transpose step of parallel FFT implementations [25].

MPI-IO

The MPI-2 specifications [28] provide a natural approach to file I/O that is consistent with the overall design of MPI. It supports basic file operations such as open, close, read, write, delete, and resize, in addition to enhanced I/O capabilities.

The concept of the *view* of a file is central to the design of MPI-IO. It is defined as a repetitive pattern of noncontiguous blocks of the file data that is accessible and appears contiguous to a process. A view is created using a *displacement*, an *etype*, and a *filetype*. An etype is an elementary unit of transfer in MPI-IO. All the file I/O is done in units that are multiples of an etype. An etype may be defined as a fixed sized array of a basic MPI data-type, or a user-defined structure. A filetype defines a template for accessing a file. If could either be a contiguous collection of etypes, a strided vector of etypes with holes in between, or an arbitrary organization of etypes with holes. A view is defined by a repetitive pattern of a filetype, beginning at a displacement, where each element of the filetype is a specified etype.

Different MPI processes may have different views of a file and can read and write the data available in their own views. Thus, by suitably defining a filetype and displacements for different MPI processes, a file may be partitioned into disjoint sections for reading by different MPI processes. The views of different MPI processes may also overlap, and therefore, have common data. The views can be dynamically created and destroyed during runtime.

MPI-IO allows independent access to file data using individual file pointers. It also allows *collective access* using shared file pointers. It supports blocking, as well asynchronous nonblocking I/O. For efficient access to the data, MPI-IO allows users to specify several different types of *hints* regarding data access patterns. It also provides portability across multiple systems by automatically carrying format conversion when needed.

5.5 Other Programming Models

MPI has been a highly successful parallel programming model because it strikes the "right" balance between its level of abstraction, its performance, and its portability. For parallel programmers, several other parallel programming models and tools are also available that are worth mentioning. While some of these models and tools are complementary to MPI and are often used along with MPI to enhance its functionality, some others offer alternate methods to program parallel systems. Some of these tools are early research prototypes, exploring different points in the productivity-portability-performance trade-off curves. It is difficult to predict if any of these research prototypes will evolve into a programming model as successful as MPI. Nevertheless, these models offer novel and innovative ways to program parallel systems and are interesting in their own right. Some of these are models and tools, which are briefly described next. These include OpenMP [6, 9], optimized general purpose libraries such as Basic Linear Algebra Subprograms (BLAS) [5], Parallel BLAS (PBLAS), Linear Algebra Package (LAPACK) [1], Scalable LAPACK (ScaLAPACK) [4], Portable, Extensible Toolkit for Scientific computation (PETSc) [3], Parallel Virtual Machine (PVM) [17, 30], Unified Parallel C (UPC) [11, 12], and the X10 programming language [10].

OpenMP

OpenMP [6, 9, 31] complements MPI by providing a shared-memory abstraction to its users. High-performance parallel systems comprising multiple shared-memory nodes are often programmed using a combination of OpenMP and MPI.

Since MPI provides a distributed memory abstraction, the memory of a parallel system composed of shared-memory nodes, needs to be partitioned into private memories among the processors sharing the memory. This significantly reduces the amount of memory available to each MPI process. The ideal solution should have one process per processor, sharing its memory with other processes running on other processors of the same shared-memory node. OpenMP along with MPI achieves exactly the same goal.

OpenMP provides the abstraction of multiple *threads* sharing the same address space. These threads may be dynamically created and destroyed using the fork/join or other OpenMP primitives. A typical MPI/OpenMP application has one MPI process per node, which may create multiple shared-memory threads, running on the other processors of the node, using the OpenMP primitives.

For performance reasons, OpenMP provides a "relaxed-consistency" model of the shared memory where each thread is free to maintain a locally cached copy of the memory. It provides primitives to "flush" the memory, thereby making it consistent with rest of the threads at the node.

Most of the OpenMP primitives are supplied using compiler directives. Thus, to enable OpenMP on a system, the compiler needs to be suitably modified.

OpenMP provides data-parallel as well as task-parallel abstraction. In the data-parallel mode, different threads run the same code, but act on different parts of the data. An example of a data-parallel mode is a *parallel for loop*, where different threads execute the same code-block with different values of the loop index variable. In the task-parallel mode, it is possible to define different "sections" of the code that are run by different threads.

Primitives are provided to identify variables in a parallel code section as either *private* or *shared*, depending on whether they are replicated or shared across the threads. It is also possible to set the default data-type as either private or shared. Several synchronization primitives for *critical sections*, *atomic operations*, and *barrier synchronization* are available. It is also possible to have a better control over thread scheduling, using primitives for *static*, *dynamic*, and *guided* scheduling. Various *reduction operations* are also available in a manner similar to MPI.

MPI with OpenMP provides a very effective model to program the current generation of parallel systems.

Optimized General-Purpose Libraries

Optimized general-purpose libraries provide a simple and attractive method to write portable, high-performance parallel programs. These libraries contain code, carefully optimized for specific architectures. The optimizations typically performed on these codes: (1) optimally utilize the cache hierarchy, (2) make the best use of instruction-level parallelism (ILP) and pipelining capabilities of the underlying hardware, (3) utilize single instruction multiple data (SIMD) units in the hardware, if any, and (4) possibly make use of multiple processors in a shared-memory node.

If an application spends most of its time in a code that is available as a standard library function (such as FFT), then the application may be modified to use the library function. This usually gives very good performance improvements with little efforts. Some commonly used general purpose libraries are:

- *BLAS:* The Basic Linear Algebra Subprograms (BLAS) [5] provide basic matrix and vector operations.

- *PBLAS:* A Parallel version of BLAS.
- *LAPACK:* The Linear Algebra Package (LAPACK) [1] provides routines for solving systems of simultaneous linear equations, least-squares solutions of linear systems of equations, eigenvalue problems, singular value problems, matrix factorizations routines such as LU, Cholesky, QR, SVD, Schur, and generalized Schur. The package works on dense and banded matrices. Real and complex matrices in single and double precision are supported.
- *ScaLAPACK:* The Scalable LAPACK library (ScaLAPACK) [4] includes a subset of LAPACK routines, redesigned for distributed memory parallel computers that support MPI or PVM. It assumes matrices are laid out in a two-dimensional block cyclic decomposition.
- *PETSc:* The Portable, Extensible Toolkit for Scientific computation (PETSc) [2, 3] provides a set of tools for the parallel numerical solution of partial differential equations that require solving large-scale, sparse nonlinear systems of equations. PETSc employs a variety of Newton techniques and Krylov subspace methods. It provides several parallel sparse matrix formats, including compressed row, block compressed row and block diagonal storage.

Parallel Virtual Machine (PVM)

PVM was the predominant parallel programming environment before the advent of MPI. PVM provides a simple and portable programming model for developing and running parallel applications on a cluster of (possibly heterogeneous) machines, in a manner similar to MPI. There are however, some important key differences between PVM and MPI.

MPI has evolved as an extensive user-level communication library for writing parallel applications. PVM, on the other hand, as the name "virtual machine" suggests, contains much more functionality of a distributed operating system. It includes functions for resource management, dynamic process creation and termination and interoperability with multiple heterogeneous systems. While later versions of MPI contain support for some of these features, it is not as rich and widely used as PVM.

On the other hand, PVM also provides primitives for point-to-point communication and group communication, but the functions provided are not as extensive as in MPI. As a result, most high-performance applications running on homogeneous massively parallel systems typically end up using MPI. On the other hand, on heterogeneous distributed environments such as in Grid computing systems [16], PVM may provide certain benefits over MPI.

Unified Parallel C

Unified Parallel C (UPC) is an extension of the C programming language, designed for programming large parallel systems. Similar to MPI, it offers a Single Program Multiple Data (SPMD) model of computing. The basic execution unit in UPC is a *thread*, which is akin to a process in MPI. The number of threads in a program is fixed, as in MPI and decided at the time of launching the job. Each thread is typically executed on a single processor. UPC provides shared-memory as well as distributed-memory abstractions to the programs.

A key element of UPC is the concept of *shared* and *private* address space. Variables in a thread may be defined to be either shared or private. A private variable, as the name suggests, is a variable that is accessible only to the thread declaring it. If multiple threads declare a private variable, each of them will have an individual copy of the variable. A shared variable, on the other hand, refers to a single memory location accessible to every thread. Each thread can read and modify a shared variable. The shared as well as private variables can be used transparently in the programs as if they were normal variables. Seamless integration with the C programming language is provided by using a new C-type qualifier `shared` to declare shared variables. The compiler is responsible for generating the code to get a variable and store it back, in case it is located at a remote memory location.

The language also defines the concept of *affinity*, using which shared variables are associated with threads that "own" them. Typically, a shared variable is stored in the physical memory of the processor that runs the thread owning the variable.

UPC also defines few data distribution primitives for specifying affinity of arrays. These may be *cyclic*, for distributing successive elements of an array to successive threads, *blocked-cyclic*, for combining successive elements of an array into blocks and distributing the blocks cyclically among the threads, and *blocked* for dynamic data distribution ensuring that the array is "evenly" distributed among the threads.

The language also provides certain synchronization primitives using which the threads may coordinate their activities. These primitives include *locks*, *barriers*, and *memory fences*.

Two user-controlled consistency models for accessing the shared data are supported. In the *strict* model, the program executes under a strict sequential consistency semantics [27], whereas in the *relaxed* model the memory accesses are locally consistent at the thread level (i.e., it appears to the issuing thread that all shared references in that thread occur in the order they were written). The consistency model can be specified on a "per-variable" basis or on a "per-statement" basis. It is also possible to set the "default" consistency model to be either relaxed or strict.

The UPC language evolved from other parallel extensions (AC [8] and Split-C [24]) to C proposed earlier. UPC is an attempt to take the best characteristics of each of the proposed extensions and provide the simplicity of the shared-memory abstraction, while giving the control over data layout, and a performance of the message passing programming paradigm.

The X10 Programming Language

X10 is a new object-oriented language for high-performance computing with emphasis on the programmer productivity [10]. The X10 effort is part of the IBM PERCS project (Productive Easy-to-use Reliable Computer Systems) whose goal is to design adaptable scalable systems for the 2010 timeframe.

X10 is a strongly typed language based on the Java programming language and has been called an "extended subset" of Java. It provides a partitioned global address space model, along with support for arrays and concurrency.

X10 is based on the concept of a *place*, defined as a collection of lightweight threads (called activities) and the associated data. A place is typically resident on a shared-memory node with multiple processors. The number of activities in a place can vary dynamically and so can the memory needed to store the place. Multiple places can be created, which reside on different nodes of a parallel distributed memory system. The concept of an *atomic section* in X10 provides the functionality of a lock; the concept of *clocks* may replace barriers; and the concept of *asynchronous operation* is used to handle concurrent activity within a node or a place.

5.6 Conclusions

Trends in semiconductor technologies point to the growing importance of parallel computing at every level. As the semiconductor technologies evolve, the transistor densities are expected to increase at a rate comparable to the historic rate of Moore's law. However, unlike the past, the microprocessor clock frequencies are expected to stabilize in the 1- to 10-GHz range. Future systems will have more and more processors (cores) on a single chip. Therefore, the future applications will have to make use of parallelism for performance improvements. For the microscopy and imaging applications, the volume of data and the computational needs are expected to continue increasing at the historical rates. Therefore, parallel computation will be of immense importance in this domain.

The basic elements of parallel computing have been discussed in this chapter. The first step in writing a parallel application is to gain the understanding of the architecture of the systems where the application is expected to run. The second step is the development of a parallel algorithm, suitable for the target architecture. The final step is the implementation of the algorithm for the target system using a suitable parallel programming model.

This chapter has introduced MPI—the most widely used parallel programming model for programming distributed memory systems. Although MPI specifications [14] contain over 90 functions, in this chapter only nine basic MPI functions were explained in detail. Using a series of simple programs, some of the common issues faced while writing parallel programs were illustrated. It should be possible to write fairly complex MPI programs using these nine basic functions. Other MPI features were also discussed briefly so that the reader is aware of the functionality available in MPI and can look up suitable references, should there be a need.

Finally, a quick overview of some other parallel libraries and parallel programming models was provided. Some of these models such as OpenMP and libraries such as BLAS, SCALAPACK, and PETSc are often used in conjunction with MPI to write high-performance parallel programs.

This chapter should provide sufficient details to enable the reader to begin experimenting with parallel programming using MPI. The references given should be adequate to explore advanced issues in parallel programming using MPI or other parallel programming models and tools.

References

[1] E. Anderson, Z. Bai, C. Bischof, S. Blackford, J. Demmel, J. Dongarra, J. Du Croz, A. Greenbaum, S. Hammarling, A. McKenney, and D. Sorensen. *LAPACK Users' Guide*. Society for Industrial and Applied Mathematics, Philadelphia, PA, third edition, 1999.

[2] Satish Balay, Kris Buschelman, William D. Gropp, Dinesh Kaushik, Matthew G. Knepley, Lois Curfman McInnes, Barry F. Smith, and Hong Zhang. PETSc Web page, 2001. http://www.mcs.anl.gov/petsc.

[3] Satish Balay, William D. Gropp, Lois Curfman McInnes, and Barry F. Smith. "Efficient management of parallelism in object oriented numerical software libraries," in E. Arge, A. M. Bruaset, and H. P. Langtangen, editors, *Modern Software Tools in Scientific Computing*, pages 163–202. Birkhauser Press, 1997.

[4] L. S. Blackford, J. Choi, A. Cleary, E. D'Azevedo, J. Demmel, I. Dhillon, J. Dongarra, S. Hammarling, G. Henry, A. Petitet, K. Stanley, D. Walker, and R. C. Whaley. *ScaLAPACK Users' Guide*. Society for Industrial and Applied Mathematics, Philadelphia, PA, 1997.

[5] L. S. Blackford, J. Demmel, J. Dongarra, I. Duff, S. Hammarling, G. Henry, M. Heroux, L. Kaufman, A. Lumsdaine, A. Petitet, R. Pozo, K. Remington, and R. C. Whaley. "An updated set of basic linear algebra subprograms (BLAS)," *ACM Trans. Math. Softw.*, 28(2):135–151, 2002.

[6] OpenMP Architecture Review Board. *OpenMP specifications (version 3.0)*, 2008. http://openmp.org/wp/openmp-specifications/.

[7] Greg Burns, Raja Daoud, and James Vaigl. "LAM: An open cluster environment for MPI," in *Proceedings of Supercomputing Symposium*, pages 379–386, 1994.

[8] William W. Carlson and Jesse M. Draper. "Distributed data access in AC," *SIGPLAN Not.*, 30(8):39–47, 1995.

[9] Rohit Chandra, Ramesh Menon, Leonardo Dagum, Dave Kohr, Dror Maydan, and Jeff McDonald. *Parallel Programming in OpenMP*. Morgan Kaufmann Publishers Inc., San Francisco, CA, 2001.

[10] Philippe Charles, Christian Grothoff, Vijay Saraswat, Christopher Donawa, Allan Kielstra, Kemal Ebcioglu, Christoph von Praun, and Vivek Sarkar. "X10: An object-oriented approach to non-uniform cluster computing," *OOPSLA '05: Proceedings of the 20th annual ACM SIGPLAN conference on Object oriented programming, systems, languages, and applications*, pages 519–538, New York, NY, 2005. ACM.

[11] The UPC Consortium. *UPC language specifications, V 1.2,* June 2005.

[12] Tarek El-Ghazawi, William Carlson, Thomas Sterling, and Katherine Yelick. *UPC: Distributed Shared Memory Programming*. Wiley-Interscience, June 2005.

[13] Brian K. Flachs, Shigehiro Asano, Sang H. Dhong, H. Peter Hofstee, Gilles Gervais, Roy Kim, Tien Le, Peichun Liu, Jens Leenstra, John S. Liberty, Brad W. Michael, Hwa-Joon Oh, Silvia M. Müller, Osamu Takahashi, Koji Hirairi, Atsushi Kawasumi, Hiroaki Murakami, Hiromi Noro, Shoji Onishi, Juergen Pille, Joel Silberman, Suksoon Yong, Akiyuki Hatakeyama, Yukio Watanabe, Naoka Yano, Daniel A. Brokenshire, Mohammad Peyravian, VanDung To, and Eiji Iwata. "Microarchitecture and implementation of the synergistic processor in 65-nm and 90-nm SOI," *IBM Journal of Research and Development*, 51(5):529–544, 2007.

[14] Message Passing Interface Forum. "MPI: A message-passing interface standard (version 1.1)," Technical report, June 1995.

[15] Ian Foster. *Designing and Building Parallel Programs: Concepts and Tools for Parallel Software Engineering*. Addison-Wesley Longman Publishing Co., Inc., Boston, MA, 1995.

[16] Ian Foster and Carl Kesselman. "The Globus toolkit," *The Grid: Blueprint for a New Computing Infrastructure*, pages 259–278, 1999.

[17] Al Geist, Adam Beguelin, Jack Dongarra, Weicheng Jiang, Robert Manchek, and Vaidyalingam S. Sunderam. *PVM: Parallel Virtual Machine: A Users' Guide and Tutorial for Network Parallel Computing*. MIT Press, 1994.

[18] W. Gropp, E. Lusk, N. Doss, and A. Skjellum. "A high-performance, portable implementation of the MPI message passing interface standard," *Parallel Computing*, 22(6):789–828, September 1996.

[19] William Gropp, Ewing Lusk, and Anthony Skjellum. *Using MPI (2nd ed.): Portable Parallel Programming with the Message-Passing Interface*. MIT Press, Cambridge, MA, 1999.

[20] William D. Gropp and Ewing Lusk. *User's Guide for MPICH, A Portable Implementation of MPI*. Mathematics and Computer Science Division, Argonne National Laboratory, 1996. ANL-96/6.

[21] HPC challenge benchmarks. http://icl.cs.utk.edu/hpcc/index.html.

[22] James A. Kahle, Michael N. Day, H. Peter Hofstee, Charles R. Johns, Theodore R. Maeurer, and David Shippy. "Introduction to the Cell multiprocessor," *IBM Journal of Research and Development*, 49(4-5):589–604, 2005.

[23] Nicholas T. Karonis, Brian Toonen, and Ian Foster. "MPICH-G2: A grid-enabled implementation of the message passing interface," *J. Parallel Distrib. Comput.*, 63(5):551–563, 2003.

[24] A. Krishnamurthy, D. E. Culler, A. Dusseau, S. C. Goldstein, S. Lumetta, T. von Eicken, K. Yelick, and K. Yelick. "Parallel programming in Split-C," *Proceedings of the 1993 ACM/IEEE conference on Supercomputing*, pages 262–273, New York, NY, 1993. ACM.

[25] Sameer Kumar, Yogish Sabharwal, Rahul Garg, and Philip Heidelberger. "Performance analysis and optimization of all-to-all communication on the Blue Gene/L supercomputer," in *International Conference on Parallel Processing (ICPP)*, 2008.

[26] Leslie Lamport. "Time, clocks, and the ordering of events in a distributed system," *Commun. ACM*, 21(7):558–565, 1978.

[27] Leslie Lamport. "The temporal logic of actions," *ACM Transactions on Programming Languages and Systems*, 16(3):872–923, May 1994.

[28] Message Passing Interface Forum MPIF. "MPI-2: Extensions to the Message-Passing Interface," Technical Report, University of Tennessee, Knoxville, 1996.

[29] MPI toolbox for Octave (MPITB). http://atc.ugr.es/javier-bin/mpitb.

[30] PVM parallel virtual machine. http://www.csm.ornl.gov/pvm/.

[31] Michael J. Quinn. *Parallel Programming in C with MPI and OpenMP*. McGraw-Hill Education Group, 2003.

[32] Marc Snir and Steve Otto. *MPI-The Complete Reference: The MPI Core*. MIT Press, Cambridge, MA, 1998.

[33] Jeffrey M. Squyres and Andrew Lumsdaine. "A Component Architecture for LAM/MPI," in *Proceedings, 10th European PVM/MPI Users' Group Meeting*, number 2840 in Lecture Notes in Computer Science, pages 379–387, Venice, Italy, September/October 2003. Springer-Verlag.

[34] Top500 supercomputers site. http://www.top500.org.

[35] V. M. Eguiluz, et al. "Scale-free functional brain networks," *Phys. Rev. Lett.*, 94(018102), 2005.

[36] John von Neumann. "First draft of a report on the EDVAC." University of Pennsylvania, June 1945.

CHAPTER 6

Parallel Feature Extraction

A. Ravishankar Rao and Guillermo A. Cecchi

6.1 Introduction

In order to motivate the need for parallel feature extraction techniques, we consider the following problem, which has been receiving increasing attention. One of the grand challenges identified by the National Academy of Engineering in 2008 (http://www.engineeringchallenges.org) was to reverse-engineer the brain. This is indeed a complex problem, requiring insights from multiple disciplines such as neuroscience, physics, electrical engineering, mathematics, and computer science [1–41].

One approach suggested by Sporns et al. [38] and others is to build a detailed structural model that captures the connectivity among the different neurons in the brain. Having such a model would help in simulating interactions between neural units, and understand signaling pathways such as feedback. As noted by Kasthuri and Lichtman [23], a connectivity map provides an anatomical basis for understanding how information is processed within a brain. An example of a computational model of brain function built with such knowledge of anatomical connectivity is the work of Seth et al. [31]. Shepherd et al. [35] have analyzed the relationship between structure and function in cortical circuits.

We now consider the challenges in creating such a detailed structural model of the brain. We examine a specific problem in order to make the discussion concrete. However, many of the solution components are quite general, such as 3D image filtering tasks, and can be applied to similar problems, irrespective of the data source.

6.2 Background

We consider the problem of imaging neural tissue. There are many imaging mechanisms are available, ranging from lower resolution optical imaging to high resolution scanning electron microscopy (SEM). If SEM is used to image neural tissue, 1 petabyte of storage is required for imaging one cubic millimeter [23]. Due to this large volume of data, researchers have to select a given resolution and then determine the tissue mass that can be imaged. If a high resolution is used (e.g., SEM), then it is feasible to scan only certain neural structures (e.g., retina, hippocampus, or the olfactory bulb). If lower resolution is used (e.g., the knife-edge scanning microscope [26] or ultra-microscopy [11]), then it is possible to obtain whole-brain maps. There are trade-offs to be made for working at each resolution level. With SEM techniques, it is possible to image individual synapses, thus determining precise signaling connectivity at the neuronal level. With whole-brain imaging, connections between different functional areas of the brain can be determined. Blow discusses [5] further issues that arise in the creation of a detailed connectivity diagram of neural tissue. As technology advances, it is possible that high-resolution maps of an entire brain could be obtained.

Regardless of the imaging method and resolution used, the task of 3D reconstruction is faced with significant computational challenges, which we term the *computational bottleneck*. The computational tasks that need to be performed include filtering the images to reduce noise, extract relevant features such as contours, track contours over multiple slices, create a unified representation, and visualize the structure.

In order to provide a concrete illustration, we examine images gathered from a technique called serial block-face scanning, introduced by Denk [10].

6.2.1 Serial Block-Face Scanning

In this method, pioneered by Denk and Horstmann [10], the neural tissue is mounted in a rigid substrate, called a block. The top surface of this block is directly imaged with an SEM, giving rise to a 2D image. A diamond knife is used to cut away a thin section (approximately 60 nanometers), which is discarded. The newly exposed top surface of the block is imaged to give another 2D image at a different depth. The advantage of this technique is that the successive 2D images are nearly perfectly co-registered. This overcomes a major difficulty with prior approaches that required each section to be collected, mounted, and imaged, thus introducing registration and distortion errors.

The use of this technique gives rise to a 3D *image stack*, as shown in Figure 6.1. Once such an image stack is assembled, the next task is to isolate neural structures,

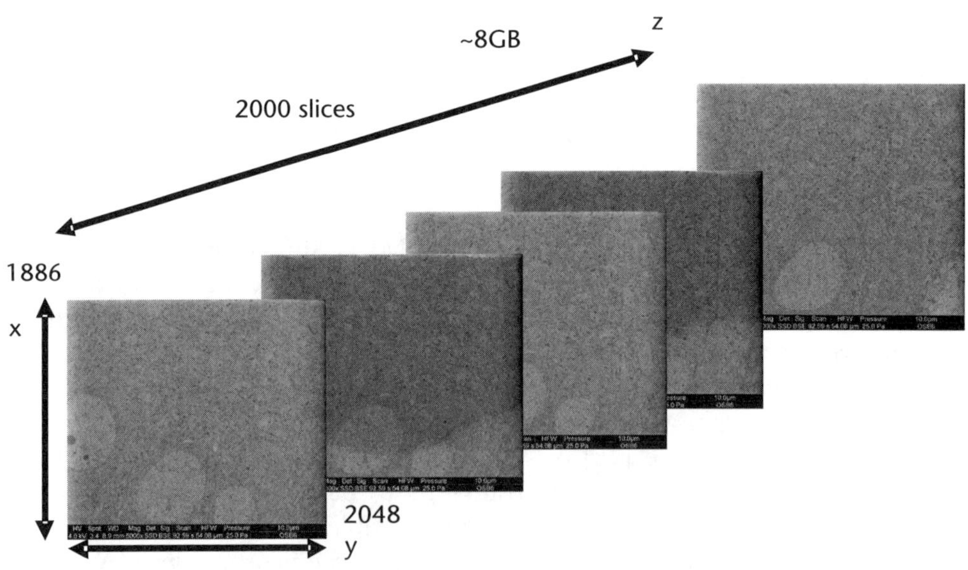

Figure 6.1 A 3D image stack created from the method of Denk and Horstmann [10]. Each 2D grayscale image is of size 188 × 2,048 pixels. The dataset available online in [10] contains 2,000 such slices. This gives rise to almost 8 GB of image data that needs to be processed.

and trace their contours. A set of such contours across all the slices can be combined to create a detailed 3D model of the neural structures. This is described in more detail in the following section.

6.3 Computational Methods

We briefly review computational methods that are useful in processing the serial section microscopy data, gathered as described in Section 6.2.1. The purpose of this section is to review useful techniques and examine their parallel implementation. Our primary objective is to examine algorithm requirements that impact parallel processing, rather than to recommend a specific solution to the problem.

6.3.1 3D Filtering

As a preprocessing step, 3D filtering is typically applied to remove noise and enhance the features of interest. Their utility has been demonstrated by several studies [34]. Median filters are specially effective at removing impulse or speckle noise. In contrast to linear filters which smooth across edges, median filters preserve edges, which constitute desirable details in images. It is advantageous to take the full multidimensional nature of the input into account when applying the median filtering. This results in superior performance when a 3D median filter is applied to 3D data rather than using a 2D or 1D median filter [2, 29]. Furthermore, repeated applications of a median filter may be necessary to obtain a stable, or invariant representation of the input image. This representation is known as the root of the image, and does not change with subsequent applications of the median filter [6]. The repeated application of a median filter is known as iterative median filtering [16] and results in improved image quality.

Though there are specific hardware implementations of 3D median filters available using field-programmable gate arrays [21], we focus on implementations on general purpose computers. The use of 3D median filtering requires that successive image slices are properly registered. If this is not the case, then errors are introduced due to misregistration, and it is better to apply 2D median filtering to each slice. In the case of serial block-face scanning, misregistration between successive slices is minimized, allowing 3D median filtering techniques to be used effectively.

6.3.2 3D Connected Component Analysis

A standard image analysis technique is to perform binarization using a threshold, which segments an image into foreground and background regions. The foreground region is analyzed to determine the number of connected foreground clusters. This operation is termed connected component analysis, which assigns two voxels to a single component if they are connected. Connectedness implies that two voxels either share a common face (termed 6-connectedness) or that they share a vertex (termed 26-connectedness). Once such connected component analysis is performed, several relevant measures can be computed, such as the number of clusters in the image, their shape, size distribution, and Euler numbers. The results of connected component analysis can also be used to perform shape matching of 3D objects [19].

Apostol and Peyrin [3] discuss the acquisition of high-resolution 3D images through X-ray microtomography for the analysis of bone samples. This gives rise to very large image datasets, typically 2,048 × 2,048 × 2,048 voxels, or 8 GB of data. A similar situation arises when one considers the analysis of large quantities of data as gathered in serial block-face scanning. A standard operation that needs to be performed on this microtomography dataset is 3D connected component analysis, which is carried out as described above.

6.3.3 Mathematical Morphological Operators

Mathematical morphological operators are useful to perform 3D shape filtering operations. Commonly used operators such as *erode* and *dilate* are useful for removing small 3D shapes due to noise, and merging 3D structures that are close together. It may be necessary to iterate over these operations a few times before the desired filtering is achieved. One of the challenges in applying morphological techniques is their computational complexity. This can be overcome through specific fast implementations and parallel computation [28].

Gu et al. [18] present techniques to perform medical image segmentation that utilize 3D mathematical morphology operators applied recursively. This operation eliminates the connectivity between a region-of-interest and neighboring tissues, leading to an effective segmentation. Benini et al. [4] devised an interactive environment to perform segmentation of biomedical images using 3D morphological operators. In order for the environment to be truly interactive, it is essential to provide fast computation. Another type of operation that can be performed with mathematical morphology techniques is edge detection [41], which is a preprocessing step that aids segmentation. Cai et al. [8] have used mathematical morphology to filter images of neuronal axons in 3D. This operation removes small areas while preserving the integrity of the axonal cross-sections. Shan, Yue, and Liu [33] report on the usage of 3D mathematical morphology operators to perform segmentation of 3D brain images gathered via functional magnetic resonance. This segmentation operation is a prerequisite for many types of processing such as brain registration, warping, and tissue classification.

6.3.4 Contour Extraction

One of the goals in processing the 3D slice data is to be able to extract 3D contours of neural structures such as axons and dendrites, which form the basis of a structural model of connectivity. This could be tackled as a single 3D image problem or a succession of 2D image problems. There are many algorithms for tackling this problem of contour extraction from images. One strategy is to first extract edge segments from the image based on kernel operators like the Sobel operator or Canny edge operator [9]. These edge segments are then grouped to form closed contours of objects. However, it is not always possible to achieve the desired results due to the presence of noise, which causes the formation of discontinuous contours.

An alternate strategy is to start with a continuous boundary which forms an estimate of the true contour, and deform the boundary so it conforms to shape features of the object of interest. This strategy is guaranteed to produce continuous boundaries. Such deformable contours are known as snakes [22] and form part of a general class of algorithms based on energy minimization principles [25, 27, 32]. Briefly, the snake is an energy-minimizing deformable shape, where the energy is a function of its internal elastic energy and an external force field. The force field is computed from the image data. A popular choice for the force field is to use the gradient of the image.

6.3.5 Requirements

It is desirable to process 3D images arising in biological applications such as serial block-face scanning by using the operations described above such as 3D filtering, connected component analysis, mathematical morphology, segmentation, and contour extraction. A complete system dedicated to handle biological imaging tasks will need to meet the following computational requirements:

- *Data collection:* The system needs to gather and store the images acquired. Appropriate tagging of information is required in order to facilitate search and retrieval at a later stage.
- *Preprocessing:* The acquired images may contain distortions such as blur that occur in the front end optics. This can be overcome through techniques such as deconvolution. In addition, the images may contain noise that needs to be filtered out.
- *Segmentation and feature extraction:* The raw images need to be segmented into regions of interest and processed to extract domain-specific features that aid in classification.
- *Analysis and interpretation:* The segmented regions are interpreted as parts of a larger biological structure such as a neuron or organ. It may be necessary to incorporate domain knowledge to aid the interpretation task, through techniques such as machine learning.
- *Modeling and prediction:* Models for function (e.g., neural connectivity models) can be built to predict the behavior of entities of interest. This phase may require the use of simulation and optimization techniques.

Typically it is not possible for a machine to perfectly perform the above tasks such as segmentation, analysis and interpretation, and human assistance is required in the loop. Hence it is important to have the ability to interact with the data in real time.

Though this list of requirements has been derived from the viewpoint of utilizing image-related biological data, it is similar to requirements in other fields, such as systems biology [39]. Research in systems biology can be considered to proceed in four stages comprising measurement, mining, modeling, and manipulation [39]. The understanding of complex biological systems proceeds in an iterative fashion, beginning with the measurement of data, and proceeding towards building quantitative models. These models can be used to make testable hypotheses, leading to experimental manipulation to verify the hypotheses. This requires further measurement, and refinement of the model. Eventually the model evolves to become sufficiently explanatory.

From the list of tasks above, the following core computational requirements can be identified:

- The ability to handle large datasets, including storage and retrieval;
- High throughput processing is required.

6.4 Parallelization

One way to address the core computational requirements identified above is to use parallel processing. We prefer to use general purpose computing platforms such as clusters of workstations based on open software environments, rather than special hardware solutions such as field programmable gate arrays [21]. The general purpose platforms permit open-ended experimentation, whereas special hardware solutions are useful when the precise algorithm has been determined, and replication cost is an issue.

With parallelization, a large computational task is broken down into smaller chunks which are performed by multiple processors. Based on the typical 3D image processing tasks that need to be carried out, we examine their parallelization from the viewpoint of computation involved per processor, communication patterns between processors, and memory issues.

6.4.1 Computation Issues

An important issue to consider when a computational task is distributed amongst multiple processes is that of load balancing. Ideally we would like the work to be distributed and executed in such a way that each process finishes its allocated work at approximately the same time. However, it may not be easy to predict how long a process may take to complete a given task. Consider median filtering, where a sort operation is required. The time to sort a sequence depends on the algorithm used, and also the arrangement of numbers in the sequence. For a sequence of length N, the quicksort algorithm requires on the average $O(N\log(N))$ comparisons to sort the items. In the worst case, $O(N^2)$ comparisons are required.

There are various scheduling algorithms that address load-balancing, such as the self-scheduling or boss-worker scheduling algorithms discussed in [17]. In these algorithms, a boss process distributes work amongst worker processes. When a worker finishes its allocated task, the boss process assigns it the next available chunk of work. This continues until the entire task is completed. The advantage of this algorithm is that processes are not kept idle, waiting for work. In general, load balancing is a complex task, and optimum performance management requires detailed knowledge of the task, data, and machine architecture. The purpose of this discussion is to make the reader aware of these issues, and suggest approaches that can be reasonably effective.

6.4.2 Communication Issues

We use the iterated 3D median filter as a concrete example to understand communication issues. Let us assume that a large 3D dataset has been partitioned into 3D cubes, such that each process is responsible for performing the filtering operation on the data within a given cube. Assume that we need to sweep an $N \times N \times N$ median filter across the 3D dataset, such that the voxel at the center of the $N \times N \times N$ region is replaced by the median of the values in that region. The computation of the median at voxels close to the boundary of a given cube requires data from the adjacent cube, which may be obtained through interprocess communication. A worst-case scenario is that half the data in the adjacent cube is required. This may require several megabytes of image data to be transmitted by each process to its neighbors. Furthermore, this communication cost may need to be incurred at each step of the iterative filtering operation. Fortunately, a characteristic of this communication pattern is that the bulk of the communication takes place between nearest neighbors. This is a common situation in many computational models of the physical world, such as computational fluid dynamics [13] or finite element analysis of materials [14].

Not all types of 3D image operations may require such intensive communication between processes. For instance, image thresholding requires no communication. However, we are examining the parallel processing requirements from a general viewpoint, which would facilitate the broadest range of image operations.

6.4.3 Memory and Storage Issues

Algorithms for 3D connected component analysis are typically run on a single processor system with local memory. Landis et al. [24] discuss the implementation of an efficient algorithm for 3D connected component analysis. However, the large image datasets pose a challenge for running the connected component analysis on single processor systems with limited memory. Apostol and Peyrin [3] present a method to overcome this challenge by partitioning the image data into subblocks. The 3D connected component analysis is run on each subblock separately. This is followed by a merging operation, as it is possible that objects are disconnected in a given subblock may extend to other subblocks where they may be connected.

The implementation of Apostol and Peyrin's technique can be easily carried out on a parallel processing architecture, say by using message passing interface (MPI) as described in the prior chapters on MPI and domain decomposition techniques. Each MPI process reads in a subblock and computes the 3D connected components. A single process can then merge the results computed over the subblocks. The advantage of the parallel implementation is that it distributes the memory usage and the computation load, rendering the problem tractable and also speeding up the computation. The computational speedup can prove very beneficial if the user wants to interactively process the high-resolution 3D image, say by adjusting the threshold gradually to observe its effect on the binarization process, and the metrics computed from 3D connected component analysis.

The above analysis motivates the following domain decompositions for different types of image-processing problems.

6.4.4 Domain Decomposition for Filtering Tasks

We propose a Cartesian grid decomposition for the 3D filtering tasks such as iterative median filtering. Figure 6.2 illustrates this decomposition which divides

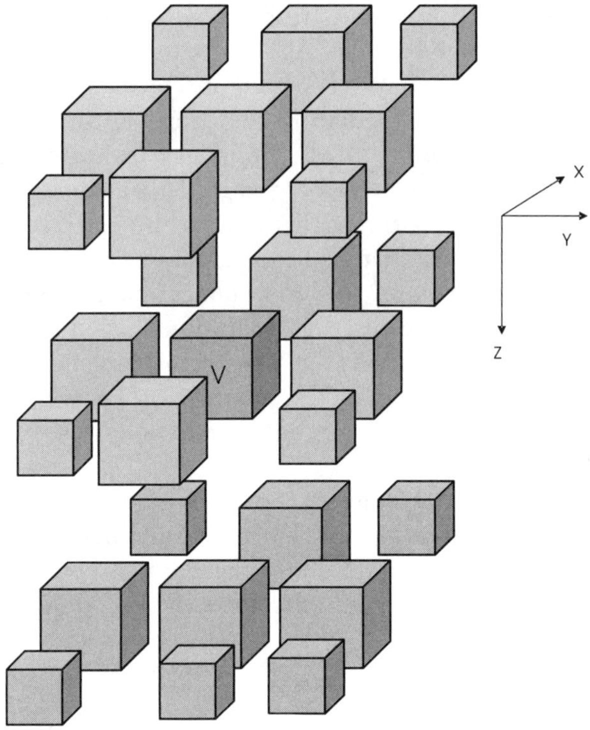

Figure 6.2 Domain decomposition for a 3D image using a 3D Cartesian grid. Each unit communicates with 26 neighbors. Shown here is a typical unit V, which communicates with 26 neighbors. Up to half the data within each unit may need to be sent to its neighbors. Each unit also receives such data from its neighbors. Thus, both send and receive commands are required. This communication cost may need to be incurred at every iteration of an algorithm such as in iterated median filtering, or iterated 3D morphological filtering.

the 3D dataset into cubes (or rectangular prisms), and each process is responsible for filtering the data within a given cube. A given process communicates with other processes that are responsible for processing data within neighboring cubes.

Note that this decomposition is independent of the actual multiprocessor architecture and connectivity. It applies equally well whether we consider clusters of workstations connected via switches, or a dedicated architecture such as Blue Gene [1] that utilizes special 3D interconnection mechanisms. However, if the above domain decomposition is combined with a mapping to a specific processor topology with enhanced nearest neighbor connectivity, then the overall algorithm performance will be superior. This is because the bulk of the communication is between nearest neighbors in both the image space and the physical processor space. As an example, the Blue Gene machine provides special high-speed interconnections for the nearest neighbors in a 3D torus [1]. Studies have shown that this architecture is beneficial for the types of 3D filtering tasks discussed above [30].

The MPI parallel programming environment provides special directives to allow the program to take advantage of specific processor configurations. In particular, the Cartesian grid communicator is used to map MPI process numbers sequentially in a 3D grid that corresponds to physical process locations [37]. This capability is evoked through the MPI_CART_CREATE and MPI_CART_GRID commands. In general, MPI assigns numbers to processes, and there is no restriction on the physical location of a process.

6.4.5 Domain Decomposition for Morphological Operators

A similar domain decomposition to the one used in 3D median filtering can be deployed. Each process handles a 3D partition and performs the morphological operation on its local data. After this operation, the relevant data are sent to the 26 neighbors, and the procedure is repeated. Finally, a merge operation is required to collate results.

6.4.6 Domain Decomposition for Contour Extraction Tasks

In the case of contour extraction, there are several possible decompositions that can be used. The first decomposition uses a fixed partitioning scheme to divide the large 3D image into subimages. A given contour may lie in different subimages as illustrated in Figure 6.3. This requires communication between the different processes that handle this contour at every step of an interactive algorithm, such as the use of snakes to compute the energy of the contour.

An alternate decomposition is contour-centric, in that a process is responsible for handling the entire contour in successive 3D slices. The process loads the image data surrounding the contour in the first slice, and acquires the next relevant portions of successive slices as illustrated in Figure 6.4. This procedure minimizes the need for communication between processes. The separate contours gathered are merged to create a single 3D representation over the volume of interest. This may require the elimination of duplicate contour information if two neural structures merge together.

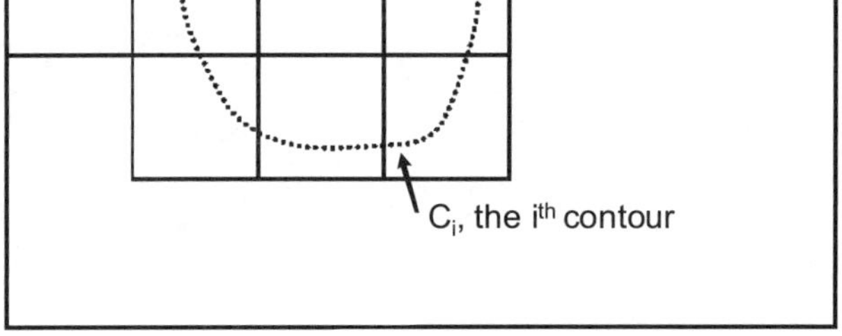

Figure 6.3 If a fixed partitioning scheme is used, a continuous contour may need to be processed by separate processes.

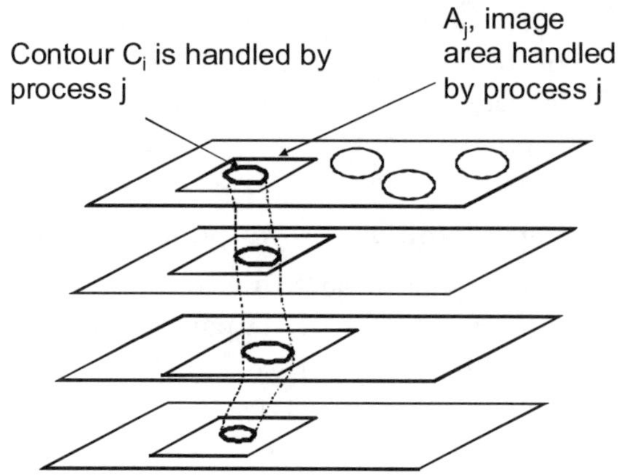

Figure 6.4 In a contour-centric model, each process is responsible for handling the entire contour, and acquires the relevant image data from successive slices.

6.5 Computational Results

We present preliminary results obtained by implementing some of the image processing algorithms described above on parallel architectures. This constitutes research work that is in progress, and significant effort remains towards building a general-purpose parallel image processing library. These results show the significant computational speedup that can be achieved through parallelization.

6.5.1 Median Filtering

We implemented the iterated median filter operation on the Blue Gene/L supercomputer, using MPI. We found that a $5\times 5\times 5$ filtering operation takes 38 seconds for 64 slices of $2{,}048 \times 1{,}872$ grayscale images on 512 processors in an $8\times8\times8$ cartesian grid. The same computation running in MATLAB on a single Intel 2-GHz processor took 143 minutes, or about 250 times slower. This illustrates that the use of high-performance computing dramatically improves the throughput of image processing tasks. Figure 6.8 depicts the variation in processing time with the number of processors used. Further details of the processing times on different architectures are provided in [30].

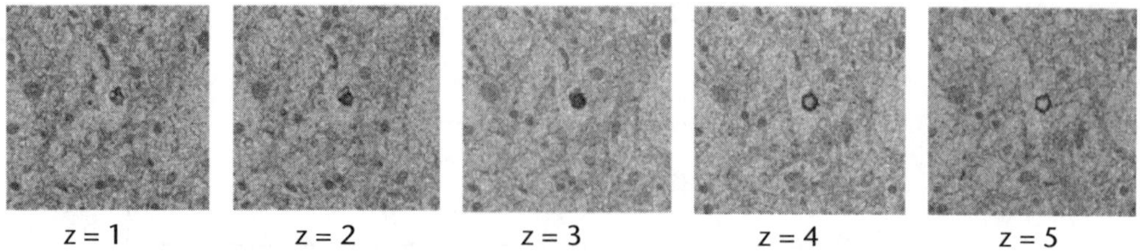

Figure 6.5 Image data from five successive slices represented by a *z* index ranging from 1 to 5. This image data constitutes the input to the 3D median filter.

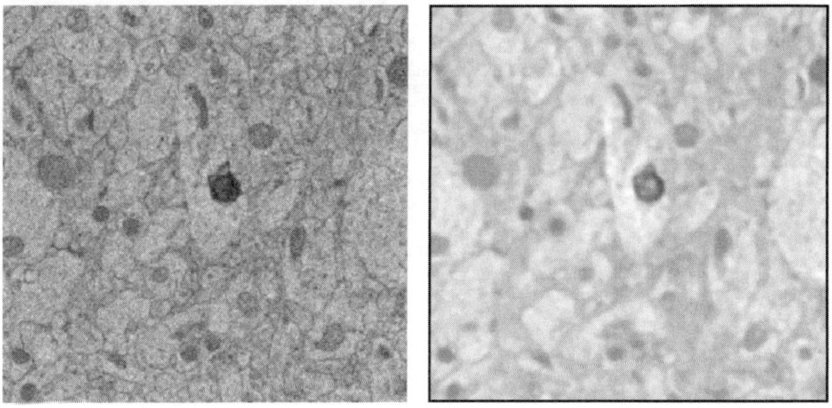

(a) (b)

Figure 6.6 (a) Image data for the index $z = 3$. (b) The 3D median filtered data at $z = 3$.

Detailed results are shown in Figures 6.5 to 6.7. A 3D median filter of size 5×5×5 voxels was used. Figure 6.6 shows a sequence of five slices taken at increasing depth, indexed by increasing z values. In Figure 6.6(a) the image in the 2D plane at $z = 3$ is shown for the sake of comparison. Figure 6.6(b) contains the result of applying the 3D median filter, where only the $z = 3$ plane is shown for clarity. Figure 6.7 shows enlarged views around a circular neural structure in Figure 6.6. Note that the high-frequency noise has been filtered out. This is a useful operation, because the subsequent stages of contour extraction rely on image gradient information. Since the gradient involves differentiation, which is well known to amplify noise, it is important to first apply a smoothing filter.

6.5.2 Contour Extraction

The next processing step is to extract contours corresponding to neural structures, such as dendrites and axons. Though, there are many possible solutions to this

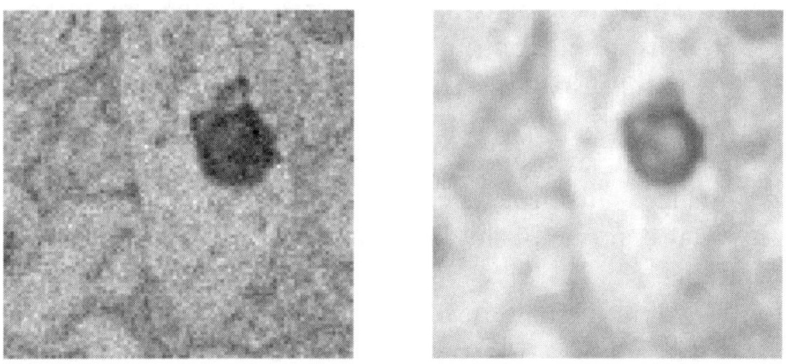

(a) (b)

Figure 6.7 (a) Image data for the index $z = 3$, shown enlarged. (b) The 3D median filtered data at $z = 3$, shown enlarged.

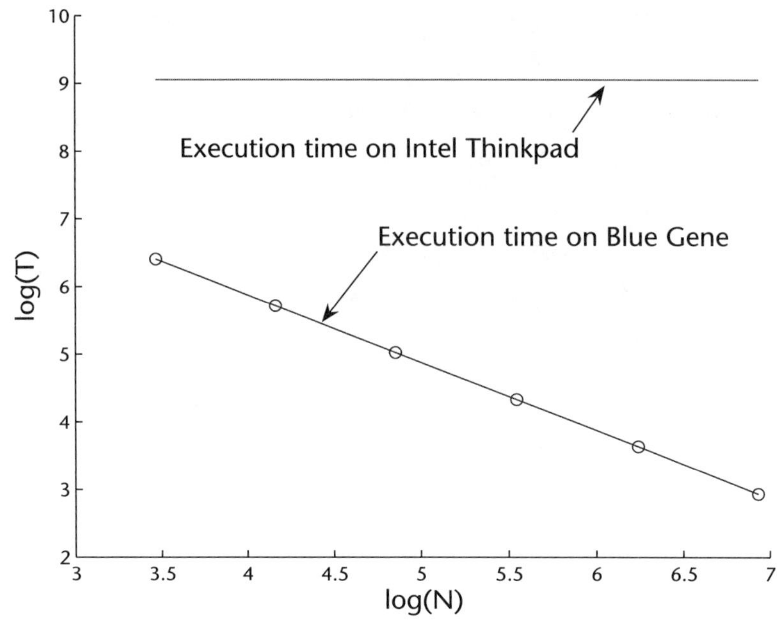

Figure 6.8 Comparison of execution times for median filtering. A 246-MB dataset was used, comprising images from [10]. The horizontal line on the top shows the computation time for this task when an IBM Thinkpad machine is used, with a 2-GHz Intel processor. The curved line at the bottom indicates the computation time for 3D median filtering on the IBM Blue Gene supercomputer as a function of the number of processors used. The computation time decreases from 606 to 18.8 seconds as the number of processors increases from 32 to 1,024; A logarithmic scale is used for plotting. This plot shows that several orders of magnitude speedup can be achieved with multiple processors.

problem, we implemented a technique based on the concept of deformable templates, or snakes. This choice is based on widespread usage of this technique, and the availability of detailed algorithmic descriptions. We implemented the technique described by Xu and Prince [40].

The contour of the first slice, at depth $z = 1$, was initialized by manual intervention, which is shown by the dotted ellipse around the neural structure in Figure 6.9. This initial contour is iteratively refined with the snake algorithm with image data within this slice to produce a final 2D contour that conforms to the precise boundary of the structure. This final contour at depth $z = 1$ is used to initialize the contour at the next depth $z = 2$, and then iteratively refined with data within this slice. This process continues, with subsequent contours being initialized with the final contour of the previous slice, and then deformed to conform to the image data.

We show intermediate results for the contour extraction process in Figures 6.10 and 6.11. This application of this technique over multiple slices results in a sequence of contour estimates, as shown in Figure 6.12. This sequence of contours can be used to fit a more detailed 3D contour model through interpolation, if desired.

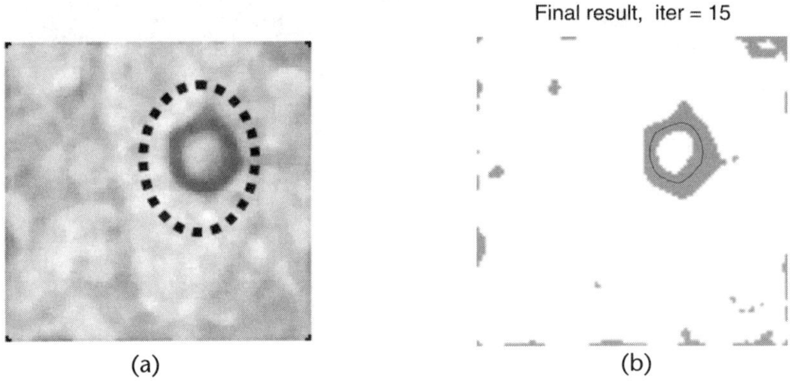

Figure 6.9 (a) An initial guess contour is indicated by the dotted line. (b) The initial contour is deformed to fit the precise features in the image, producing a final contour. The final contour is then used to automatically initialize the search for the corresponding contour in the next slice.

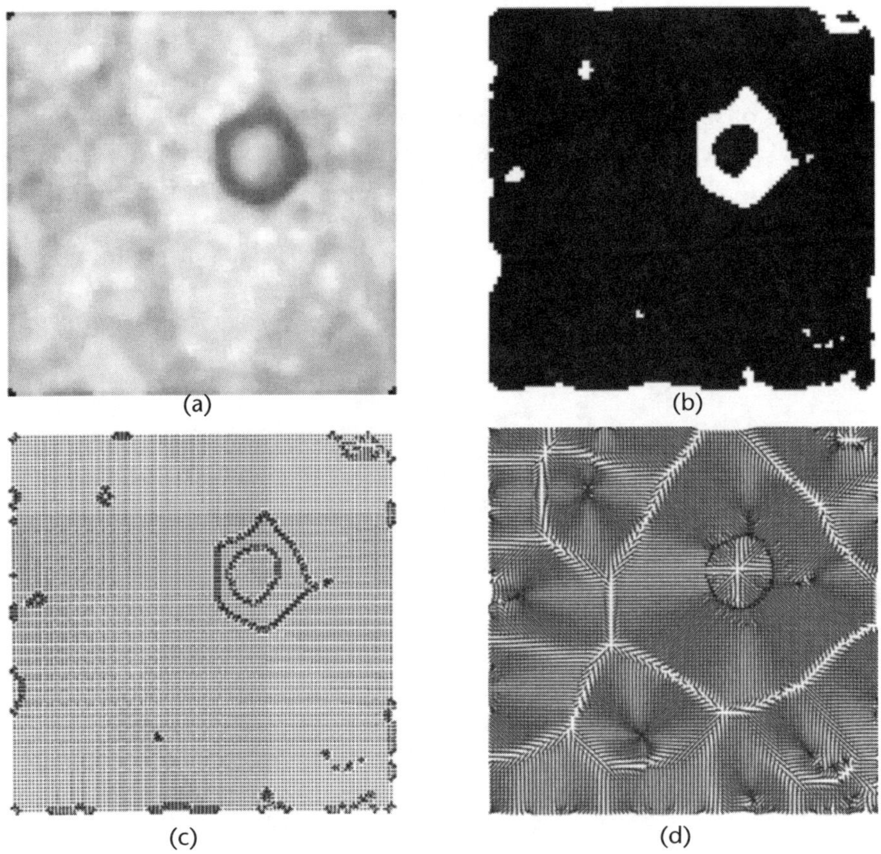

Figure 6.10 (a) The original image. (b) The edge map. (c) Gradient of the edge map. (d) The normalized gradient vector flow field.

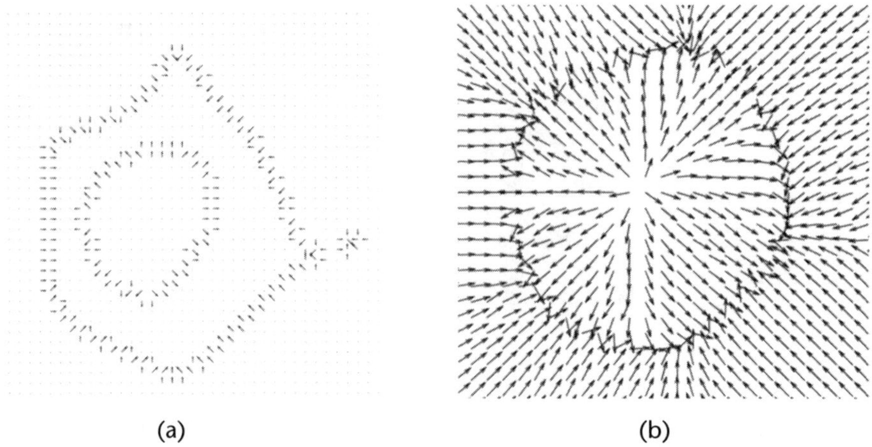

Figure 6.11 (a) Enlarged image of the edge gradient map. (b) Enlarged image of the normalized gradient vector flow field.

To illustrate this process, we used the alpha-shapes model [12] to render 3D views created from the data in Figure 6.12. This is depicted in Figure 6.13.

6.5.3 Related Work

The use of snake models to extract contours from images remains an active research topic. Cai et al. [7] have improved the gradient vector flow snake model [40] to address problems when two axonal contours are close together.

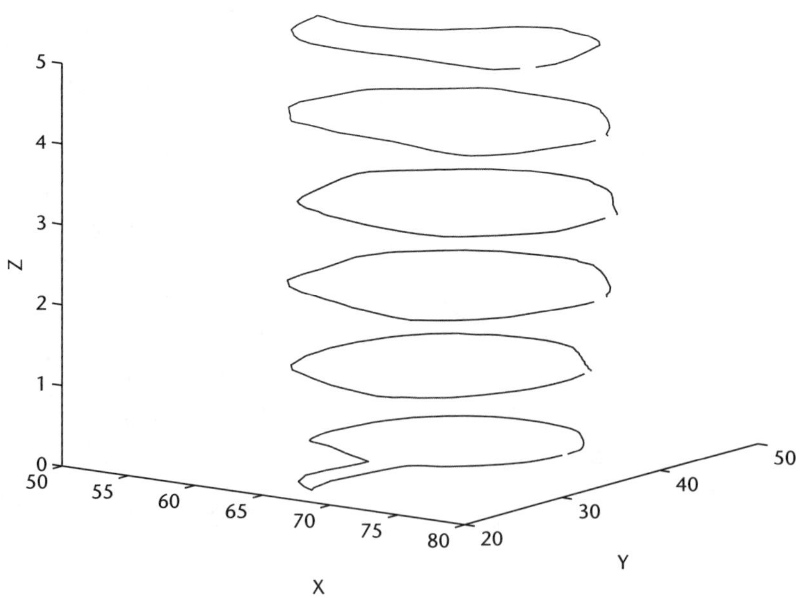

Figure 6.12 A partial 3D reconstruction shown by estimating the contour over successive slices.

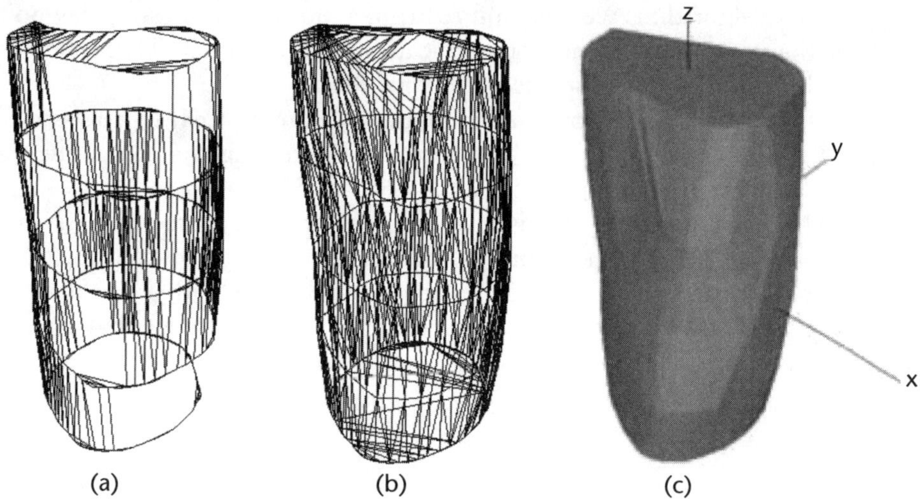

Figure 6.13 (a) The slice data visualized with a 3D graphics package. A coarse wire-frame model is used. (b) The slice data visualized with a fine wire-frame model. (c) The slice data visualized with a continuous surface.

Jain et al. [20] use an alternate approach to process brain slice data, and segment the image pixels into two types of classes: those that fall within a cell and those that fall outside. Human experts generate suitable training data by labeling different image regions according to these classes. A nonlinear filtering step is first applied to reduce noise. A convolutional network is trained using back-propagation to perform such a classification automatically.

Fiala [15] has created a tool for facilitating the viewing and processing of serial slice data. Features provided include alignment, region growing, and volume visualization.

Smith [36] reviews current techniques for neural circuit reconstruction. Smith observes that "progress in electron microscopy-based circuit analysis will depend heavily on the development of schemes for robust and efficient automated segmentation." Promising emerging directions are the use of machine learning techniques, and the use of complementary imaging techniques such as immunofluorescence.

6.6 Conclusion

The use of large 3D image datasets is growing rapidly in the biological sciences, driven by problems such as neural circuit reconstruction through serial slice data. In this chapter, we reviewed several basic 3D image processing algorithms that are useful in handling such serial slice data. Some of the algorithms considered were median and morphological filtering, and 3D connected component analysis. We also examined the problem of contour extraction between successive slices.

These algorithms were examined from a parallel processing viewpoint, to highlight the computation and communication requirements. These factors affect the way domain decomposition is performed.

Our results demonstrate a dramatic speedup in image processing tasks on a parallel platform. This will be essential in processing the enormous datasets that are being gathered, ranging in trillions of voxels.

The main message of this chapter is to encourage researchers in image processing to write parallel programs to take advantage of increasing computational resources in order to tackle the problem of dealing with large datasets. This requires a paradigm shift in the field, which is an investment that pays off through increased productivity.

References

[1] R. Adiga, et al., "Blue gene/l torus interconnection network," *IBM J. Research and Development*, 49(2), 2005.

[2] M.B. Alp and Y. Neuvo, "3-dimensional median filters for image sequence processing," *IEEE International Conference on Acoustics Speech and Signal Processing*, pp. 2917–2920, 1991.

[3] A. Apostol and F. Peyrin, "Connectivity analysis in very large 3d microtomographic images," *IEEE Trans. on Nuclear Science*, 54(1):167–172, 2007.

[4] S. Benini, E. Boniotti, R. Leonardi, and A. Signoroni, "Interactive segmentation of biomedical images and volumes using connected operators," *International Conference on Image Processing*, volume 3, pp. 452–455, 2000.

[5] N. Blow, "Following the wires," *Nature Methods*, 4(11):975–980, November 2007.

[6] W.W. Boles, M. Kanefsky, and M. Simaan, "Recursive two-dimensional median filtering algorithms for fast image root extraction," *IEEE Transactions on Circuits and Systems*, 35(10):1323–1326, 1988.

[7] H. Cai, et al., "Repulsive force based snake model to segment and track neuronal axons in 3d microscopy image stacks," *Neuroimage*, 32(4):1608–1620, 2006.

[8] H. Cai, et al., "Use mean shift to track neuronal axons in 3d," *Life Science Systems and Applications Workshop*, pp. 1–2, 2006.

[9] J. Canny, "A computational approach to edge detection," *IEEE Transactions on Pattern Analysis and Machine Intelligence*, 8(6):679–698, 1986.

[10] W. Denk and H. Horstmann, "Serial block-face scanning electron miscroscopy to reconstruct three-dimensional tissue nanostructure," *PLOS Biology*, 2(11), 2004.

[11] H. Dodt, et al., "Ultramicroscopy: three-dimensional visualization of neuronal networks in the whole mouse brain," *Nature Methods*, 4(11):331–336, April 2007.

[12] H. Edelsbrunner and E. P. Mucke, "Three-dimensional alpha shapes," *ACM Trans. Graphics*, 13:43–72, 1994.

[13] E. W. Evans, et al., "Automatic and effective multi-dimensional parallelisation of structured mesh based codes," *Parallel Computing*, 26(6):677–703, 2000.

[14] M. W. Fahmy and A. H. Namini, "A survey of parallel nonlinear dynamic analysis methodologies," *Computers and Structures*, 53(4):1033–1043, 1994.

[15] J. C. Fiala, "Reconstruct: a free editor for serial section microscopy," *Journal of Microscopy*, 218:52–61, 2005.

[16] A. R. Forouzan and B. N. Araabi, "Iterative median filtering for restoration of images with impulsive noise," *Proceedings of the IEEE International Conference on Circuits, Electronics and Systems*, volume 1, pp. 232–235, 2003.

[17] W. Gropp, E. Lusk, and A. Skjellum, *Using MPI: Portable Parallel Programming with the Message-Passing Interface*, MIT Press, Cambridge, MA, 1999.

[18] L. Gu, J. Xu, and T.M. Peters, "Novel multistage three-dimensional medical image segmentation: Methodology and validation," *IEEE Transactions on Information Technology in Biomedicine*, 10(4):740–748, 2006.

[19] N. Iyer, et al., "Three-dimensional shape searching: state-of-the-art review and future trends," *Computer-Aided Design*, 37(5):509–530, 2005.

[20] V. Jain, et al., "Supervised learning of image restoration with convolutional networks," in *IEEE International Conference on Computer Vision*, pp. 1–8, 2007.

[21] M. Jiang and D. Crookes, "High-performance 3d median filter architecture for medical image despeckling," *Electronics Letters*, 42(24):1379–1380, 2006.

[22] M. Kass, A. Witkin, and D. Terzopoulos, "Snakes, active contour model," *International Journal of Computer Vision*, 1988.

[23] N. Kasthuri and J.W. Lichtman, "The rise of the 'projectome'," *Nature Methods*, 4(4):307–308, April 2007.

[24] E. N. Landis, T. Zhang, E. N. Nagy, G. Nagy, and W. R. Franklin, "Cracking, damage and fracture in four dimensions," *Materials and Structures*, 40(4):357–365, 2007.

[25] R. Malladi and J. Sethian, "Image processing via level set curvature flow," *Proc. of National Academy of Science*, pp. 7046–7050, 1995.

[26] B. H. McCormick, et al., "Construction of anatomically correct models of mouse brain networks," *Neurocomputing*, 58-60:379–386, 2004.

[27] H. Nguyen, M. Worring, and R. Boomgaard, "Watersnakes: Energy-driven watershed segmentation," *IEEE Transactions on Pattern Analysis and Machine Intelligence*, 25: 330–342, 2003.

[28] N. Nikopoulos and I. Pitas, "A fast implementation of 3-d binary morphological transformations," *IEEE Transactions on Image Processing*, 9(2), 2000.

[29] H. Rantanen, M. Karlsson, P. Pohjala, and S. Kalli, "Color video signal processing with median filters," *IEEE Transactions on Consumer Electronics*, 38(3):157–161, 1992.

[30] A. R. Rao, G. A. Cecchi, and M. O. Magnasco, "High performance computing environment for high throughput microscopy," *BMC Cell Biology*, (Suppl 1):S9, 2007.

[31] A. K. Seth, J. L. McKinstry, G. M. Edelman, and J. L. Krichmar, "Visual binding through reentrant connectivity and dynamic synchronization in a brain-based device," *Cerebral Cortex*, 14(11):1185–1199, 2004.

[32] J. Sethian, *Level Set Methods*, Cambridge University Press, 1996.

[33] Z. Shan, G. Yue, and J. Liu, "Automated histogram-based brain segmentation in t1-weighted three-dimensional magnetic resonance head images," *NeuroImage*, 17(3): 1587–1598, 2002.

[34] R. Shekhar and V. Zagrodsky, "Mutual information-based rigid and nonrigid registration of ultrasound volumes," *IEEE Trans. Med. Imaging*, 21(1):9–22, 2002.

[35] G. Shepherd, A. Stepanyants, I. Bureau, D. Chklovskii, and K Svoboda, "Geometric and functional organization of cortical circuits," *Nature Neuroscience*, 8:782–790, 2005.

[36] S. J. Smith, "Circuit reconstruction tools today," *Current Opinion in Neurobiology*, 17(5):601–608, 2007.

[37] M. Snir, et al., *MPI: The Complete Reference, Vol. 1*, MIT Press, 1998.

[38] Olaf Sporns, Giulio Tononi, and Rolf Kotter, "The human connectome: A structural description of the human brain," *PLoS Computational Biology*, 1(4), September 2005.

[39] B. Tadmor and B. Tidor, "Interdisciplinary research and education at the biology-engineering-computer science interface: a perspective," *Drug Discovery Today*, 10(23-24), December 2005.

[40] C. Xu and J. L. Prince, "Snakes, shapes, and gradient vector flow," *IEEE Transactions on Image Processing*, 7:359–369, 1998.

[41] Z. Yu-qian, G. Wei-hua, C. Zhen-cheng, T. Jing-tian, and L. Ling-yun, "Medical images edge detection based on mathematical morphology," *27th Annual International Conference of the Engineering in Medicine and Biology Society*, pp. 6492–6495, 2005.

CHAPTER 7

Machine Learning Techniques for Large Data

Elad Yom-Tov

The field of machine learning is devoted to the study and development of algorithms that attempt to learn from data; that is, they extract rules and patterns from data provided to them. In this chapter we survey different types of machine learning algorithms, with a focus on algorithms for pattern classification and analysis. We further emphasize algorithms that are useful for processing large data, such as that prevalent in computational biology.

7.1 Introduction

Machine learning (ML) methods are algorithms for identifying patterns in large data collections and for applying actions based on these patterns. ML methods are data driven in the sense that they extract rules from given data. ML algorithms are being successfully applied in many domains, including analysis of Internet data, fraud detection, bioinformatics, computer networks analysis, and information retrieval.

Watanabe [1] described a pattern as "the opposite of chaos; it is an entity, vaguely defined, that could be given a name." Examples of patterns are DNA sequences that may cause a certain disease, human faces, and behavior patterns. A pattern is described by its features. These are the characteristics of the examples for a given problem. For example, in a bioinformatics task, features could be the genetic markers that are active or nonactive for a given disease.

Once data is collected, ML methods are usually applied in two stages: training and testing. During the training phase, data is given to the learning algorithms which then construct a model of the data. The testing phase consists of generating predictions for new data according to the model. For most algorithms, the training phase is the computationally intensive phase of learning. Testing (or applying) the model is generally orders of magnitude faster than the learning phase.

ML algorithms are commonly divided according to the data they use and the model they build. For example, generative algorithms construct a model for generating random samples of the observed data, usually assuming some hidden parameters. In contrast, discriminative methods build a model for differentiating between examples of different labels, where all parameters of the model are directly measurable. In this chapter, we focus on discriminative algorithms.

Discriminative algorithms are often used to classify data according to labeled examples of previous data. Therefore, the training data is provided with both inputs and outputs. This is known as supervised learning. Alternatively, unsupervised learning algorithms are applicable when only inputs are provided, and the ML algorithm can, for example, cluster the data into distinct partitions.

ML algorithms can also be classified according to their training mode. Off-line learning refers to a learning process where all training data is collected before learning commences. In online learning the ML model is built and updated as the data is collected. The latter is more appropriate when large datasets are involved or when the environment changes over time, but is usually less accurate than off-line learning.

Recently, an important distinction was made between small-scale and large-scale learning methods. According to Bottou and Bousquet [2], a small-scale learning problem is constrained by the maximal number of training examples, whereas a large-scale learning problem is limited by the maximal computing time (and resources) available. Traditionally, most ML algorithms were developed for small-scale learning. In recent years many of them have been adapted to large-scale data. The latter are the focus of this chapter.

Generally, large-scale ML has been tackled using one of three approaches (or a combination thereof): online, where each data item is examined only once; parallel or distributed algorithms, where the computational load is partitioned between several processing nodes; and efficient data structures, which can speed up computation through either exact or approximate data representation.

In the following sections we describe three main classes of ML algorithms: feature selection and/or feature reduction algorithms, clustering algorithms, and classification algorithms. We do not discuss data collection and representation, as these are mostly problem-specific and are therefore difficult to generalize. In broad terms, one should try to find invariant features that are as informative as possible about the task at hand.

7.2 Feature Reduction and Feature Selection Algorithms

Often it is the case that many features can be used to describe an example, but it is unknown how informative each feature is to the classification task at hand. In these cases one may ask if all available features should be used for classification or clustering, or only those which are known to be highly indicative. Several studies (for example, [3, 4]) have shown that uninformative features can be detrimental, especially when they are highly correlated to each other, noisy, or completely irrelevant. Conversely, if too few features are collected, it will be impossible to distinguish between the examples [1]. Therefore, it is usually advantageous to use feature selection or feature reduction algorithms, which remove useless features and keep informative features.

Two approaches can be used to process features. One is to choose a subset of the features, a process known as feature selection. Alternatively, a good combination of the features can be found through a process of feature reduction.

Feature selection is a difficult combinatorial optimization problem for even a reasonably small set of features. Finding the best subset of features by testing all possible combinations of 100 input features will require testing 10^{30} combinations.

Consequently numerous methods have been proposed for finding a (suboptimal) solution by testing a fraction of the possible combinations.

Feature selection methods can be divided into three main types [4]:

1. *Wrapper methods:* The feature selection is performed around (and with) a given classification algorithm. The classification algorithm is used for ranking possible feature combinations.
2. *Embedded methods:* The feature selection is embedded within the classification algorithm.
3. *Filter methods:* Features are selected for classification independently of the classification algorithm.

The simplest wrapper algorithms are exhaustive search (which, as noted above is practical only when the number of features is small), sequential forward feature selection (SFFS), and sequential backward feature selection (SBFS). More advanced wrapper algorithms include simulated annealing and genetic algorithms [5]. Sequential forward feature selection works by first selecting the feature with which the lowest classification error is reached and then greedily adding features that provide the largest reduction in error. This process is continued until the classification error starts to increase. The underlying assumption of this method is that features can be evaluated individually. While this assumption often works well in practice, it may cause this method to miss combinations of features that are useful only when selected together.

Because wrapper algorithms use a classification algorithm as the scoring function, it is easy to parallelize them if the classification algorithm can be parallelized. Alternatively, methods such as SFFS and SBFS can be parallelized by exploiting the fact that features are evaluated independently and thus, different features can be evaluated by different processors.

One example of the filtering approach to feature selection is that developed by Koller and Sahami [6]. Here, information theoretic considerations are used to decide on the most informative features, without the need for a classification algorithm. This method estimates the cross-entropy between every pair of features and discards those features that have a large cross-entropy with other features, thus removing features that add little additional classification information. However, because the pair-wise estimate of the cross-entropy between features needs to be computed, this algorithm scales quadratically with the number of features. Thus, only data with a small to medium number of features can be evaluated using this method. We note, however, that this method is easy to parallelize because of the independence between the estimation procedure for each pair of features.

An altogether different approach to reducing dimensions is through feature reduction, that is, finding a low-dimensional combination of features. This combination can be linear or nonlinear.

Arguably the most well-known of these methods is principal component analysis [PCA, or Karhunen-Loeve Transform (KLT)]. PCA reshapes the data along directions of maximal variance. This is done by finding the eigenvectors corresponding to the largest eigenvalues of the covariance matrix of the data, and projecting the data along these eigenvectors. Thus, the low-dimensional features are linear combinations of the high-dimensional features. PCA is an unsupervised method; that is, it does not take into account the labels of the data. A similar, albeit supervised, linear method is Fisher discriminant analysis, which projects the data on a single dimension, while maximizing the separation between the classes of the data.

Because of the usefulness of PCA, much research has been devoted to making it applicable to large-scale data. One approach is to use a simple neural network, trained stochastically with an appropriate learning rule so that it converges to the principal components [7]. This method has also been used in a parallel setting using batch training [8]. Another approach to PCA computation is through an expectation-maximization algorithm (see also below) [9], an algorithm which can also be parallelized.

In the information retrieval field, a similar approach to PCA is used to transform high-dimensional term-document matrices into lower dimensional ones by applying singular value decomposition (SVD) to the term-document matrix. This process is known as Latent Semantic Indexing (LSI). The difference between PCA and LSA is that the former applies eigenvector decomposition to the covariance matrix, while the latter does the same to the term-document matrix. Thus, the results of the two analyses are different, while the methods are similar. Several methods for parallel computation of SVD are reviewed in Letsche and Berry [10].

Using different basic assumptions, independent component analysis (ICA) [11] finds a linear mixture of the data such that each of the projections will be as independent as possible from the other projections. Versions of this algorithm can be modified for online training and parallel versions of the algorithm have been developed in [12].

Recently, nonnegative matrix factorization (NMF) [13] methods have been developed for feature reduction. Given a data matrix X of size $N \times D$, where N are the number of examples and D the number of features, NMF finds two matrices W and H, such that $X \approx W \cdot H$. The number of columns of W is smaller than D, so that W represents a lower dimensional projection of X. NMF has been used, for example, in text and image clustering. For large data, a parallel NMF algorithm is shown in [14] and an online algorithm in [15].

Many feature reduction algorithms project high-dimensional data via a nonlinear projection, frequently through modifications of the above-mentioned linear methods. Examples of such methods are nonlinear component analysis [16], nonlinear FDA [17], and Kernel PCA [18]. The latter method works by remapping data by way of a kernel function into feature space where the principle components of the data are found.

7.3 Clustering Algorithms

Clustering algorithms are (usually) unsupervised methods for finding a partitioning of the data points into clusters. The clusters can be represented either as a list of their members or by computing a representative sample of each class, known as the centroid. Clusters can be defined [19] as partitions where patterns are more similar to each other in a given cluster than they are to patterns outside it, or as a volume of high-density points separated from other clusters by relatively low density volumes.

Clustering algorithms require a similarity function by which points can be compared and (often) a criterion for joining points into clusters or the number of clusters to be found.

The computational cost of finding an optimal partition is usually prohibitively high. Therefore, in all but the most simple cases, clustering algorithms find what may be a suboptimal partition, albeit in a reasonable number of computations. This is done using top-down (or partitional) algorithms and bottom-up (or hierarchical) algorithms.

A simple example for bottom-up algorithms is the agglomerative hierarchical clustering algorithm (AGHC). This algorithm starts by assuming that each data point is a cluster. At each iteration two clusters are merged until a preset number of clusters is reached. The decision on which clusters are to be merged can be done using the distance between cluster centers, distance between the two nearest points in different clusters, and so forth. AGHC is a very simple, intuitive scheme. However, it is computationally intensive and thus impractical for medium and large datasets.

Top-down methods are the type more frequently used for clustering (especially of large data) due to their lower computational cost. The k-Means algorithm [20] is probably the most popular amongst top-down clustering algorithms. The algorithm attempts to minimize the squared error between points labeled to be in a cluster and the centroid.

A pseudo-code of the k-Means algorithm is given in Algorithm 1. k-Means proceeds by iteratively finding the closest centroid to each data point and then recomputing centroids as the average of all points associated with a centroid. Variants of the k-Means algorithm include fuzzy k-Means [21] and k-Medians [22].

There are several ways in which k-Means has been modified to be used with large data. KD-trees [23] in conjunction with an analysis of the geometry of clusters have been shown to reduce the computational load to sublinear dependencies on the number of data points. An online version of k-Means has been developed in [24]. Finally, parallelization of k-Means is extremely easy because at each step the assignment of each point to the centroids, as well as the update to the centroids, can be conducted independently for each point. Therefore, subsets of points can be allocated to different processors.

The advent of kernel-based methods has led to the development of clustering algorithms that use kernels functions in their operation (e.g., [25]). The basic idea behind these algorithms is to map the data into a higher dimension using

Algorithm 1 Pseudo-Code of the k-Means Clustering Algorithm

1: Initialize K random cluster centers.
2: **while** not converged **do**
3: Assign each of the data points the nearest of the K cluster centers.
4: Recompute the cluster centers by averaging the points assigned to each cluster.
5: **end while**
6: Return the cluster centers.

a nonlinear function of the input features and to cluster the data using simple clustering algorithms at the higher dimension. (More details regarding kernels are given in Section 7.4.) One of the main advantages of kernel methods is that simple clusters (for example, ellipsoid clusters) formed in a higher dimension correspond to complex cluster geometries in the input space.

The spectral clustering methods [26, 27] are a related class of clustering algorithms. These methods first map the data into a matrix representing the distance between the input patterns. The matrix is then projected onto its k largest eigenvectors and the clustering is performed on this projection. Because of the need to compute the pair-wise kernel function values, there has been limited application of these methods to large data [8].

7.4 Classification Algorithms

The goal of classification algorithms is to label examples based on their features. Classification algorithms use training data to learn how feature patterns translate into labels. Formally, the goal of a classifier is to find a functional mapping between input data X and a class label Y so that $Y = f(X)$. The mapping f should give the smallest possible error in the mapping; that is, the smallest number of examples where Y will be the wrong label, as measured on previously unseen (and untrained-on) test data. We note that many classification algorithms can generate a classifier that will classify all the training data correctly. This does not imply that they will perform well on unseen test data. Indeed, it can be a sign that the classifier has been overfitted to the training data.

Given full knowledge about the distribution of the data, an optimal classification rule (with regard to the classification error) can be derived. This is known as the Bayes rule. If one assumes a zero/one loss (i.e., a wrong decision entails a penalty of one and a correct decision results in a penalty of zero), the Bayes rule simplifies to the maximum aposteriory (MAP) rule, which requires that we label an input sample with the most probable label according to the data distribution.

It is usually impossible to use the Bayes rule because it requires full knowledge of the class-conditional densities of the data. Thus, one is frequently left with one of two options. If a model for the class-conditional densities is known (for example, if it is known that the data is derived from a mixture of two Gaussians) one can use *plug-in* rules to build a classifier. Plug-in classifiers estimate the parameters of the distribution and use the MAP rule to classify data. When the parametric structure of the data is unknown, one can use nonparametric classifiers that estimate the density of the data or the decision boundaries between different classes.

One of the most widely used methods for estimating the parameters of distributions for use in plug-in classifiers is the expectation-maximization (EM) algorithm [28]. In order to operate the EM algorithm, the number of components (e.g., the number of Gaussians) in each class must be known in advance, or at least estimated using methods such as ML-II [29] or MDL [30].

As its name implies, the EM algorithm works in two stages. During the expectation stage an expectation of the likelihood of each example, given the current parameter estimates, is computed. In the maximization step, the maximum likelihood estimates of the distribution parameters are estimated by maximizing the expected likelihood of the examples found in the expectation. The algorithm begins with random values of the distribution parameters and iterates between the E and M steps until it converges. The most demanding computational part of the algorithm is the expectation step, where each sample in the data must be accessed. However, this is also a task that is trivial to parallelize; each node in a parallel system can observe some of the data and compute the statistics independently of other nodes. The total update can then be aggregated at a central location (a master node) and the M step executed. Alternatively, stochastic methods have been suggested for online training of the EM algorithm [31].

When the underlying distribution of data cannot be modeled using parametric distributions, or when these distributions are unknown, one is forced to use other classification approaches. One of these is to model the density of data for each of the classes and classifying test points according to the maximum posterior probability. Frequently, this is done using the Parzen windows estimator [32]. This method is, however, expensive both computationally and memory-wise. Additionally, a large number of training points is required for an accurate estimation of density. Therefore, this method is suitable mainly for small to medium datasets.

A different method for classification uses a naive Bayes classifier. This classifier works under the assumption that each feature is independent of other features. Under this assumption, using Bayes' theorem, the probability that a sample drawn from a given class is simply the prior probability of this class multiplied by the probability that the value of the sample for each feature is derived from the class. The probabilities for each feature are approximated with relative frequencies found in the training set.

In spite of their naive design and underlying assumptions, naive Bayes classifiers often work well on real-world problems. This has been shown to have a theoretic justification [33].

Speeding up a naive Bayes classifier (i.e., computing the frequencies of feature values) can be done by either sending different features to different processing nodes or by partitioning the data to different nodes and aggregating the frequencies returned from each node.

Most classification algorithms directly try to construct a boundary between classes to minimize the test error. The k-nearest neighbor classifier is probably the simplest classification method. This classifier labels test points according to a majority vote amongst the k closest training points. This method frequently achieves remarkably low classification errors but it suffers from two main drawbacks. The first is that it is computationally expensive because of the need to find the k closest training points to each test point. The second drawback is that a large memory is required for storing the training data. One way to overcome the first drawback is through efficient data structures such as the KD-tree [34] or, more recently, cover trees [35].

A different approach to classification uses decision trees. In a decision tree, each node corresponds to a variable and each arc to a range of possible values of that variable. The leaves of the tree represent predictions for values of the labels.

Decision trees are built to create leaves that are as pure as possible; that is, they contain the largest possible fraction of points from a single class. Classification of

Algorithm 2 High-Level Pseudo-Code for Building a Decision Tree

```
Given: Training set T
function Decision-Tree-Building (Set T)
if T meets the stopping criteria (Required purity reached) then
   Return
else
   for each attribute in T do
      Find the best values to split T along the i-th attribute
   end for
   Use the attribute that gives the best split among all the attributes to partition T into T1
   and T2
   Decision-Tree-Building (T1)
   Decision-Tree-Building (T2)
end if
```

a new test point is performed by traversing the tree from its root to the leaves, taking the right arcs according to the values of the example's features.

The most well-known algorithms for building decision trees are CART [36], ID3 [37], and C4.5 [38]. These algorithms construct the decision tree in a greedy fashion, starting from the root. A schematic representation of a tree building algorithm is shown in Algorithm 2. At each node, the data is split according to a decision criterion based on the statistics of each feature. This process is repeated recursively until the nodes reach a desired purity.

Decision trees are a simple yet effective classification. One of their main advantages is that they provide human-readable rules of classification. Decision trees have several drawbacks, especially when trained on large data, where classifiers tend to result in complicated trees, which in turn require either pruning or else a large memory for storage. There is considerable literature on methods for simplifying and pruning decision trees (for example, [39]). Techniques for handling large data on a single node include transformation and sorting (SPRINT, [40]), sorting and using data appropriate data structures (SLIQ, [41]), sampling [38], and approximation (SPEC, [42], with or without parallelization). Direct use of parallelization has been demonstrated, for example, in [43].

Many classification algorithms build a functional mapping from input patterns to output labels, optimizing an objective function that aims to minimize the training error.

A simple linear mapping can be performed using the Moore-Penrose pseudo-inverse, which links the input patterns to the labels via the weighted sum of the input patterns to optimize the mean square error. If the training patterns are represented in a matrix of size $N \times D$ where D is the input dimension and N the number of examples, and the corresponding labels in a $N \times 1$ vector $T \in \{-1, +1\}$, the optimal weights for the classifier are given by

$$w = \left(P^T \cdot P\right)^{-1} \cdot P^T \cdot T \tag{7.1}$$

and the classification rule is

$$\hat{t} = sign\left(W^T \cdot x\right) = \begin{cases} +1 & if\ w^T \cdot x > 0 \\ -1 & if\ w^T \cdot x < 0 \end{cases} \tag{7.2}$$

Several methods exist for parallelization of this algorithm, for example, [44]. Stochastic gradient methods can be used for online approximations for least squares, as described later in this chapter.

Neural networks, first suggested by Alan Turing [45], are a computational model inspired by the connectivity of neurons in animate nervous systems. A schematic diagram of a neural network with a single hidden layer is shown in Figure 7.1. Each circle denotes a computational element referred to as a neuron. A neuron computes a weighted sum of its inputs and possibly performs a nonlinear function on this sum. The sign function and the hyperbolic tangent are usually chosen as the nonlinear functions.

Many variations on the simple layout shown in Figure 7.1 (known as a feed-forward architecture) have been developed. Some examples include: recurrent neural networks, which feed the output back as an input to the network; neural networks with multiple outputs, useful for multiclass classification; and networks with a single neuron. The latter is known as a perceptron.

Perceptrons find a linear separating hyperplane, optimizing a cost function such as mean least-squares. Many algorithms have been developed for training (i.e., finding the weight vector for the perceptron): batch and stochastic, online and offline, with and without memory [46]. The perceptron is a good choice for an online linear classifier.

Multilayered neural networks are generally trained using the backpropagation algorithm or one of its variants. This algorithm propagates errors in the classification of training examples from the output layer, through the hidden layers, to the input layer, whilst modifying the weights of the neurons to decrease the error. Second-order backpropagation methods such as conjugate-gradient descent (CGD) [47] and Quickprop [48], can be used to achieve faster convergence.

Neural networks are very amenable to training with large data using stochastic gradient descent, offline as well as online [49]. Alternatively, parallel methods (both stochastic and batch) have been used for such training [50].

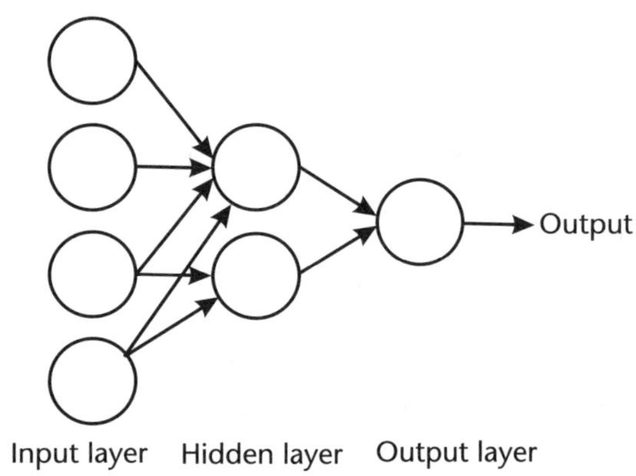

Figure 7.1 A schematic diagram of a neural network.

In recent years, the development of kernel-based learning methods has put support vector machine (SVM) classifiers at the forefront of classification algorithms. SVMs, first introduced by Boser, Guyon, and Vapnik [51], find a linear hyperplane that separates the data with the largest margin; that is, the largest distance between the hyperplane and the closest data points. Maximizing the margin has been shown to improve generalization performance [18]. SVMs further improve classification by transforming the data into a high dimension, known as the feature space.

A linear separating hyperplane is a decision function in the form

$$f(\mathbf{x}) = sign\left(\langle \mathbf{w}, \mathbf{x} \rangle + b\right) \tag{7.3}$$

where $\mathbf{x} \in \Re^N$ is an input pattern, $\mathbf{w} \in \Re^N$ is a weight vector, b a bias term, and $\langle \cdot, \cdot \rangle$ denotes an inner product.

For data to be classified correctly, the hyperplane defined by $\mathbf{w}$ should ensure that

$$y_i \cdot \left(\langle \mathbf{w}, \mathbf{x_i} \rangle + b\right) > 0 \quad for\ all\ i = 1, \ldots, m \tag{7.4}$$

assuming that $y \in \{-1, +1\}$.

There exists a single such hyperplane that maximizes the margin, which can be found by maximizing

$$\|\mathbf{w}\|^2$$

Taking both requirements into account, the SVM problem can be written as

$$\begin{aligned} &minimize\ \ \tfrac{1}{2}\|\mathbf{w}\|^2 \\ &subject\ to\ \ y_i \cdot \left(\langle \mathbf{w}, \mathbf{x_i} \rangle + b\right) \geq 1 \quad for\ all\ i = 1, \ldots, m \end{aligned} \tag{7.5}$$

Note that the right-hand side of the bottom equation was modified to one instead of zero to avoid a trivial solution to $\mathbf{w}$.

This constrained minimization problem is solved using Lagrange multipliers. Transformed into the dual optimization form, the problem can be rephrased as

$$\begin{aligned} &maximize\ \ W(\alpha) = \textstyle\sum_{i=1}^{m} \alpha_i - \frac{1}{2}\sum_{i,j=1}^{m} \alpha_i \alpha_j y_i y_j \langle \mathbf{x_i}, \mathbf{x_j} \rangle \\ &s.t.\ \ \alpha_i \geq 0, \textstyle\sum_{i=1}^{m} \alpha_i y_i = 0 \end{aligned} \tag{7.6}$$

Classification of new samples is performed using the following equation:

$$y = sign\left(\sum_{i=1}^{m} y_i \alpha_i \langle \mathbf{x}, \mathbf{x_i} \rangle + b\right) \tag{7.7}$$

As this equation shows, any coefficients α_i that are close to zero correspond to input patterns that are not used to construct the resulting classifier. The remaining coefficients are known as support vectors. This quadratic optimization problem can be solved in many ways, including: a perceptron, which finds the largest margin hyperplane separating the data [46]; quadratic programming optimization algorithms, which solve the optimization problem [46]; or through other efficient optimization algorithms such as the sequential minimal optimization (SMO) algorithm [52].

As noted earlier, much better classification is achieved by mapping the input patterns into a high dimensional space. This can be done by mapping each input pattern using a function such $\tilde{\mathbf{x}} = \varphi(\mathbf{x})$. However, in practice, if the function maps the data into a very high dimension, it would be problematic to compute and to store the results of the mapping.

Note, however, that in (7.6) only pairwise inner products of examples are used. Therefore, instead of mapping each sample to a higher dimension and then performing the inner product, it is possible (for certain classes of mapping functions) to first compute the inner product between patterns and only then compute the mapping for the resulting scalar. This is known as the kernel trick [51].

Thus, in (7.6) we now replace the inner products $\langle \mathbf{x}, \mathbf{x}' \rangle$ with $k(\mathbf{x}, \mathbf{x}')$, where k is the mapping function, also known as the kernel function. Kernel functions should conform to conditions known as the Mercer conditions [53]. Examples of such functions include polynomials, radial basis functions (Gaussian functions), and hyperbolic tangents.

SVMs have been studied extensively [18] and have been extended to many directions. Some notable examples include cases where the classes are not separable, through the introduction of a penalty term that allows some training patterns to be incorrectly classified [54], and for feature selection [55, 56].

As noted above, the solution to an SVM requires the solution of a quadratic optimization problem. Such problems have $O(n^3)$ time complexity and $O(n^2)$ space complexity [57]. In practice, the computational complexity is usually closer to $O(n^2)$ [58]. Recently, some approximation algorithms have claimed to further reduce the computational complexity of SVMs to $O(n)$ complexity [57].

Some research has been devoted to the solution of SVMs on parallel computing nodes. In [58] an SVM solver is parallelized by training multiple SVMs, each on a subset of the training data, and aggregating the resulting classifiers into a single classifier. The training data is then redistributed to the classifiers according to their performance and the process is iterated until convergence is reached. A more low-level approach is taken in [59], where the quadratic optimization problem is divided into smaller quadratic programs, each of which is solved on a different node. The results are aggregated and the process is repeated until convergence. Graf et al. [60] partition the data and solve an SVM for each partition. The support vectors from each pair of classifiers are then aggregated into a new training set, for which an SVM is solved. The process continues until a single classifier remains. The aggregation process can be iterated using the support vectors of the final classifier in the previous iteration to seed the new classifiers.

An alternative approach [61] to parallelization of SVM solvers is to compute the full kernel matrix in a distributed memory so that each node holds a slice of the full kernel matrix, and execute a gradient descent-like algorithm to find the Lagrange multipliers. This approach is advantageous when it is hard to separate between the data from different classes.

Online SVMs have been extensively researched (see, for example, [62]). A simple though effective idea is the Forgetron algorithm [63]. Here, a "budget" of support vectors is prespecified. As each new example is received, it is first classified according to the support vectors held in memory. If it is misclassified, it is accepted into the learner's memory as a new support vector and the weight of all previous support vectors is reduced. The support vectors with the lowest weight are discarded to limit the memory required for the algorithm.

The last approach to classification we discuss in this chapter is based on combining multiple classifiers to form a single, improved classifier. In *bagging*, multiple instances of the dataset are created by drawing subsets of examples from the training set and building a classifier based on each of these subsets. Each of these classifiers is known as a component classifier. Classification of a test example is performed by taking the majority vote of all the component classifiers [46]. Bagging can be parallelized trivially by training the component classifiers on different nodes [64].

Boosting is a more sophisticated approach to combining classifiers. Here, a simple (or weak) classifier is trained on the data, as shown in Algorithm 3. Then, those points of the train-set that are incorrectly classified are identified and given a higher weight. Another weak classifier is trained on the weighted data to improve the classification of these incorrectly labeled points. This process is repeated until a sufficiently low training error is reached.

The training of the weak classifiers can be performed by either drawing points from the training data proportionally to error or by selecting a subset of the incorrectly trained points. The former algorithm is known as AdaBoost [65], which is arguably the most popular boosting algorithm. The latter algorithm is the local boosting algorithm [66].

Several approaches exist for training boosting classifiers with large data. Online [67] and semi-online (also known as filtering) methods [68] have been demonstrated in the literature. In the latter, it is assumed that a very large number of examples are provided to the learner, one at a time. Instead of sampling a fixed dataset according to a distribution as in Algorithm 3, each new example is accepted with a probability relative to this distribution. Once enough samples have been accepted, a weak learner is trained and the examples are discarded. This process continues until enough weak learners have been trained. Thus, very large data can be handled because only a small chunk is held in memory at any one time.

Similar to bagging, boosting can be parallelized trivially by parallel training of the weak learner. Alternatively, each node can train a weak learner on a subset of the data and share the resulting hypothesis with all other nodes [64]. The nodes

Algorithm 3 Pseudo-Code for the AdaBoost Algorithm

Given: Training set T, comprised of n patterns $\mathbf{x}$ and corresponding labels $y \in \{-1, +2\}$.
Initialize: $W_1(i) = 1/n$
for k = 1 to M **do**
  Train weak classifier C_k using T sampled according to W_i
  Estimate E_k, the error of C_k measured on T sampled according to W_i.
  Let $\alpha_k = 1/2 \ln((1 - E_k)/E_k)$

$$W_{k+1}(i) = \frac{W_k(i)}{Z_k} \times \begin{cases} e^{-\alpha_k} & \textit{if the ith sample is correctly classified} \\ e^{+\alpha_k} & \textit{if the ith sample is incorrectly classified} \end{cases}$$

end for

then weigh each sample assigned to them according to a combination of all the weak learners.

7.5 Material Not Covered in This Chapter

The field of machine learning spans a wide range of methods and practices, making it difficult to cover all of them in a single chapter. The following paragraphs list some useful approaches that are not described in this chapter.

In addition to supervised and unsupervised learning, semi-supervised learning uses both labeled and unlabeled examples to learn a classifier. A related field is that of transductive learning [69]. In contrast to supervised learning, where the goal is to learn a general rule for classifying examples, transductive learning aims at learning rules for classifying specific test examples. This makes transductive learning more accurate for these examples.

Graphical models [70] are learning methods that represent dependencies between random variables using a graph, where nodes correspond to random variables and edges denote independencies between the variables. Graphs can be directed or undirected. Examples of graphical models include Bayesian networks, Gaussian Belief Propagation, Markov random fields, hidden Markov models, as well as neural networks discussed earlier.

Reinforcement learning [71] is a learning paradigm where algorithms learn how to act, based on observations of the world. Every action taken by the algorithms has an impact on their environment. The environment provides feedback to the learning algorithm, which uses this feedback to improve its model (or policy).

Reinforcement learning algorithms aim to maximize some long-term reward. Reinforcement learning differs from supervised learning in that a training set does not explicitly exist. As the learning algorithm performs actions in its environment, it receives feedback but there is no "correct" action as such.

References

[1] W. Watanabe, *Pattern Recognition: Human and Mechanical*. Wiley, 1985.

[2] L. Bottou and O. Bousquet, "The tradeoffs of large scale learning," in *Advances in Neural Information Processing Systems*, vol. 20, Cambridge, MA: MIT Press, 2008.

[3] G. Trunk, "A problem of dimensionality: A simple example," *IEEE Transactions on Pattern Analysis and Machine Intelligence*, vol. 1, no. 3, pp. 306–307, 1979.

[4] A. Blum and P. Langley, "Selection of relevant features and examples in machine learning," *Artificial Intelligence*, vol. 97, pp. 245–271, 1997.

[5] E. Yom-Tov and G. Inbar, "Feature selection for the classification of movements from single movement-related potentials," *IEEE Transactions on Neural Systems and Rehabilitation Engineering*, vol. 10, no. 3, pp. 170–177, 2001.

[6] D. Koller and M. Sahami, "Toward optimal feature selection," *Proceedings of the 13th International Conference on Machine Learning*, (Bari, Italy), pp. 284–292, Morgan Kaufmann, 1996.

[7] E. Oja, "Simplified neuron model as a principal component analyzer," *Journal of Mathematical Biology*, vol. 15, no. 3, pp. 267–273, 1982.

[8] E. Yom-Tov, U. Aharoni, A. Ghoting, E. Pednault, D. Pelleg, H. Toledano, and R. Natarajan, "An introduction to the IBM parallel machine learning toolbox," IBM developerWorks, 2007.

[9] S. Roweis, "Em algorithms for PCA and SPCA," in *Advances in Neural Information Processing Systems*, vol. 10, Cambridge, MA: MIT Press, 1997.

[10] T. A. Letsche and M. W. Berry, "Large-scale information retrieval with latent semantic indexing," *Information Science*, vol. 100, no. 1-4, pp. 105–137, 1997.

[11] J.-F. Cardoso, "Blind signal separation: Statistical principles," *Proceedings of the IEEE*, vol. 9, no. 10, pp. 2009–2025, 1998.

[12] D. Keith, C. Hoge, R. Frank, and A. Malony, "Parallel ICA methods for EEG neuroimaging," in *20th International Parallel and Distributed Processing Symposium (IPDPS)*, pp. 25–29, IEEE, 2006.

[13] S. H. S. Daniel D. Lee, "Algorithms for non-negative matrix factorization," in *Advances in Neural Information Processing Systems*, vol. 13, pp. 556–562, Cambridge, MA: MIT Press, 2000.

[14] S. A. Robila and L. G. Maciak, "A parallel unmixing algorithm for hyperspectral images," *Intelligent Robots and Computer Vision XXIV: Algorithms, Techniques, and Active Vision*, vol. 6384, no. 1, p. 63840F, 2006.

[15] B. Cao, D. Shen, J.-T. Sun, X. Wang, Q. Yang, and Z. Chen, "Detect and track latent factors with online nonnegative matrix factorization," *Proceedings of International Joint Conferences on Artificial Intelligence (IJCAI)*, pp. 2689–2694, 2007.

[16] S. Mika, B. Schölkopf, A. J. Smola, K.-R. Müller, M. Scholz, and G. Rätsch, "Kernel pca and de-noising in feature spaces," *Advances in Neural Information Processing Systems 11* (M. Kearns, S. Solla, and D. Cohn, eds.), (Cambridge, MA, USA), MIT Press, 1999.

[17] S. Mika, G. Rätsch, J. Weston, B. Schölkopf, and K.-R. Müller, "Fisher discriminant analysis with kernels," in *Neural Networks for Signal Processing IX* (Y.-H. Hu, J. Larsen, E. Wilson, and S. Douglas, eds.), pp. 41–48, IEEE, 1999.

[18] B. Schölkopf and A. Smola, *Leaning with Kernels: Support Vector Machines, Regularization, Optimization, and Beyond*, Cambridge, MA: MIT Press, 2002.

[19] A. Jain, R. Duin, and J. Mao, "Statistical pattern recognition: A review," *IEEE Transactions on Pattern Analysis and Machine Intelligence*, vol. 22, no. 1, pp. 4–37, 1999.

[20] S. Lloyd, "Least squares quantization in PCM," *IEEE Transactions on Information Theory*, vol. IT-2, pp. 129–137, 1982.

[21] J. Bezdek, *Fuzzy mathematics in pattern classification*, PhD thesis, Cornell University, Applied mathematics center, Ithaca, NY, 1973.

[22] A. Juan and E. Vidal, "Fast k-means-like clustering in metric spaces," *Pattern Recogn. Lett.*, vol. 15, no. 1, pp. 19–25, 1994.

[23] D. Pelleg and A. Moore, "Accelerating exact k-means algorithms with geometric reasoning," *KDD '99: Proceedings of the Fifth ACM SIGKDD International Conference on Knowledge Discovery and Data Mining*, (New York, NY), pp. 277–281, ACM, 1999.

[24] S. Zhong, "Efficient online spherical k-means clustering," *Proceedings of the IEEE International Joint Conference on Neural Networks (IJCNN)*, pp. 3180–3185, 2005.

[25] A. Ben-Hur, D. Horn, H. Siegelmann, and V. Vapnik, "Support vector clustering," *Journal of Machine Learning Research*, vol. 2, pp. 125–137, 2001.

[26] A. Ng, M. Jordan, and Y. Weiss, "On spectral clustering: Analysis and an algorithm," in *Advances in Neural Information Processing Systems* (T. Dietterich, S. Becker, and Z. Ghahramani, eds.), vol. 14, (Cambridge, MA), pp. 849–856, MIT Press, 2002.

[27] N. Cristianini, J. Shawe-Taylor, and J. Kandola, "Spectral kernel methods for clustering," in *Advances in Neural Information Processing Systems* (T. Dietterich, S. Becker, and Z. Ghahramani, eds.), vol. 14, (Cambridge, MA), pp. 649–655, MIT Press, 2002.

[28] A. Dempster, N. Laird, and D. Rubin, "Maximum-likelihood from incomplete data via the em algorithm (with discussion)," *Journal of the Royal Statistical Society, Series B*, vol. 39, pp. 1–38, 1977.

[29] D. MacKay, "Bayesian model comparison and backprop nets," in *Neural Networks for Signal Processing 4* (J. Moody, S. Hanson, and R. Lippmann, eds.), (San Mateo, CA), pp. 839–846, Morgan Kaufmann, 1992.

[30] A. Barron and T. Cover, "Minimum complexity density estimation," *IEEE Transactions on information theory*, vol. IT-37, no. 4, pp. 1034–1054, 1991.

[31] A. Same, C. Ambroise, and G. Govaert, "An online classification em algorithm based on the mixture model," *Statistics and Computing*, vol. 17, no. 3, pp. 209–218, 2007.

[32] E. Parzen, "On estimation of a probability density function and mode," *Annals of mathematical statistics*, vol. 33, no. 3, pp. 1065–1076, 1962.

[33] H. Zhang, "The optimality of naive bayes," *Proceedings of Florida Artificial Intelligence Research Society (FLAIRS)*, 2004.

[34] J. L. Bentley, "Multidimensional binary search trees used for associative searching," *Commun. ACM*, vol. 18, no. 9, pp. 509–517, 1975.

[35] A. Beygelzimer, S. Kakade, and J. Langford, "Cover trees for nearest neighbor," *ICML '06: Proceedings of the 23rd International Conference on Machine Learning*, (New York, NY), pp. 97–104, ACM, 2006.

[36] L. Breiman, J. Friedman, R. Olshen, and C. Stone, *Classification and Regression Trees*, New York: Chapman and Hall, 1993.

[37] J. Quinlan, "Learning efficient classification procedures and their application to chess end games," in *Machine Learning: An Artificial Intelligence Approach* (R. Michalski, J. Carbonell, and T. Mitchell, eds.), (San Francisco, CA), pp. 463–482, Morgan Kaufmann, 1983.

[38] J. Quinlan, *C4.5: Programs for Machine Learning*, San Francisco: Morgan Kaufmann, 1993.

[39] M. Mehta, J. Rissanen, and R. Agrawal, "Mdl-based decision tree pruning," *Proceedings of the First International Conference on Knowledge Discovery and Data Mining*, pp. 216–221, 1995.

[40] J. C. Shafer, R. Agrawal, and M. Mehta, "Sprint: A scalable parallel classifier for data mining," *VLDB'96, Proceedings of 22th International Conference on Very Large Data Bases, September 3-6, 1996, Mumbai (Bombay), India* (T. M. Vijayaraman, A. P. Buchmann, C. Mohan, and N. L. Sarda, eds.), pp. 544–555, Morgan Kaufmann, 1996.

[41] M. Mehta, R. Agrawal, and J. Rissanen, "Sliq: A fast scalable classifier for data mining," *EDBT '96: Proceedings of the 5th International Conference on Extending Database Technology*, (London, UK), pp. 18–32, Springer-Verlag, 1996.

[42] E.-H. Han, G. Karypis, and V. Kumar, "Scalable parallel data mining for association rules," *SIGMOD '97: Proceedings of the 1997 ACM SIGMOD International Conference on Management of Data*, (New York, NY), pp. 277–288, ACM, 1997.

[43] N. Amado, J. Gama, and F. M. A. Silva, "Parallel implementation of decision tree learning algorithms," *EPIA '01: Proceedings of the 10th Portuguese Conference on Artificial Intelligence on Progress in Artificial Intelligence, Knowledge Extraction, Multi-agent Systems, Logic Programming and Constraint Solving*, (London, UK), pp. 6–13, Springer-Verlag, 2001.

[44] S. Chunguang, "Parallel solution of sparse linear least squares problems on distributed-memory multiprocessors," *Parallel Computing*, vol. 23, no. 13, pp. 2075–2093, 1997.

[45] A. Turing, "Intelligent machinery," in *Collected Works of A.M. Turing: Mechanical Intelligence* (D. Ince, ed.), (Amsterdam, The Netherlands), Elsevier Science Publishers, 1992.

[46] R. Duda, P. Hart, and D. Stork, *Pattern Classification*, New York: John Wiley and Sons, Inc., 2001.

[47] M. Muller, "A scaled conjugate gradient algorithm for fast supervised learning," *Neural Networks*, vol. 6, pp. 525–533, 1993.

[48] S. Fahlman, "Faster-learning variations on back-propagation: An empirical study," in *Connectionist Models Summer School* (T. Sejnowski, G. Hinton, and D. Touretzky, eds.), (San Mateo, CA), Morgan Kaufmann, 1988.

[49] L. Bottou and Y. LeCun, "Large-scale online learning," in *Advances in Neural Information Processing Systems*, vol. 15, Cambridge, MA: MIT Press, 2004.

[50] P. Sundararajan, and N. Saratchandran, *Parallel Architectures for Artificial Neural Networks: Paradigms and Implementations*, Wiley-IEEE Computer Society Press, 1998.

[51] B. Boser, I. Guyon, and V. Vapnik, "A training algorithm for optimal margin classifiers," *Proceedings of the 5th Annual ACM Workshop on Computational Learning Theory* (D. Haussler, ed.), (Pittsburgh, PA), pp. 144–152, ACM Press, 1992.

[52] J. Platt, "Fast training of support vector machines using sequential minimal optimization," in *Advances in Kernel Methods - Support Vector Learning* (A. Smola, P. Bartlett, B. Schölkopf, and D. Schuurmans, eds.), (Cambridge, MA), pp. 185–208, MIT Press, 1999.

[53] S. Haykin, *Neural Networks: A Comprehensive Foundation,* 2nd ed., Prentice-Hall, 1999.

[54] C. Cortes and V. Vapnik, "Support vector networks," *Machine Learning*, vol. 20, pp. 273–297, 1995.

[55] I. Guyon, J. Weston, S. Barnhill, and V. Vapnik, "Gene selection for cancer classification using support vector machines," *Machine Learning*, vol. 46, no. 1-3, pp. 389–422, 2002.

[56] J. Weston, A. Elisseeff, B. Schölkopf, and M. Tipping, "Use of the zero-norm with linear models and kernel methods," *Journal of Machine Learning Research*, vol. 3, pp. 1439–1461, 2003.

[57] I. W. Tsang, J. T. Kwok, and P.-M. Cheung, "Core vector machines: Fast SVM training on very large data sets," *Journal of Machine Learning Research*, vol. 6, pp. 363–392, 2005.

[58] R. Collobert, S. Bengio, and Y. Bengio, "A parallel mixture of SVMs for very large scale problems," in *Advances in Neural Information Processing Systems*, MIT Press, 2002.

[59] G. Zanghirati and L. Zanni, "A parallel solver for large quadratic programs in training support vector machines," *Parallel Computing*, vol. 29, pp. 535–551, 2003.

[60] H. P. Graf, E. Cosatto, L. Bottou, I. Durdanovic, and V. Vapnik, "Parallel support vector machines: The cascade SVM," in *Advances in Neural Information Processing Systems*, 2004.

[61] E. Yom-Tov, "A distributed sequential solver for large-scale SVMs," in *Large Scale Kernel Machines* (O. Chapelle, D. DeCoste, J. Weston, and L. Bottou, eds.), (Cambridge, MA), pp. 141–156, MIT Press, 2007.

[62] A. Bordes and L. Bottou, "The huller: a simple and efficient online SVM," in *Machine Learning: ECML 2005*, Lecture Notes in Artificial Intelligence, LNAI 3720, pp. 505–512, Springer Verlag, 2005.

[63] O. Dekel, S. Shalev-Shwartz, and Y. Singer, "The forgetron: A kernel-based perceptron on a budget," *SIAM J. Comput.*, vol. 37, no. 5, pp. 1342–1372, 2008.

[64] C. Yu and D. Skillicorn, "Parallelizing boosting and bagging," Technical report, Queen's University, Kingston, 2001.

[65] Y. Freund and R. Schapire, "A decision-theoretic generalization of online learning and an application to boosting," *Journal of Computer and System Sciences*, vol. 55, pp. 119–139, 1995.

[66] R. Meir, R. El-Yaniv, and S. Ben-David, "Localized boosting," *Proceedings of the 13th Annual Conference on Computer Learning Theory*, (San Francisco), pp. 190–199, Morgan Kaufmann, 2000.

[67] N. Oza, "Online bagging and boosting," *2005 IEEE International Conference on Systems, Man and Cybernetics*, vol. 3, pp. 2340–2345, 2005.

[68] J. K. Bradley and R. E. Schapire, "Filterboost: Regression and classification on large datasets," in *Advances in Neural Information Processing Systems*, vol. 20, Cambridge, MA: MIT Press, 2007.

[69] V. N. Vapnik, *Statistical Learning Theory*, Wiley, 1998.

[70] E. M. Airoldi, "Getting started in probabilistic graphical models," *PLoS Computational Biology*, vol. 3, p. e252, Dec. 2007.

[71] R. S. Sutton and A. G. Barto, *Reinforcement Learning: An Introduction.* MIT Press, 1998.

PART III

Specific Applications of Parallel Computing

CHAPTER 8

Scalable Image Registration and 3D Reconstruction at Microscopic Resolution

Lee Cooper, Kun Huang, Manuel Ujaldon, and Antonio Ruiz

8.1 Introduction

Characterizing the phenotypes associated with specific genotypes is critical for understanding the roles of genes and gene interactions. Three-dimensional morphologies of cellular and tissue structure are one aspect of phenotype that provides information key to the study of biological processes such as the initiation of cancer in the tumor microenvironment, the development of organs via complex induction processes, or the formation of neural networks in the brain cortex. Nevertheless, existing techniques for obtaining high-magnification 3D information from biomedical samples are rather limited. While confocal and multiphoton imaging offer 3D capability, both are limited in field and depth and only structures with fluorescent labeling are visible. Therefore, another fundamental approach for 3D acquisition is to perform reconstruction from multiple 2D images obtained from tissue sectioning. The basis for this approach is the process of automatically aligning images of serial sections using *image registration* [1–22] to compensate for misalignments and distortions.

Image registration has been extensively studied in biomedical imaging, geological survey, and computer vision [7, 8]. It can be considered as an optimization problem for finding the optimal transformation T between two images I_1 and I_2 to maximize their similarity. Commonly used similarity metrics include mutual information [23], cross-correlation, and summed square difference. The transformation spaces include rigid transformation, which deals with only rotation and translation, and nonrigid transformation which compensates for deformations such as bending, stretching, shearing, and warping [11, 24, 25]. Like any optimization process a good initialization is critical for a global optimum outcome. In many cases, a good rigid registration result serves as an ideal initialization for nonrigid registration [10]. For large images with conspicuous deformations, hierarchical multiresolution registration methods have also been widely used in medical imaging applications [26, 27].

There are four primary issues for the registration of section images for tissue reconstruction:

1. *Heavy computational requirements.* High-resolution slide scanners are capable of generating images with resolutions of 0.46μm/pixel (with 20X objective lens), often producing images with hundreds of millions or

even billions of pixels. The reconstruction of an individual tissue sample may involve hundreds of slides, and a full study may contains several samples with image data easily ranging in the terabytes. Commonly a multiscale approach is taken to registering images of this size; however, this requires transformation of the free-floating image at each scale which is computationally nontrivial. For instance, in this paper we are dealing with images with sizes up to 23K × 62K pixels. With a scale factor of two, this implies transformation of an image containing around 12K × 31K pixels prior to the final stage.

2. *Feature-rich environment.* The textural quality of microscopic image content provides a unique challenge to the problems of feature selection and matching. When applied to microscopic image content, traditional feature detection schemes such as corner detection generate an overwhelming abundance of features that are regular in appearance making matching prone to error. In addition, at submicron resolutions a shift of 1 mm corresponds to thousands of pixels. The search for corresponding features is therefore infeasible without good initialization.
3. *Nonrigid distortion and local morphological differences.* The key challenge for image registration of section images is to compensate for distortion between consecutive images that is introduced by the sectioning process. Tissue sections are extremely thin (3 to 5 μm) and delicate as a result. The preparation process (i.e., sectioning, staining, and coverglass application) can introduce a variety of nonrigid deformations including bending, shearing, stretching, and tearing. At micron resolutions, even minor deformations become conspicuous and may prove problematic when accuracy is critical to the end application. In order to compensate for such deformations, a *nonrigid registration* is essential and success depends on establishing a large number of precise *feature correspondences* throughout the extent of the image. This precision requires comparison of intensity information and is very time consuming with popular comparison measures such as mutual information (MI).
4. *Preservation of 3D morphology of microscopic structures.* A specific issue for 3D reconstruction is the preservation of the 3D morphology of important microanatomical structures. Most image registration algorithms focus on "optimally" aligning two images. In principle, they do not take the integrity of the 3D structure into account. Therefore, the 3D reconstruction process has to go beyond traditional pairwise image registration by integrating 3D structure integrity constraint into its overall consideration.

This chapter discusses research efforts in both algorithm design and high-performance computing (HPC) implementation to address the challenges described above. A two-stage image registration pipeline designed for aligning large histological images is presented. The first stage is a fast rigid registration algorithm based on the matching of high-level features. This approach circumvents the issue of the presence of numerous and ambiguous local features, providing effective initialization. In the second stage, nonrigid registration is achieved by precisely matching a large number of local intensity features using cross-correlation. To address computing requirements, implementation is presented in several HPC environments including CPU clusters and graphical processing units (GPU). Additionally, a scheme is proposed for preserving the integrity of 3D duct-like structures in tissue reconstruction. This work is motivated by several large scale biomedical studies including developmental biology and breast cancer research.

This chapter is organized in the following sections. In Section 8.2, existing approaches for image registration are reviewed with emphasis on several large scale research projects that require the alignment of 2D microscopic slides for 3D reconstructions. Existing HPC solutions related to image registration and 3D reconstruction of microscopic structures are also discussed. In Section 8.3, the two-stage registration algorithm is presented along with a discussion on the preservation of 3D ductal structures in reconstruction applications. Section 8.4 discusses HPC solutions for image registration based on the two-stage algorithm, including parallel computing using CPU and GPU clusters. In order to evaluate the implementations described in the previous section, the two stage registration algorithm was applied in a high performance computing environment to a benchmark of images taken from biological studies in Section 8.5. The results for the algorithms and HPC implementation are presented in Section 8.6.

8.2 Review of Large-Scale Image Registration

8.2.1 Common Approaches for Image Registration

Image registration has been extensively studied in many applications including radiology, geological survey, and computer vision. It can be framed as an optimization problem; that is, finding the transformation T between two images I_1 and I_2 to maximize their similarity with respect to some measure,

$$T = \arg\max \mathit{Similarity}(I_1, T(I_2)) \tag{8.1}$$

In this context, a registration algorithm needs to specify the following conditions:

1. *The similarity metric which determines the cost function for the optimization process.* The commonly used similarity metrics include mutual information (MI) [23], normalized cross-correlation (NCC), and summed square difference (SSD) [7, 8]. Among them, MI was originally designed for registering images of different modalities such as magnetic resonance image (MRI) with positron emission tomography image (PET) for human brain scanning. MRI images provide structural information while the PET images supply functional information. By maximizing the mutual information between the two types of images, the user obtains an overlay of the functional map on a structural atlas of the brain. NCC and SSD are frequently used for matching images of the same modality. In our work, we use NCC given that we are dealing with histological images with the same type of staining. However, we also observed that for histological images with different types of staining, a variation of NCC can still be used [28] due to similar anatomical structure between the histological

sections. This provides a computationally efficient solution since NCC can be implemented efficiently using fast fourier transform (FFT).

2. *The space of transformation between two images.* Transformation spaces include rigid or Euclidean transformation (i.e., rotation and translation), and nonrigid transformations include scaling, affine transformations, projective transformations, and nonlinear transformations which compensate for deformations such as bending, stretching, shearing, and warping [11, 24, 25]. Since Euclidean transformation, scaling, and affine transformations can also be expressed using a single matrix, they are also referred to as linear transformations, while other deformable types are referred to as nonlinear transformations. Commonly used nonlinear transformations include piece-wise linear, polynomial, local-weighted mean, and thin-plate spline. Given the fragility of tissue sections, nonrigid transformations are necessary in microscopic image registration to compensate for deformation. However, since the computational cost for applying nonlinear transformations is in general much higher than for linear types, care has to be taken in selecting the transformation type so as not to compromise quality or introduce unnecessary computational burden.
3. *The search strategy for optimization.* Many image registration algorithms including maximum mutual information (MMI) use iterative approaches such as Levenberg-Marquardt to search for optima in the transformation space. At each iteration, transformations are applied to one image and the results are used compare to the second image to calculate costs and identify the next position in the transformation space. However, this approach is not computationally feasible for large images since applying transformations is such a demanding task. Additionally, for complex transformation types with relatively greater freedom, the dimensionality of the search space becomes very large. For large images with conspicuous deformations, hierarchical multiresolution registration methods have also been widely used in medical imaging applications [26, 27]. Another type of approach is to use feature matching to generate a set of control points. Once the correspondences of control points are established through feature matching, the nonrigid transformations can be computed directly without iteration.

8.2.2 Registering Microscopic Images for 3D Reconstruction in Biomedical Research

There have been many works focusing on acquiring the capability for analyzing large microscopic image sets in 3D space. In [6] and [15], the authors used stacks of confocal microscopic images to develop a 3D atlas for brains of various insects including honeybee and fruit fly. Both research groups focus on developing a consensus 3D model (atlas) for all key functional modules of insect brains. In [29], gene expression patterns in whole fly embryos are studied in 3D space using stacks of confocal microscopic images. In the Edinburgh Mouse Atlas Project (EMAP), 2D and 3D image registration algorithms have been developed to map the histological images with 3D optical tomography images of the mouse embryo [12]. Similarly, in [9] the authors presented a workflow for observing the 3D distribution of expression patterns of genes in mouse embryo. In [10], the authors built 3D models for human cervical cancer samples using stacks of histological images in clinical settings. A complete study on registering large microscopic images of mouse brain sections was presented in [22].

8.2.3 HPC Solutions for Image Registration

Large scale image registration has many applications in both biomedical research [10, 22, 28] and geophysics [30]. However, there are currently few works addressing image registration algorithms intended to run efficiently on HPC environments.

The work on parallel image registration on multicomputers is limited [22] and is restricted to either large computer clusters [31–33] or IBM cell clusters [34]. Clusters of GPUs have been used to implement other heavy workload tasks [35], mostly within the simulation and visualization fields. For example, numerical methods for finite element computations used in 3D interactive simulations [36], and nuclear, gas dispersion, and heat shimmering simulations [37].

On the other hand, commodity graphics hardware has become a cost-effective parallel platform to implement biomedical applications in general [38]. Applications similar to ours such as the registration of small radiological images [39] and computation of joint histogram for image registration [40], and others within the fields of data mining [41], image segmentation, and clustering [42] have applied commodity graphics hardware solutions. Those efforts have reported performance gains of more than 10 times [43], but they were mostly implemented using shaders with the Cg language [43].

8.3 Two-Stage Scalable Registration Pipeline

To address the specific challenges of nonrigid distortion, large image size, and feature rich content, we have proposed an algorithm that consists of two stages: rigid initialization and nonrigid registration. Rigid initialization estimates the rough alignment of the base-float image pair from the consensus of correspondences between anatomical features. The nonrigid stage seeks to establish pixel-precision correspondences by precisely matching areas of intensity information. The initialization reduces the search for matching in the nonrigid stage, resulting in a lower likelihood of erroneous matches and less computation. This combination provides an approach to the problem of automatic sectioned image registration and reconstruction that is robust, easily parallelizable, and scalable.

8.3.1 Fast Rigid Initialization

The basis of the rigid initialization stage is the matching of *high-level features* or image regions that correspond to anatomically significant features such as blood vessels, mammary ducts, or small voids within the tissue boundary. This is a natural choice for features that has several advantages over the more primitive features generated by commonly used methods such as corner detection. First, the amount of high level features is relatively limited keeping the total number of possible matches reasonable. Second, the descriptions used to match these features such as shape, size, and type are invariant under rotation and translation and so the matching can accommodate the full range of misalignments. Third, the feature descriptions are scalars and are fast and simple to compare once computed. Finally, many choices of high level features remain mutually distinguishable when the images to be registered have distinct stain types. This permits, for example, the alignment of an hematoxylin and eosin stained image with an immunohistochemically stained image.

In contrast, performing corner detection on a typical microscopy image generates an overwhelming number of features due to the textural quality of the content. These features are relatively ambiguous, and their comparison requires the use of neighborhood intensity information and has to account for differences in orientation and also appearance if distinct stains are used.

High-Level Feature Extraction

Extraction of high-level features is a simple process as the features often correspond to large contiguous regions of pixels with a common color characteristic. Color segmentation followed by morphological operations for cleanup usually suffice. The computational cost of these operations can be significantly reduced by performing the extraction on downsampled versions of the original images without compromising the quality of the final nonrigid result. The rigid estimate only serves as an initialization for the nonrigid stage and a matter of even tens of pixels difference is insignificant to the outcome of nonrigid stage. Figure 8.1 demonstrates the extraction process, showing an example from one of our placenta test images. Descriptions for these features such as size and shape can be computed using tools like `regionprops` available in MATLAB's Image Processing Toolbox.

High-Level Feature Matching

Determining an estimate for rigid registration from a set of matches requires a method that is robust to mismatches. This is especially true in microscope images where many of the features are guaranteed to have a similar appearance. Regardless of how strict criteria for matching features is, it is inevitable that a substantial amount of coincidental mismatches will occur. The fundamental idea in the approach presented here is that, given feature sets from the base image $\{b_i\}$ and

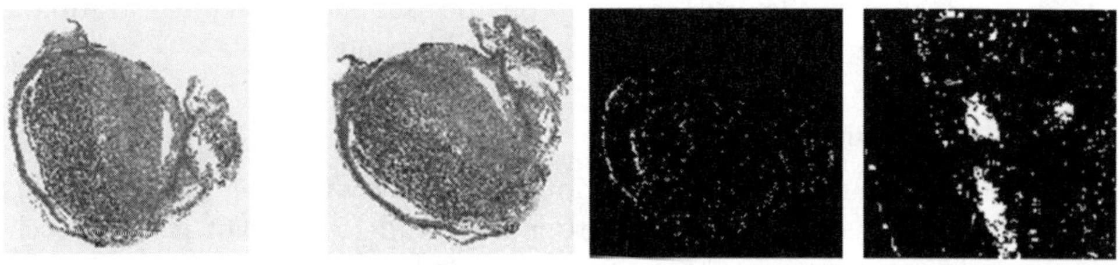

(a) (b) (c) (d)

Figure 8.1 Rigid registration for initialization using high level features. (a, b) 5x decimated placenta image (originally 20x), approximately 4,000 × 4,000 pixels. (c) Regions corresponding to blood vessels, extracted by color segmentation and morphological cleanup. Correspondences between elements in the feature sets of the base and float images are used to produce an estimate of the orientation θ and translation T between them. (d) Close-up of (c).

float image $\{f_j\}$ and recognizing that any pair of matches $(b_i, f_j), (b_k, f_l)$ define a model rigid transformation $(\tilde{\theta}, \tilde{T})$, for carefully chosen matchings and matching pairings a large portion of the model transformations will concentrate around the true parameters in the Euclidean transformation space.

Careful choice of matches and match pairs is achieved with a set of criteria that ensure consistency at two levels: between feature descriptions for matches between individual base and float features, and geometrically between pairs of matches. These criteria are depicted in Figure 8.2. The description consistency criteria is the percent difference in the similarity between description attributes like size and eccentricity. For a feature set where the base feature b_i is described by the attributes $(s_i^1, \ldots, s_i^n)$, and float feature f_j is described by $(t_i^1, \ldots, t_i^n)$, a match (b_i, f_j) is established if the set of n criteria $\{C_1, \ldots, C_n\}$

$$C_k = \frac{|s_i^k - t_j^k|}{min(s_i^k, t_j^k)} < \epsilon_k, k = 1, \ldots, n \tag{8.2}$$

are met within the tolerances ϵ_k. Generating a model rigid transformation requires a pair of matches. To identify models originating from coherent pairs, geometric consistency criteria are used to ensure similar intra-image distances and feature orientations. For a pair of matches $(b_i, f_j), (b_k, f_l)$, the intra-image distances from feature centroid to centroid are required to be similar

$$\frac{|\|\overrightarrow{b_i} - \overrightarrow{b_k}\| - \|\overrightarrow{f_j} - \overrightarrow{f_l}\||}{min(\|\overrightarrow{b_i} - \overrightarrow{b_k}\|, \|\overrightarrow{f_j} - \overrightarrow{f_l}\|)} < \tau \tag{8.3}$$

Additionally, the orientations of the feature semimajor axes must be consistent with the angle of the model transformation $\tilde{\theta}$. That is, given the orientations of the semimajor axes of the base features ϕ_i, ϕ_j, and float features ψ_j, ψ_l

$$(|\phi_i - \psi_j| < \omega) \bigwedge (|\phi_k - \psi_l| < \omega) \tag{8.4}$$

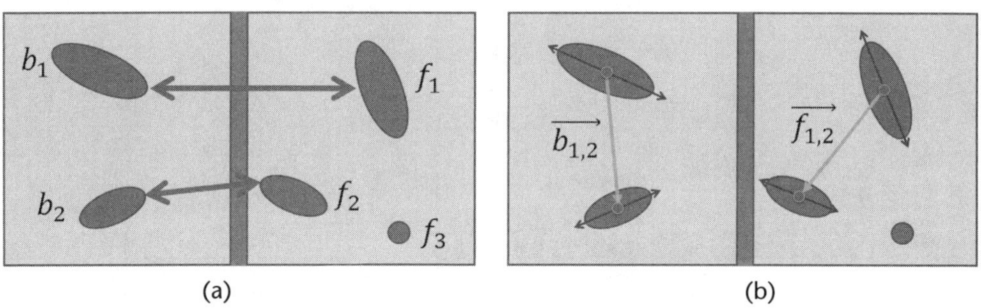

Figure 8.2 Feature characteristic and geometric constraints for high level feature matching. (a) The rotationally invariant characteristics of features such as size and eccentricity are used to form correspondences between the feature sets of the base and float images. (b) Each pair of correspondences defines a model rigid transformation $(\tilde{\theta}, \tilde{T})$. Checking for consistency between the intra-image distances of the correspondences, $\|b_{1,2}\|, \|f_{1,2}\|$, and between $\tilde{\theta}$ and the feature semimajor axis orientations helps to eliminate models based on erroneous correspondences.

Models from consistent match pairs are pooled for the voting process that is described next.

Histogram Voting

With a set of model transformations identified, a histogram voting scheme is used to estimate the initialization parameters (θ, T_x, T_y). First, θ is estimated by counting the models in the w_θ-interval centered at at each $\tilde{\theta}$, taking θ as the $\tilde{\theta}$ with the largest w_θ-interval count. The models that fall within this maximum count w_θ-interval are then selected and used to estimate the translation parameters. Interval counting is then applied with w_T-intervals centered at each of $\tilde{T}_x, \tilde{T}_y$ from the selected models to identify T. The use of intervals centered at each model adds computation; however, the use of arbitrary boundaries as in simple histograms may effectively split the maximum count bin allowing another spurious bin to prevail. An example of histogram voting is presented in Figure 8.3. Typical values for percent difference tolerances ϵ_k, τ are 0.1 to 0.2, and 5° to 10° for the orientation threshold ω. Interval sizes for histogram voting range from 0.5° to 1° for θ and from 30 to 50 pixels for T_x, T_y.

While it is possible to parallelize most parts of the rigid initialization stage, execution times are reasonable even with a modest serial implementation. Using MATLAB the estimates for the example in Figure 8.3 were calculated in less than 4 seconds with a serial implementation on a system similar to the one described in Section 8.5.

8.3.2 Nonrigid Registration

With the rigid initialization stage described, we turn our focus to the nonrigid stage. Correcting for nonrigid distortion to the accuracy necessary for end applications in quantitative phenotyping requires establishing a large number of precise correspondences between the base and float images. The desired pixel-level precision suggests that assignment of correspondences between representative features, such as high level features, is not enough. Instead, direct comparison of intensity information is needed which introduces the problem of computational burden.

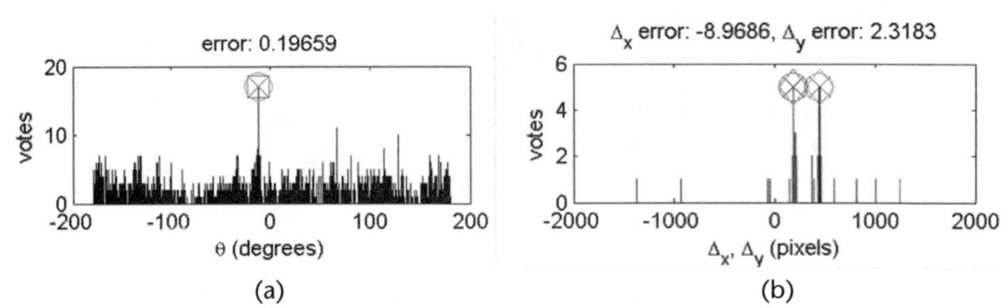

Figure 8.3 Model transformation voting to estimate rigid initialization. Models that satisfy the geometric constraints are pooled and the parameters (θ, T) are estimated using histogram voting. (a) The difference between the voting result estimate and a manual registration less than 0.2 degree. (b) For T_x, T_y the differences between manual and voting results are within 10 pixels.

These considerations are addressed in an approach to extraction and matching of *intensity features* that compares small tile regions between the base and float images in an efficient manner, using the rigid initialization parameters and fast fourier transform to compute their cross-correlations.

Intensity Feature Extraction

The first issue in intensity feature extraction is the selection of unambiguous regions that are likely to produce accurate matches. This issue is especially important for matching in nonrigid registration since using the aggregate of matching results to make inference about the quality of any single match is difficult due to the freedom and subtlety of nonrigid distortions. In this sense the matchings at this stage are local: the only information available to judge their quality comes from the individual intensity regions themselves. Good candidate regions for matching have rich content, a mixture of different tissues or tissue and background that forms a distinctive appearance. Often these regions will coincide with blood vessels, ducts, or other content with distinctive shape. Regions containing uniform tissue are not good candidates for matching, as accurate matchings are unlikely due to the textural quality of content and the morphological differences between sections. A simple way to enforce this is to select tiles whose variance meets a certain minimum threshold. That is, for any feature point p with coordinates $\left[\begin{smallmatrix} x \\ y \end{smallmatrix}\right]$ centered in the $W \times W$-pixel window, we require

$$\frac{1}{W^2 - 1} \sum_{i,j} (t(i, j) - \bar{t})^2 \geq \sigma^2 \tag{8.5}$$

where t is the *template*, a grayscale representation of the p-centered pixel window with mean value $\bar{t}$, and σ^2 a significance threshold. There are cases where the variance threshold can be met and an ambiguous matching result can occur (consider matching a two templates with upper half white and lower half black) although these cases are uncommon in natural images.

Another important issue in intensity feature matching is the spatial distribution of features. The correspondences resulting from the matching of intensity features form the basis for a nonrigid transformation of the float image. These correspondences should be fairly distributed throughout areas of interest in order to produce a result that conforms in the areas where further analysis will take place on the registered result. To keep the total number of features reasonable and attempt an even spatial distribution, we sample features uniformly over the image with a $W \times W$ tiling. For example, in the 16K × 16K placenta images we would typically tile in the range of 150 to 350 pixels to generate a total of 2,025 to 11,236 possible features, the large majority of which are discarded due to insufficient variance.

Intensity Feature Matching

For any selected feature point p_1 with coordinate $\left[\begin{smallmatrix} x_1 \\ y_1 \end{smallmatrix}\right]$ in the base image, we first select a $B \times B$-pixel window centered at p_1. This window is converted to grayscale and rotated by the angle θ obtained from the initialization stage. The central $W_1 \times W_1$-pixel patch is then used as the $p1$ template for identifying p_2, the correspondence point of p_1 in the float image. B is calculated from θ and W_1, taken large enough to accommodate the template.

The coordinate p_2 is estimated using

$$p_2' = \begin{bmatrix} x_2 \\ y_2 \end{bmatrix} = \begin{bmatrix} \cos(\theta) & -\sin(\theta) \\ \sin(\theta) & \cos(\theta) \end{bmatrix} \begin{bmatrix} x_1 \\ y_1 \end{bmatrix} + \mathbf{T} \tag{8.6}$$

with $\mathbf{T}$ being the translation vector obtained from the initialization stage. A $W_2 \times W_2$-pixel tile ($W_2 > W1$) centered at p_2' designated as the *search window* is taken from the float image. The NCC between the template and search window is computed, and p_2 is taken as the center of the area corresponding to the maximum NCC value. If this maximum value exceeds a threshold (usually 0.8 or greater), then the match is considered successful and $p1$, $p2$ are recorded as a correspondence. The choice of W_1 and W_2 is based on the severity of the deformation as well as computational capacity. Empirically we set $W_2 = 2W_1$, however cases with large deformation may require a larger search area. Demonstration of the effect of tile size on execution time is demonstrated in Section 8.6.

Commonly used similarity measures for intensity information other than NCC include summed square of difference and mutual information. SSD is not a good choice for microscopic images since the content tends to be discrete (e.g., sharp boundaries between cell nucleus and cytoplasm and cell membrane). MI is commonly used as a metric in gradient search strategies but the cost of joint-histogram computation makes its use in exhaustive search prohibitively expensive. We choose NCC since it is not only robust in identifying structural similarity but also highly efficient when implemented with fast fourier transform. Furthermore, NCC values have an intuitive interpretation, making threshold parameter selection easier.

The large number of features that exist within a typical dataset makes efficient computation of NCC critical. Additionally, rather than using a search strategy we compute NCC between template and search window pairs at all spatial offsets to avoid the problem of local minima.

Given a template t size $W_1 \times W_1$ with mean $\bar{t}$, and search window s size $W_2 \times W_2$, $W_2 > W_1$, the NCC between t and s is the quotient of covariance and individual variances

$$\rho(u, v) = \sum_{x,y} \frac{\{t(x-u, y-v) - \bar{t}\}\{s(x,y) - \bar{s}_{u,v}\}}{(\{t(x-u, y-u) - \bar{t}\}^2\{s(x,y) - \bar{s}_{u,v}\}^2)^{\frac{1}{2}}} \tag{8.7}$$

where $\bar{s}_{u,v}$ is the mean of the search window portion overlapping the template at offset (u, v).

For calculating normalizing factors in the denominator we use the method of running sums presented in [44]. This avoids the expensive local calculations of search window mean and variance for the template overlap region as the template is shifted through each of $(W_1 + W_2 - 1)^2$ positions, reducing the operation count from $3W_2^2(W_1 - W_2 + 1)^2$ to approximately $3W_1^2$.

The unnormalized cross-correlation from the numerator is calculated via the convolution theorem of the discrete fourier transform that relates the product of DFT spectra to circular convolution in the spatial domain. For cross-correlation we are interested in ordinary convolution so t and s are padded with zeros to size $W_1 + W_2 - 1$ prior to forward transform to ensure that the circular overlap portions of the convolution result are null.

8.3.3 Image Transformation

The collection of point correspondences generated by the precise matching process provides the information needed to form a mapping that transforms the float image into conformation with the base. A variety of nonrigid mappings are used in practice, differing in computational burden, robustness to erroneous correspondences, and existence of inverse form. High-performance implementation of the transformation process is not a simple matter and is a separate issue not treated in this work.

The Polynomial Transformation

In choosing a transformation type we seek something capable of correcting complex distortions, that is robust to matching errors, that admits a closed inverse form, and that is computationally reasonable to calculate and apply. Of the commonly used nonrigid mapping types such as thin-plate spline, local weighted mean, affine, polynomial, and piece-wise variations, we choose polynomial mapping. Thin-plate spline provides a minimum energy solution which is appealing for problems involving physical deformation; however, perfect conformity at correspondence locations can potentially cause large distortion in other areas and excess error if an erroneous correspondence exists. The lack of an explicit inverse form means the transformed image is calculated in a forward direction, likely leaving holes in the transformed result. Methods such as gradient search can be used to overcome the inverse problem but at the cost of added computation which can become astronomical when applied at each pixel in a gigapixel image. Kernel-based methods such as local weighted mean require a uniform distribution of correspondences. Given the heterogeneity of tissue features, this distribution cannot always be guaranteed.

Polynomial warping admits an inverse form, is fast in application, and from our experience is capable of satisfactorily correcting the distortion encountered in sectioned images. Polynomial warping parameters can be calculated using least squares or least squares variants which can mitigate the effect of matching errors. Affine mapping offers similar benefits but is more limited in the complexity of the warpings it can represent.

In our algorithm, we use second degree polynomials. Specifically, for a point (x, y) in the base image, the coordinate (x', y') of its correspondence in the float image is

$$\begin{cases} x' = a_1x^2 + b_1xy + c_1y^2 + d_1x + e_1y + f_1 \\ y' = a_2x^2 + b_2xy + c_2y^2 + d_2x + e_2y + f_2 \end{cases} \tag{8.8}$$

Since each pair of matched point correspondences provides two equations, we need at least six pairs of point correspondences to solve for the coefficients in (8.8). In practice, a much larger number of point correspondences is obtained.

8.3.4 3D Reconstruction

For 3D tissue reconstruction applications, where a sequence of images is to be registered together, the matching process is applied successively to each ordered pair in the sequence. Images are transformed starting at one end of the sequence, and at each step the transformations from prior image pairs are propagated through the match coordinates in order to achieve a coherent set of transformed images. Figure 8.10 in Section 8.6 shows a reconstruction result generated from a sample set of mouse placenta images. The improvement with respect to reconstruction quality that is provided by nonrigid registration is demonstrated in Figure 8.10. In some cases however, there are caveats for the use of nonrigid registration for 3D reconstruction, as discussed next.

The Stacking Phenomenon

Consider a volume of tissue containing meandering vessel-like structures with trajectories that are more or less transverse to the sectioning planes. Each section of this tissue contains an arrangement of vessel cross-sections. Registering this sequence nonrigidly and exactly, using the vessel centroids as correspondences, produces an arrangement of straight cylindrical structures absent of any the trajectory components that lie in the sectioning plane. This example demonstrates what we refer to as the *stacking phenomenon*, a type of overfitting that can occur in certain reconstruction scenarios where nonrigid registration is applied. This phenomenon illustrates a key point: the difference between reconstruction and registration.

In cases where intensity features are used to form correspondences this type of severe structural distortion is mitigated since each correspondence is based upon the consensus of a neighborhood of pixels, rather than imposed as a geometric concept as described above. There may be cases, however, where a different solution is sought. Perhaps intensity features are deemed to be unsatisfactory due to excessive morphological differences, too expensive in terms of computation, or another type of feature is readily available. For example, in mouse mammary images such as the one in Figure 8.4, the content is characterized by sparse adipose tissue interspersed with cross-sections of duct-like structures that are the focus of the post-reconstruction phenotypical analysis. At some point the duct regions will have to be segmented so the segmented ducts could be used as features for registration as well.

Methods exist to address the stacking phenomenon by imposing a structural constraint during the registration process. Malandain et al. [45] propose an iterative method for reconstruction from brain autoradiograph sections where the registered autoradiograph images are fused with a magnetic resonance volume standard that is obtained prior to sectioning. This provides a result that is structurally sound; however, the acquisition of a volume standard is not practical in many cases. For histological images that contain vessel or duct-like structures, [17] presents a method that uses trajectory smoothing. In this case the images are rigidly registered and the trajectories of duct centroids are tracked from one image to the next. These trajectories are smoothed, by application of splines or low-pass filtering, and the duct centroids are then nonrigidly registered to the smoothed trajectories. This approach is illustrated in Figure 8.4 along with rendered reconstruction results.

(a) (b) (c)

Figure 8.4 Stacking phenomenon. (a) Mammary images contain sparse adipose tissue interspersed with duct structures. (b) Nonrigid registration of mammary images by corresponding duct centroids results in severe structural deformation as all components of duct trajectory within the sectioning planes are eliminated, resulting in vertical columnar structures. (c) This can be corrected by rigidly registering the sequence, tracking the duct centroid trajectories, smoothing these trajectories, and nonrigidly registering the duct centroids to the smoothed trajectories.

8.4 High-Performance Implementation

The size of high-resolution microscope image datasets presents a computational challenge for automatic registration. Scanning slides with a 20X objective lens produces images with submicron pixel spacing, often ranging in the gigapixel scale with tens of thousands of pixels in each dimension. At this scale the amount of data necessary in quantitative phenotyping studies can easily extend into the terabytes. This motivates the development of a high-performance computing approach to registration.

This section discusses high-performance implementation of the two-stage registration algorithm and introduces solutions at both the hardware and software layers. At the hardware layer two areas are pursued: *parallel systems* based on clusters of multisocket multicore CPUs, and GPUs for hardware acceleration. At the software layer, performance libraries are used for computing the normalized correlations by fast Fourier transform (FFT) on both CPU and GPU. Performance results from the implementation varieties discussed here are presented in further detail in Section 8.6.

8.4.1 Hardware Arrangement

Both single and multiple node implementations are described below. In the multiple node implementations, a simple head/workers organization is assumed. Communication on single node implementations is accomplished with Pthreads, and message passing interface (MPI) is used for internode communication in multiple node implementations. The details of division of work between CPU and GPU at the level of individual nodes are discussed further in Section 8.4.3.

8.4.2 Workflow

The workflow for the two-stage registration algorithm is summarized in Figure 8.5. With the exception of computing nonrigid transformations, the CPU-bound

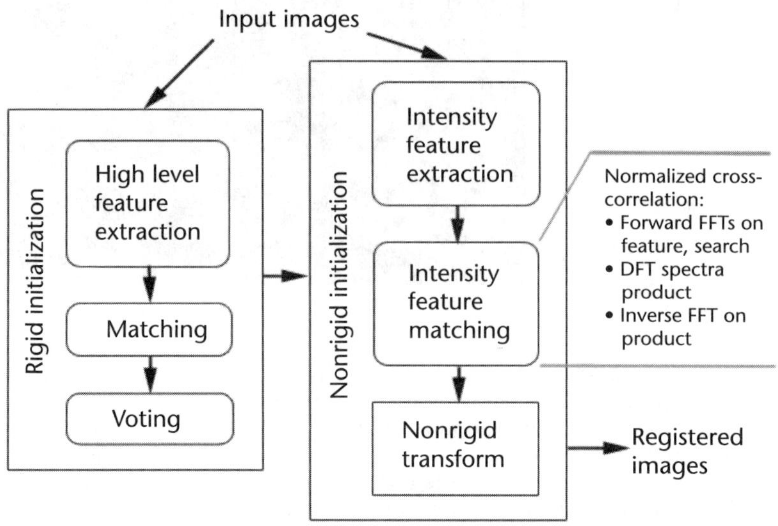

Figure 8.5 Workflow of the two-stage nonrigid registration algorithm. Rounded items indicate operations that can be carried out simply in parallel. The most computationally demanding phase is the intensity feature matching portion, consisting of two forward FFTs and one inverse.

operations of each stage are simply parallelizable. The overwhelming majority of computation, however, takes place in the normalized cross-correlation operations of intensity feature matching. To illustrate where to focus efforts to increase performance, consider that the rigid initialization stage for a typical 16K × 16K mouse placenta image executes in 4 seconds where the nonrigid stage executes in 60 to 90 seconds (depending on intensity feature and search window sizes). Roughly 60% of the time in the nonrigid stage is spent in computing normalized cross-correlations. Therefore, effort is focused on the nonrigid stage, primarily on distributing and accelerating intensity feature matching operations, but also on intensity feature extraction and grayscale conversion although to a lesser extent.

Rigid Initialization Stage

Due to modest computational requirements no effort is made to reduce execution times for the rigid initialization stage. If real-time response is desired, however, there is room for improvement. The first operation in this stage is the extraction of high level features by color segmentation followed with morphological operations for cleanup. Each of these are independent local operations that admit an embarrassingly parallel implementation. Additionally these operations can be pipelined with reading source images from disk. Furthermore, GPU acceleration can be employed to accelerate convolution where present within morphological operations. The second and third operations in the initialization stage are the matching of high level features and histogram voting, which consist of the simple search procedures detailed in Section 8.3. These searches can be carried out straightforwardly in parallel as well.

Nonrigid Stage

Where the primary effort is focused on improving intensity feature matching performance, simple efforts can be made to improve the performance of reading images from disk, grayscale conversion, and intensity feature extraction.

Given the large size of microscope images, some in excess of 10 GB, reading from disk and decoding can require a considerable amount of time. A parallel file system may be employed to reduce this time, although this requires distributing large amounts of data over a network and can complicate later steps since the data will be distributed among several nodes rather than a single head node. A portion of the time spent reading and decoding can be hidden, however, by overlapping reading/decoding with grayscale conversion, and using the head node to read/decode incrementally and asynchronous communication to defer grayscale conversion of incremental reads to worker nodes.

With the grayscale base and float images in memory, the next step is to determine which template regions will serve as candidates for intensity feature matching. The process is simple: the head node divides the base image among the worker nodes, which compute the variances of the $W_1 \times W_1$ template sized tiling of their portions and return the results.

With a set of candidate intensity feature regions identified, what remains is to rotate them, extract their templates, and perform the correlations between the templates and their corresponding search areas. The candidate features are evenly divided among the worker nodes, who rotate them, extract their templates, and perform the correlations between template and search, returning the maximum correlation result magnitudes and coordinates. The base image is stored in column-major format, so to keep communication to a minimum the candidate feature regions are buffered in order and the remainder of the image is discarded. Asynchronous communication is used to keep the head node busy while send operations post. The search windows, taken from the float image, are handled in a similar manner. However, since the search windows for distinct features can overlap significantly, they are not individually buffered, rather their union is buffered as a whole.

The division of work on a single node implementation of the nonrigid stage follows a similar strategy as the multiple node implementation except that no effort is made to overlap reading/decoding performance. In the case where GPU acceleration is used, intensity feature extraction proceeds sequentially, and as candidate features are identified they are passed to the GPU. This process is described in further detail in Section 8.4.3.

The discrete Fourier transforms necessary for calculating correlations on CPU are performed using the FFT library FFTW [46]. The 2D-DFT dimensions are critical for performance; ideally the size of the padded transform $W_1 + W_2 - 1$ is a power of two or a small prime number. For the cases when this size rule cannot be obeyed, FFTW provides a simple mechanism called a *plan* that specifies an optimized plan of execution for the transformation. This plan is precomputed and subsequently reused, resulting in a one-time cost. For example, with a template size $W_1 = 350$ and a search window size $W_2 = 700$, FFTW takes around 0.7 second to compute the two 1,049 $\times$ 1,049 forward transforms without planning, whereas with plan the computation takes only 0.32 second with a 6-second one-time penalty (a cost which can later be amortized by loading the plan from disk at runtime on subsequent transforms of the same size).

8.4.3 GPU Acceleration

To further speed up the demanding process of intensity feature matching, GPU can be employed to perform the FFT operations used in normalized cross-correlation calculation. The FFT operation is a good candidate for GPU acceleration since it is a highly parallel "divide and conquer" algorithm. In general, the performance of algorithms on GPUs depends on how well parallelism, closer memory, bus bandwidth, and GFLOPs are exploited.

General Purpose GPU: CUDA

GPU resources are a specialized set of registers and instructions intended specifically for common graphics operations. Programming models exist to access these resources for more general purpose tasks. One such interface, the Compute Unified Device Architecture (CUDA) [47], consists of a set of library functions which can be coded as an extension of the C language. The CUDA has an associated compiler that generates code acceptable for the GPU, which is seen by the CPU as a multicore processor. The CUDA hardware interface, shown in Figure 8.7, hides the graphics pipeline notions (see Figure 8.6), instead presenting the GPU resources as a collection of threads running in parallel.

FFT on GPU

To accelerate intensity feature matching, the forward FFTs, pointwise spectrum multiplication, and inverse FFT are implemented on GPU. On a single node implementation, parallelism with two GPUs is achieved either by having one CPU extract features sequentially and cycle them between each GPU, or by having two CPUs divide the work each with their own GPU. The multiple node implementation does not process features sequentially, and GPU parallelism is achieved by assigning one GPU to each CPU socket. In this case the GPU simply takes the place of the CPU in calculating the normalized cross-correlations.

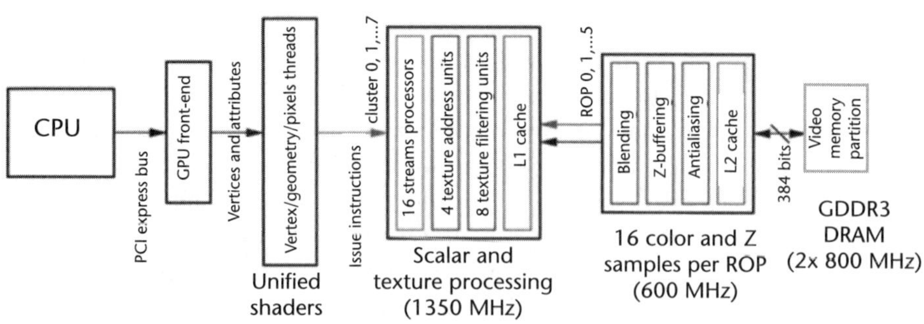

Figure 8.6 The Nvidia G80 architecture, used in implementation of intensity feature matching. Programs are decomposed as threads are executed on 128-stream processors, located in the central row. The data are stored on L1 and L2 caches and video memory is located in lower rows.

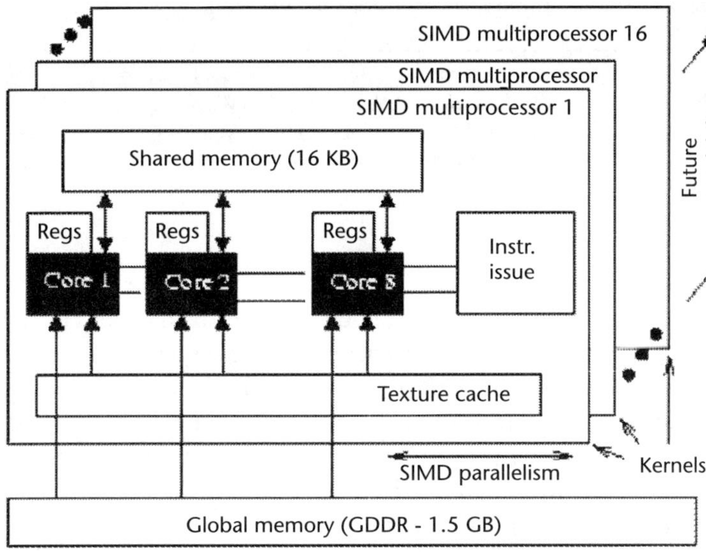

Figure 8.7 The Compute Unified Device Architecture (CUDA) hardware interface.

The remaining parts of the nonrigid stage including intensity feature extraction, and rotation of intensity features did not show any significant speedup on the GPU for three reasons:

1. They were already computationally inexpensive on the CPU.
2. More importantly, it was remarkable how much time was required to transport code and data back and forth between the CPU and the GPU through the memory bus, hypertransport link, and PCI-express controller (see Figure 8.8). This cost could not be amortized during the subsequent computation despite of the high GFLOPS rate of the GPU.
3. Most of these operations contain conditionals and are not arithmetic intensive, which makes them more appropriate for the CPU processor. Additionally, this enables our bi-processor platform to achieve a more balanced execution.

The FFTs can be efficiently implemented on GPU using the CUDA programming model. The CUFFT library [48] provides an efficient fast fourier transform implementation very similar to the FFTW library, and permits leveraging the floating-point power and parallelism of the GPU without developing a custom implementation. As with FFTW, dimensions are critical for CUFFT operations with the same power-of-two and prime convention.

8.5 Experimental Setup

8.5.1 Benchmark Dataset and Parameters

In order to evaluate the implementations described in the previous section, the two-stage algorithm was applied to a benchmark of images from: (1) mouse placenta

Table 8.1 The Benchmark Datasets for the Two-Stage Algorithm Consist of Two Image Sequences

Field of Study	*Research Area and Biomedical Goals*	*Mouse Source*	*Computational Workload*	*Image Size (pixels)*	*Slide Pairs*
Genetics	Role of a gene	Placenta	Medium	16K × 16K	100
Oncology	Breast cancer tumor	Mammary	Large	23K × 62K	4

from a morphometric study on the role of retinoblastoma gene, and (2) mouse mammary gland for studying breast cancer tumor microenvironment [16]. The goal in both sequences is the reconstruction of 3D tissue models to study microanatomy. Details of this benchmark are presented in Table 8.1.

The benchmark datasets are evaluated with a variety of values W_1 and W_2, chosen to cover both optimal and suboptimal cases for DFT size and to demonstrate the effect of parameter size on execution time performance. These parameter sets are summarized in Table 8.2.

8.5.2 The Multiprocessor System

The implementation was performed on the Visualization Cluster of the BALE system at the Ohio Supercomputer Center, a general purpose GPU equipped computing cluster (see Figure 8.8). The Visualization Cluster contains 16 nodes, each equipped with dual-socket × dual-core AMD Opteron 2218 CPUs, 8 GB DDR2 DRAM running at 667 MHz, and dual-card Nvidia Quadro FX 5600 GPUs with 2 × 1.5 GB of on-board GDDR3 DRAM running at 1,600 MHz. All nodes are connected by Infiniband and include 750 GB, 7,200 RPM local SATA II disks with 16-MB cache.

8.6 Experimental Results

The experiments from Table 8.2 were performed on the benchmark datasets of Table 8.1 using a variety of single node and multiple node configurations. For fairness in comparison between configurations, from this point forward all references to execution times are total run time without file I/O, conversion from RGB to grayscale, or rigid initialization unless explicitly stated otherwise.

8.6.1 Visual Results

A sample 3D reconstruction from 50 placenta slides is presented in Figure 8.9. Due to the absence of ground truth, evaluation of registration quality beyond visual in-

Table 8.2 Test Parameters for Template Size W_1 and Search Window Size W_2

Input Image:	*Placenta: 16K × 16K*			*Mammary: 23K × 62K*		
Window Size:	*Small*	*Medium*	*Large*	*Small*	*Medium*	*Large*
Template W_1 (pixels)	171	250	342	342	500	683
Search W_2 (pixels)	342	500	683	683	1,000	1,366
Aggregate ($W_1 + W_2 - 1$)	512	749	1,024	1,024	1,499	2,048

Note: Parameters were chosen to reflect a realistic range that demonstrates effect on performance of size and optimality with respect to FFT.

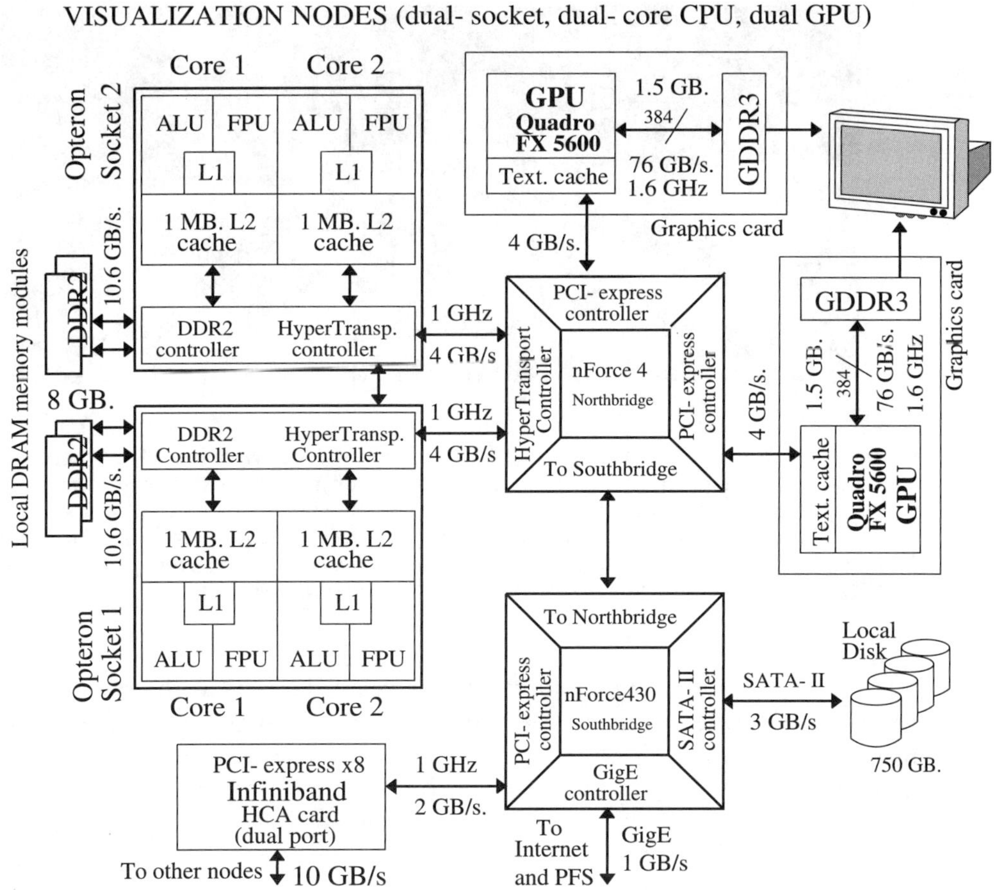

Figure 8.8 Visualization cluster node configuration on the BALE supercomputer.

spection is difficult. Differences in morphometry between adjacent sections can mask small but significant differences in quality regardless of the choice of evaluation metric. Figure 8.10 demonstrates the improvement of nonrigid registration over rigid alone, where no coherent structures are apparent in the reconstruction, preventing morphometric analyses of the volume. This improvement is also demonstrated in 2D in Figure 8.10, where difference images between the base and float are shown for the nonrigid and rigid-only cases.

8.6.2 Performance Results

The number of intensity features selected within each image varies significantly due to content. Table 8.3 summarizes the number of features extracted for each input image in the mammary set, as well as the total computation required for the registration algorithm to be computed serially on a single Opteron CPU with no GPU acceleration. The percentage of intensity features selected ranges from 4% to 30% of the total image area, with those percentages varying slightly with the value of W_1. Figure 8.11 shows the number of features extracted as a percentage of total

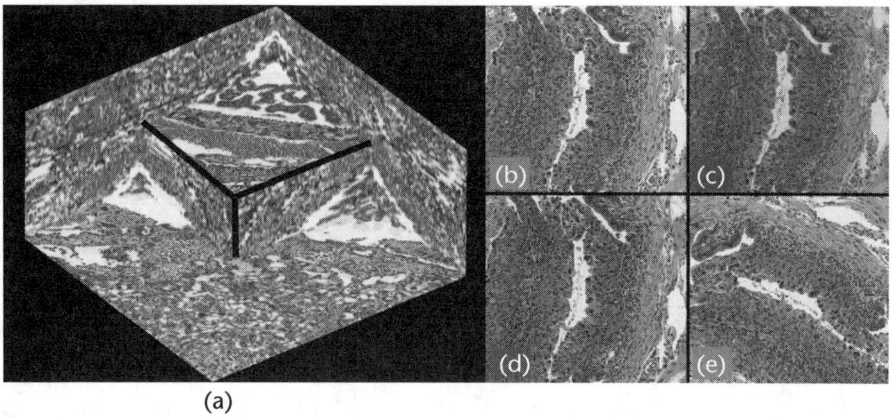

Figure 8.9 (a) A sample 3D reconstruction of mouse placenta with virtually excised corner. Only a fraction of the reconstructed volume is shown at high resolution due to the limitations of the rendering software. (b–e) Registration of mouse mammary images: (b) a 1,000 × 1,000-pixel patch from the base image; (c) corresponding 1,000 × 1,000-pixel patch taken from the float image; (d) patch from (c) after nonrigid transformation of the float image; (e) overlay between (b) and (d), with the grayscale representations embedded in the red and green channels respectively.

image area for both the mammary and placenta datasets at the respective smallest values of W_1. Due to differences in content sparsity between the two datasets, the placenta images have nearly twice the intensity feature density of the mammary dataset.

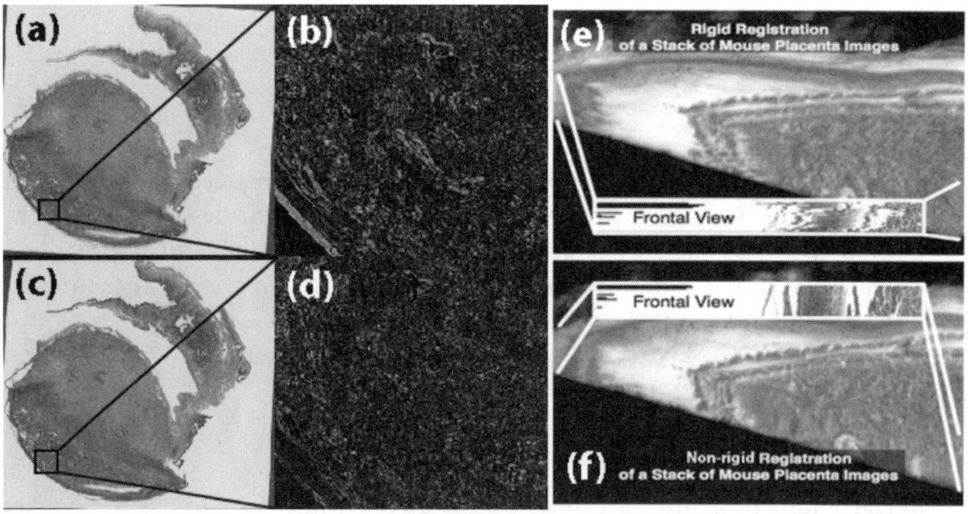

Figure 8.10 (a) Overlay of base and float placenta images after rigid registration. (b) High-resolution differenced patch from (a). (c) Overlay of the base and float images after nonrigid registration. (d) High-resolution differenced patch from the panel (c). (e, f) Rendering of an edge view of placenta reconstruction, the frontal views represent virtual cross-sections of the reconstructed tissue: (e) with rigid registration alone, no coherent structures are apparent in the frontal view; (f) nonrigid registration corrects the distortions apparent in (e) and the reconstructed volume is then suitable for further analyses.

Table 8.3 Workload Breakdown on Single CPU for Mammary Image Set

	Number of Features Extracted			*Workload on CPU (in seconds)*	
Window Size:	*Small*	*Medium*	*Large*	*Execution Time*	*Execution Time*
(template, search)	*(342, 683)*	*(500, 1000)*	*(683, 1366)*	*with I/O*	*without I/O*
Mammary 1	1,196	655	384	650.86	558.54 (85%)
Mammary 2	1,048	568	312	497.83	414.17 (83%)
Mammary 3	3,119	1,528	854	1,320.01	1,192.69 (90%)
Mammary 4	690	322	168	463.77	340.62 (73%)

Note: The number of features extracted for each input image within the mammary dataset differs due to content. Execution times in the last two columns represent the large case.

Single Node Performance

Execution times for single node implementation with a single Opteron CPU and a single CPU/GPU pair are presented in Figure 8.12. For both placenta and mammary, the small and large parameter sets both fulfill the optimal DFT size conditions, where the medium size is not compliant. This has a major effect on execution time for both CPU and GPU. For CPU with FFTW, the average time for the placenta case of 294.57 seconds for the medium size versus 57.97 seconds for the small and 91.33 seconds for the large. Doubling window sizes in the optimal cases from large to small only increments execution time by 58%, where moving from the small optimal size to the noncompliant medium size increases execution by nearly 400%. The mammary set exhibits the same behavior with increases of 26% and 213%, respectively. For GPU with CUFFT the average times for the placenta case are 19.27, 47.80, and 22.22 seconds for the small, medium, and large parameter sets, respectively. For the mammary set, the medium aggregate parameter size of 749 satisfies the DFT prime multiple condition but the overhead is still significant.

Speedup factors from CPU to GPU are also presented in Figure 8.12. For the placenta set, GPU acceleration produces a 3.00x and 4.11x speedup for intensity feature matching for the small and large parameters sets, respectively. For the mammary set the gains are more modest at 2.00x and 2.59x, respectively. In the mammary set, the speedup factors also show a significant difference depending on the input image, with more variation for larger window sizes. This is due in part to the effect of window size on heterogeneity that reveals higher disparities of feature density among images, an effect that is corroborated in Figure 8.12.

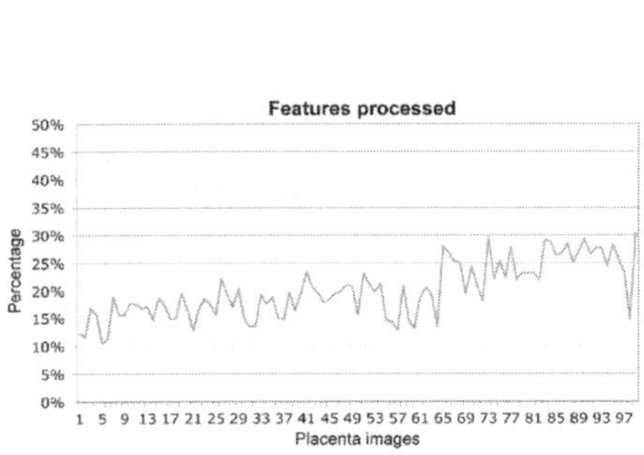

(a) Placenta image set

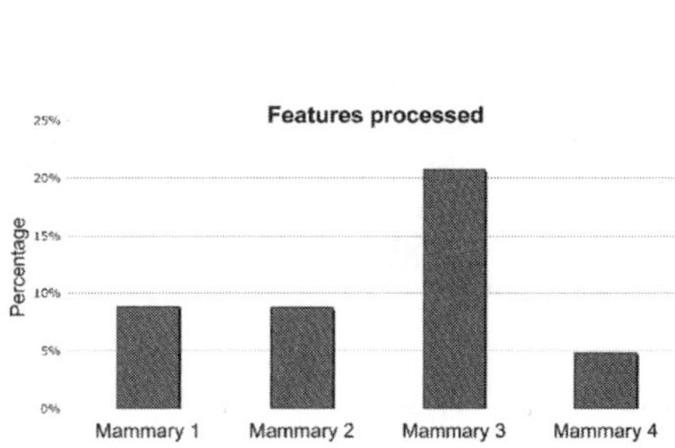

(b) Mammary image set

Figure 8.11 (a, b) Percentage of features processed per image on each input image set. We take the small size for the template and search window as the most representative.

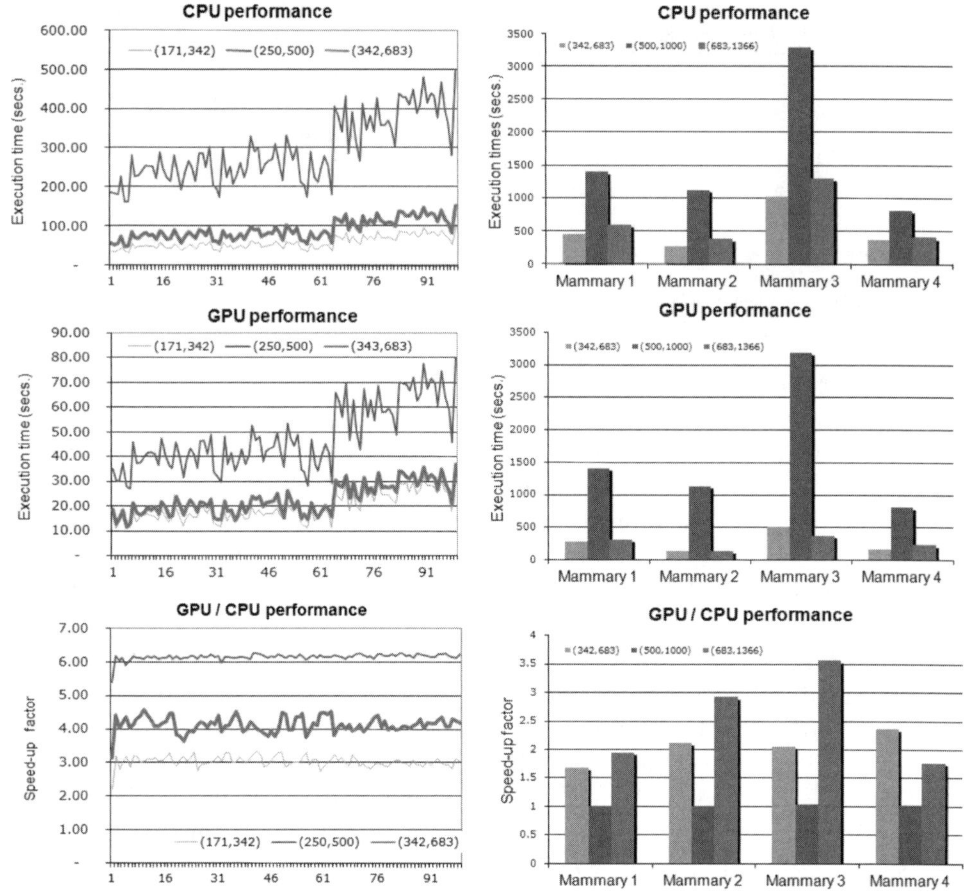

Figure 8.12 Execution times for intensity feature matching with serial implementations for both placenta (left) and mammary (right) datasets. (top) Execution times for single CPU implementation. For placenta average execution times are 57.95, 294.57, and 91.33 seconds for the small, medium, and large parameter sets, respectively. For mammary the averages are 530.41, 1,660.91, and 669.96 seconds. (middle) Single GPU accelerated execution times show marked improvement. (bottom) Speedup factor with GPU acceleration. Improvement is significant except for the non-DFT-optimal mammary parameters. The average speedup factors for placenta are 3.00x, 6.16x, and 4.11x. For mammary, the average speedup factors are 2.00x, 1.01x, and 2.59x.

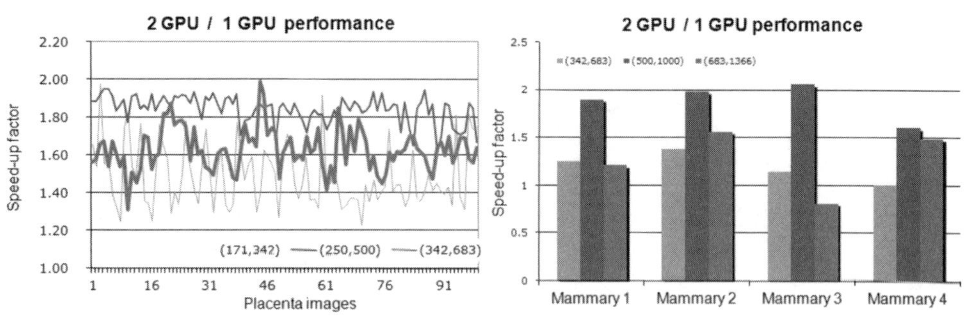

Figure 8.13 Single node GPU scalability. Improvement when enabling the second GPU using Pthreads. For placenta the average speedup factors are 1.46x, 1.83x, and 1.62x for the small, medium, and large parameter sizes, respectively. For mammary the average speedup factors are 1.17x, 1.94x, and 1.09x.

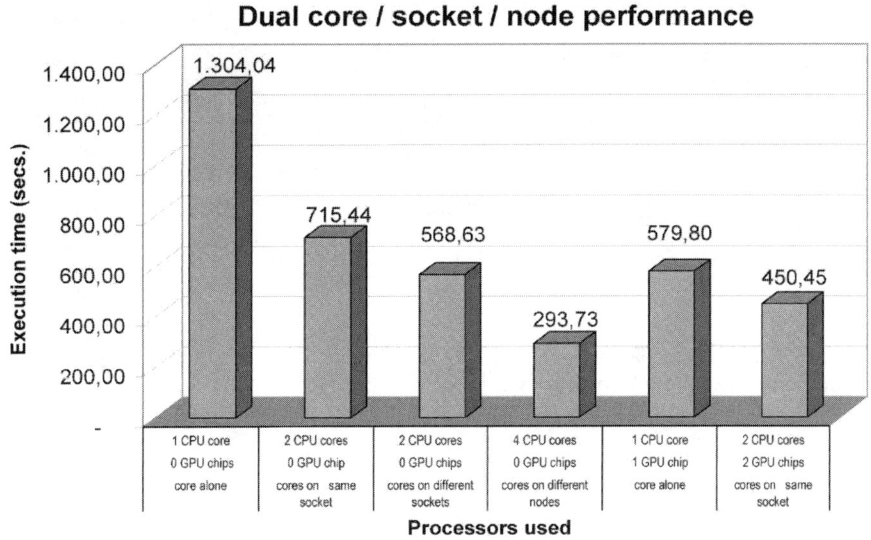

Figure 8.14 Single node performance under different node configurations for the most work-intensive image from the mammary set.

The effects of adding a second GPU to the single node configuration are presented in Figure 8.13. Gains from GPU parallelism are very diverse, starting with 30% to 50% for small window sizes, continuing with 60% for large window sizes, and finally ending with near-optimal scalability for medium sizes. These gains are proportional to computational workload, revealing that the GPU is a more scalable processor when it can exploit arithmetic intensity and GFLOPS are not limited by data shortages due to insufficient bandwidth between video memory and GPU.

Multiple Node Performance

Scalability results for CPU are shown in Figure 8.15. Gains are fairly consistent for all images and all parameter sets, since although CPU executions are slower

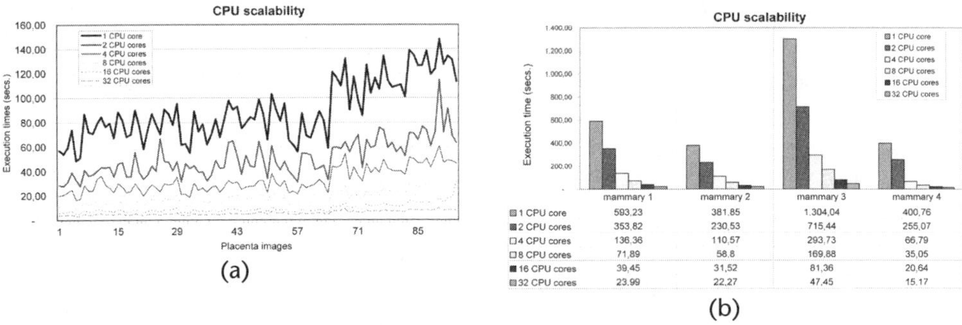

Figure 8.15 (a) CPU scalability for multiple nodes with the placenta set. (b) CPU scalability for the mammary set.

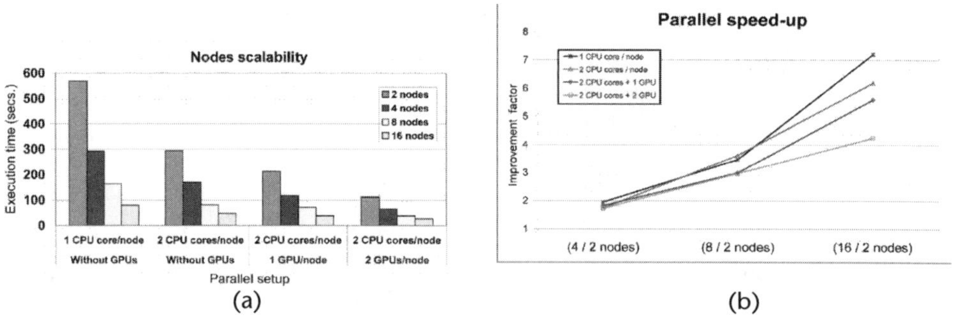

Figure 8.16 (a) Scalability and (b) speedup for increasing numbers of nodes running the most work-intensive image from the mammary set.

than GPU-assisted executions they are also more scalable on a larger number of nodes due to the communication bindings of paired CPU-GPU executions. This is confirmed in Figure 8.14, where the most work-intensive mammary image is tested for an assorted combination of CPUs and GPUs.

For increasing numbers of nodes Figures 8.15 and 8.16 show a progressive reduction in execution times. For the most work-intensive mammary image, the speedup on 16 versus 2 nodes with a 1 CPU configuration is 7.4x, where for a 2 CPU/GPU per node configuration the speedup is slightly over 4x. The less effective internode parallelism of the more aggressive configurations is due in large part to their more demanding intranode communications.

8.7 Summary

The next generation of automated microscope imaging applications, such as quantitative phenotyping, require the analysis of extremely large datasets, making scalability and parallelization of algorithms essential.

This chapter presents a fast, scalable, and simply parallelizable algorithm for image registration that is capable of correcting the nonrigid distortions of sectioned microscope images. Rigid initialization follows a simply reasoned process of matching high level features that are quickly and easily extracted through standard image processing techniques. Nonrigid registration refines the result of rigid initialization, using the estimates of rigid initialization to match intensity features using an FFT-implementation of normalized cross-correlation.

A computational framework for the two-stage algorithm is also provided along with results from sample high-performance implementations. Two hardware-based solutions are presented for nonrigid feature matching: parallel systems and graphics processor acceleration. Scalability is demonstrated on both single node systems where GPUs and CPUs cooperate, and also on multiple node systems where any variety of the single node configurations can divide the work. From a departure point of 181 hours to run 500 mammary images on a single Opteron CPU, the GPU accelerated parallel implementation is able to reduce this time to 3.7 hours (based on 26.61 seconds per slide), achieving a total speedup of 49x when all 32 CPUs and GPUs participate.

References

[1] C. Levinthal and R. Ware, "Three-dimensional reconstruction from serial sections," *Nature*, vol. 236, pp. 207–210, 1972.

[2] J. Capowski, "Computer-aided reconstruction of neuron trees from several sections," *Computational Biomedical Research*, vol. 10, no. 6, pp. 617–629, 1977.

[3] E. Johnson and J. Capowski, "A system for the three-dimensional reconstruction of biological structures," *Computational Biomedical Research*, vol. 16, no. 1, pp. 79–87, 1983.

[4] D. Huijismans, W. Lamers, J. Los and J. Strackee, "Toward computerized morphometric facilities: a review of 58 software packages for computer-aided three-dimensional reconstruction, quantification, and picture generation from parallel serial sections," *The Anatomical Record*, vol. 216, no. 4, pp. 449–470, 1986.

[5] V. Moss, "The computation of 3-dimensional morphology from serial sections," *European Journal of Cell Biology*, vol. 48, pp. 61–64, 1989.

[6] R. Brandt, T. Rohlfing, J. Rybak, S. Krofczik, A. Maye, M. Westerhoff, H.-C. Hege and R. Menzel, "A three-dimensional average-shape atlas of the honeybee brain and its applications," *The Journal of Comparative Neurology*, vol. 492, no. 1, pp. 1–19, 2005.

[7] J. Hajnal, H. Derek and D. Hawkes, *Medical Image Registration*. Boca Raton, FL: CRC Press, 2001.

[8] A. Goshtasby, *2-D and 3-D Image Registration: For Medical, Remote Sensing, and Industrial Applications*. Hoboken, NJ: Wiley-Interscience, 2005.

[9] J. Streicher, D. Markus, S. Bernhard, R. Sporle, K. Schughart and G. Muller, "Computer-based three-dimensional visualization of developmental gene expression," *Nature Genetics*, vol. 25, pp. 147–152, 2000.

[10] U. Braumann and J. Kuska and J. Einenkel and L. Horn and M. Luffler and M. Huckel, "Three-dimensional reconstruction and quantification of cervical carcinoma invasion fronts from histological serial sections," *IEEE Transactions on Medical Imaging*, vol. 24, no. 10, pp. 1286–1307, 2005.

[11] W. Crum, T. Hartkens and D. Hill, "Non-rigid image registration: theory and practice," *The British Journal of Radiology*, vol. 77, no. 2, pp. S140–S153, 2004.

[12] W. Hill, and R. Baldock, "The constrained distance transform: interactive atlas registration with large deformations through constrained distance," *Proceedings of the Workshop on Image Registration in Deformable Environments*, (Edinburgh, UK), Sept. 8, 2006.

[13] T. Yoo, *Insight into Images: Principles and Practice for Segmentation, Registration, and Image Analysis*. Wellesley, MA: AK Peters, 2004.

[14] S. Sarma, J. Kerwin, L. Puelles, M. Scott, T. Strachan, G. Feng, J. Sharpe, D. Davidson, R. Baldock and S. Lindsay, "3d modelling, gene expression mapping and post-mapping image analysis in the developing human brain," *Brain Research Bulletin*, vol. 66, no. 4-6, pp. 449–453, 2005.

[15] A. Jenett, J. Schindelin and M. Heisenberg, "The virtual insect brain protocol: creating and comparing standardized neuroanatomy," *BMC Bioinformatics*, vol. 7, p. 544, 2006.

[16] P. Wenzel, L. Wu, R. Sharp, A. de Bruin, J. Chong, W. Chen, G. Dureska, E. Sites, T. Pan, A. Sharma, K. Huang, R. Ridgway, K. Mosaliganti, R. Machuraju, J. Saltz, H. Yamamoto, J. Cross, M. Robinson and G. Leone, "Rb is critical in a mammalian tissue stem cell population," *Genes & Development*, vol. 21, no. 1, pp. 85–97, 2007.

[17] L. Cooper, K. Huang, A. Sharma, K. Mosaliganti and T. Pan,"Registration vs. reconstruction: Building 3-D models from 2-D microscopy images," *Proceedings of the Workshop on Multiscale Biological Imaging, Data Mining and Informatics*, (Santa Barbara, CA), pp. 57–58, Sept. 7-8, 2006.

[18] K. Huang, L. Cooper, A. Sharma and T. Pan, "Fast automatic registration algorithm for large microscopy images," *Proceedings of the IEEENLM Life Science Systems & Applications Workshop*, (Bethesda, MD), pp. 1–2, July 13-14, 2006.

[19] P. Koshevoy, T. Tasdizen and R. Whitaker, "Implementation of an automatic slice-to-slice registration tool," SCI Institute Technical Report UUSCI-2006-018, University of Utah, 2006

[20] J. Prescott, M. Clary, G. Wiet, T. Pan and K. Huang, "Automatic registration of large set of microscopic images using high-level features," *Proceedings of the IEEE International Symposium on Medical Imaging*, (Arlington, VA), pp. 1284–1287, April 6-9, 2006.

[21] R. Mosaliganti, T. Pan, R. Sharp, R. Ridgway, S. Iyengar, A. Gulacy, P. Wenzel, A. de Bruin, R. Machiraju, K. Huang, G. Leone and J. Saltz, "Registration and 3D visualization of large microscopy images," *Proceedings of the SPIE Medical Imaging Meeting*, (San Diego, CA), pp. 6144:923–934, Feb. 11, 2006.

[22] O. Schmitt, J. Modersitzki, S. Heldmann, S. Wirtz and B. Fischer, "Image registration of sectioned brains," *International Journal of Computer Vision*, vol. 73, no. 1, pp. 5–39, 2007.

[23] F. Maes, A. Collignon, D. Vandermeulen, G. Marchal and P. Suetens, "Multimodality image registration by maximization of mutual information," *IEEE Transactions on Medical Imaging*, vol. 16, no. 2, pp. 187–198, 1997.

[24] F. Bookstein, "Principal warps: Thin-plate splines and the decomposition of deformations," *IEEE Transactions on Pattern Analysis and Machine Intelligence*, vol. 11, no. 6, pp. 567–585, 1989.

[25] K. Rohr, *Landmark-Based Image Analysis: Using Geometric and Intensity Models*, New York: Springer, 2007.

[26] R. Bajcsy and S. Kovacic, "Multiresolution elastic matching," *Computer Vision, Graphics, and Image Processing*, vol. 46, pp. 1–21, 1989.

[27] M. Lefebure, and L. D. Cohen, "A multiresolution algorithm for signal and image registration," *Proceedings of IEEE International Conference on Image Processing*, (Washington, D.C.), 3: pp. 252–255, Oct. 26-29, 1997.

[28] L. Cooper, S. Naidu, G. Leone, J. Saltz and K. Huang, "Registering high resolution microscopic images with different histochemical stainings - a tool for mapping gene expression with cellular structures," *Proceedings of the Workshop on Microscopic Image Analysis with Applications in Biology*, (Piscataway, NY), Sept. 21, 2007.

[29] H. Peng, F. Long, M. Eisen and E. Myers, "Clustering gene expression patterns of fly embryos," in *Proceedings of IEEE International Symposium on Biomedical Imaging (ISBI)*, (Arlington, VA), pp. 1144–1147, April 6–9, 2006.

[30] T. Kim and Y.-J. Im, "Automatic satellite image registration by combination of matching and random sample consensus," *IEEE Transactions on Geoscience and Remote Sensing*, vol. 41, no. 5, pp. 1111–1117, 2003.

[31] F. Ino, K. Ooyama and K. Hagihara, "A data distributed parallel algorithm for nonrigid image registration," *Parallel Computing*, vol. 31, no. 1, pp. 19–43, 2005.

[32] S. Warfield, F. Jolesz and R. Kikinis, "A high performance computing approach to the registration of medical imaging data," *Parallel Computing*, vol. 24, no. 9–10, pp. 1345–1368, 1998.

[33] T. Rohlfing and C. Maurer, "Nonrigid image registration in shared-memory multiprocessor environments with applications to brains, breasts, and bees," *IEEE Transactions on Information Technology in Biomedicine*, vol. 7, no. 1, pp. 16–25, 1998.

[34] M. Ohara, H. Yeo, F. Savino, G. Iyengar, L. Gong, H. Inoue, H. Komatsu, V. Sheinin, S. Daijavad and B. Erickson, "Real time mutual information-based linear registration on the cell broadband engine processor," *Proceedings of the IEEE International Symposium on Medical Imaging (ISBI)*, (Washington, DC), pp. 33–36, April 12-15, 2007.

[35] Z. Fan, and F. Qiu, A. Kaufman, and S. Yoakum-Stover, "GPU cluster for high performance computing," in *Proceedings 2004 ACM/IEEE Intl. Conference for High Performance Computing, Networking, Storage and Analysis*, (Washington, DC), pp. 47–53, Nov. 11-17, 2006.

[36] W. Wu and P. Heng, "A hybrid condensed finite element model with GPU acceleration for interactive 3D soft tissue cutting: Research articles," *Computer Animation and Virtual Worlds*, vol. 15, no. 3-4, pp. 219–227, 2004.

[37] Zhao, Y., Y. Han, Z. Fan, F. Qiu, Y. Kuo, A. Kaufman, and K. Mueller, "Visual simulation of heat shimmering and mirage," *IEEE Transactions on Visualization and Computer Graphics*, vol. 13, no. 1, pp. 179–189, 2007.

[38] M. Botnen and H. Ueland, "The GPU as a computational resource in medical image processing," tech. rep., Dept. of Computer and Information Science, Norwegian Univ. of Science and Technology, 2004.

[39] F. Ino, J. Gomita, Y. Kawasaki and K. Hagihara, "A gpgpu approach for accelerating 2-D/3-D rigid registration of medical images," *Proceedings of the 4th International Symposium on Parallel and Distributed Processing and Applications (ISPA)*, (Sorrento, IT), pp. 769–780, Dec. 1–4, 2006.

[40] P. Hastreiter, C. Rezk-Salama, C. Nimsky, C. Lurig and G. Greiner, "Techniques for the Analysis of the Brain Shift in Neurosurgery," *Computers & Graphics*, vol. 24, no. 3. pp. 385–389, 2000.

[41] S. Guha, S. Krisnan, and S. Venkatasubramanian, "Data Visualization and Mining Using the GPU," *Data Visualization and Mining Using the GPU, Tutorial at 11th ACM International Conference on Knowledge Discovery and Data Mining (KDD)*, (Chicago, IL), Aug. 21-24, 2005.

[42] M. Hadwiger, C. Langer, H. Scharsach, and K. Buhler, "State of the art report on GPU-Based Segmentation," Tech. Rep. TR-VRVIS-2004-17, VRVis Research Center, Vienna, Austria, 2004.

[43] J. D. Owens, D. Luebke, N. Govindaraju, M. Harris, J. Kruger, A. E. Lefohn, and T. J. Purcell, "A Survey of General-Purpose Computation on Graphics Hardware," *Journal of Computer Graphics Forum*, vol. 26, no. 1, pp. 21–51, 2007.

[44] J. P. Lewis, "Fast normalized cross-correlation," *Vision Interface*, (Quebec City, Canada), pp. 120–123, June 15–19, 1995.

[45] G. Malandain, E. Bardinet, K. Nelissen and W. Vanduffel, "Fusion of autoradiographs with an MR volume using 2-D and 3-D linear transformations," *NeuroImage*, vol. 23, no. 1, pp. 111–127, 2004.

[46] FFTW library, "FFTW home page," http://www.fftw.org, Dec. 2007.

[47] CUDA, "Nvidia Home Page," http://developer.nvidia.com/object/cuda.html, Dec. 2007.

[48] CUFFT library, "Nvidia Home Page," http://developer.download.nvidia.com/compute/cuda/1_1/CUFFT_Library_1.1.pdf, Dec. 2007.

CHAPTER 9

Data Analysis Pipeline for High Content Screening in Drug Discovery

Daniel W. Young, Charles Y. Tao, and Yan Feng

9.1 Introduction

Automated fluorescent microscopy based high content screening (HCS) is making a major impact on many aspects of the drug discovery process by enabling simultaneous measurement of multiple features of cellular phenotype that are relevant to both therapeutic and toxic activities of compounds. HCS employs highly automated processes and typically generates immense datasets reflecting cell phenotypic information. Besides substantial research articles published in the past 10 years, three recent books were devoted entirely to HCS, encompassing the scientific and technological aspects of assay development, automation, image acquisition, and image analysis [1–3]. However, mining the large HCS dataset remains an ongoing challenge in the field. In this chapter, we will briefly summarize current HCS technology and aim to focus our discussion mainly on the integration of necessary computation and statistical methods for data handling and mining. Examples of data normalization, dose response estimation, cytometry analysis, and data driven factor analysis will be discussed in detail.

9.2 Background

Drug discovery is a complex process that entails many aspects of technology. A recent estimate put the cost of developing a new drug at more than \$1 billion, and the time for developing such a drug at 12 to 14 years [4]. How to reduce the cost and increase the efficiency of the drug discovery process has become increasingly important.

Typical drug discovery efforts employ a "targeted" process, which starts with the identification of specific disease related target, usually gene products that are either mutated or misregulated in patients. Then functional assays, such as enzyme or binding based in vitro assays, are designed to identify small molecule or biological "hits" that can be used to perturb or restore the function of the target. Such "hit" molecules are further tested for its efficacy and toxicity in cellular disease models to become a "lead" molecule. Drug leads with a desired activity profile are then tested in animals and humans before they are finally submitted to Food and Drug Administration (FDA) for approval. The targeted approach has been increasingly criticized recently due to the high failure rate of identifying active but nontoxic drug leads. Hence, there is a need for refined approaches to predict success earlier in the discovery process [5]. HCS, which can simultaneously measure multiple cellular phenotypes which are relevant to both efficacy and toxicity, became increasingly popular because it holds the promise of expediting the drug discovery process [6].

Early HCS applications were mainly focused on secondary assays for the hit-to-lead process. Evaluating the effects of hit compounds in cellular models is crucial for making decisions on the fate of the compound, as well as the direction of the drug discovery program. Hit compounds are usually tested in a variety of assays reflecting their on-target activities, off-target activities, toxicities, and physicochemical properties. Cellular imaging based HCS assays have been developed in all these areas. Successful cases include nuclear and plasma membrane translocation of transcription factors or signaling molecules, endocytosis of G-protein coupled receptor (GPCR) after activation, cell cycle, and neurite morphology assays [7–10]. With the increasing realization of the pitfalls of targeted based drug discovery, HCS has now been adopted earlier in the hit discovery stage at many major pharmaceutical companies [11]. Interestingly, phenotypic profiling of drug activities using high content analysis has been shown to be able to predict drug mechanism of action (MOA) with remarkable accuracy, comparable to the more expensive and low throughput transcription profiling technology [12, 13]. The large quantity of single cell descriptors from high content analysis might hold the key to such a result. High-throughput prediction of drug MOA and toxicity is one of the most important challenges facing drug discovery. High content analysis is predicted to have an increasingly important role in this area as the technology develops.

9.3 Types of HCS Assay

Thanks to the rapid development of knowledge on molecular and cellular functions in the past two decades, many cellular processes can be adopted into HCS assays format in rather straightforward ways and usually in a short timeframe. Current HCS assays can be roughly divided into three main formats: (1) translocation assays, (2) cytometry assays, and (3) morphology assays. Figure 9.1 illustrates examples of the three assay formats.

Translocation assays score for concentration shift of a particular protein from one cellular compartment to another. Immunocytochemistry using specific antibodies and expression of green fluorescence protein (GFP) fusion protein are among the most commonly used methods for detection. The ratio or difference of the concentration in different cellular compartments was used as a quantitative score for such events [14]. Nuclear-cytoplasmic translocation of transcription factors or signal transduction molecules, such as NFκB, NFAT, FOXO, p38MK2, and HDAC4/5 are among the most common ones of interest. Because of the simple format and robust quantitation method, nuclear-cytoplasmic translocation was also adopted in engineering biosensors to detect binding or cleavage events in cells, such as p53/MDM2 interaction or caspase3 activation [15, 16].

Cytometry assay treats each cell as a single entity and scores for population change of cells with certain phenotypic characteristics. Fluorescence activated cell

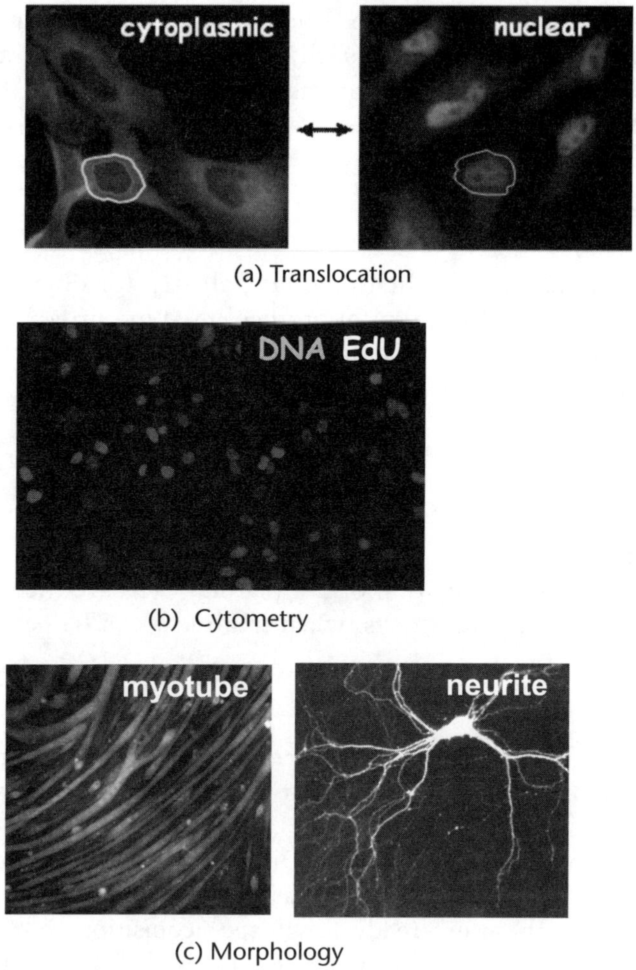

Figure 9.1 HCS assay formats. (a) Cytoplasmic-nuclear translocation of GFP-Foxo3a. The ratio of GFP-Foxo3a intensity in the nuclear mask region (circle) and in the cytoplasmic mask region (ring) was used as the quantitative measurement. (b) Cytometry assay measuring percentage of S-phase cells. S-phase cells were labeled with Ethynyl-dU incorporation and azido-rhodamine label and all cells were labeled with Hoechest 33342 dye. (c) Morphology assays on myotube formation and neurite formation. Myotube length and width, neurite length, and branches, were among the parameters of interest.

sorting (FACS) is the most common format for cytometry based assays. FACS requires a large number ($>10^5$) of cells and does not provide enough resolution to score intracellular or morphological changes in cells, whereas HCS based cytometry requires many fewer ($\sim 10^2$) cells and can be performed in high spatial resolution, thus providing higher throughput and information not accessible to conventional FACS. Cell cycle assay is a cytometry based assay widely adopted in oncology research where interfering with cell cycle and cell growth is among the main goals of therapeutic intervention. Cell cycle stages are determined by DNA content, as well as elevation of other specific cell cycle markers such as phospho-Histone H3 for mitotic cells and cleaved PARP for apoptotic cells [17].

Morphological based assays are generally more challenging for image analysis, and thus more difficult to develop into a fully automated high throughput format. But a few exceptions, such as neurite outgrowth and myotube differentiation, where no other surrogate assay formats are available, have been studied more extensively [18, 19].

9.4 HCS Sample Preparation

Sample preparation is the first (and crucial) step for obtaining high-quality image data. Here we provided a typical protocol used in our lab as the starting point for HCS assay development. In practice, we found more than 90% of the assays can be covered by this simple procedure.

9.4.1 Cell Culture

Cells are plated at $\sim 10^5$/ml, 30 μl for 384 well and 100 μl for 96 well plate, respectively. We use a Multidrop (ThermoFisher), Microfill (BioTek), or PlateMate (Matrix) to assist automation of cell plating. Cells are then grown at 37°C in 5% CO_2 from overnight to 2 days, when treatment can be performed in between.

9.4.2 Staining

We use an ELX405 plate washer (BioTek) running at the lowest dispensing/aspiration speed to assist all fixation and staining processes. In brief, an equal volume of 2× concentrated fixatives, 7.5% formaldehyde in phosphate buffered saline (PBS) or Mirsky's (National Diagnostics), were added directly to the wells. After 15 minutes, the wells were washed 1× with PBS. For cells expressing GFP fusion protein, these are ready for image acquisition. For immunocytochemistry staining, the cells were then incubated 1× with PBS-TB (PBS with 0.2% Triton X-100, 0.1% BSA) for 10 minutes to permeabilize the cell membrane. Primary antibody was then added at $\sim$0.5 to 5 μg/ml in PBS-TB and incubated at room temperature for 1 hour. The cells were then washed 2× with PBS-TB. Fluorescently labeled secondary antibody was added at 2 μg/ml together with 1 μg/ml nuclear stain Hoechst33342 in PBS-TB and incubated at room temperature for 30 minutes. The cells were then washed 2× with PBS-TB and 1× PBS before image acquisition.

9.5 Image Acquisition

Currently more than 10 commercial vendors provide automated fluorescence imaging hardware, including confocal and nonconfocal imagers. In principle, a confocal imager can provide images with reduced out-of-focus light. But in practice most of the applications use low magnification objectives on a flat monolayer of cells; a wide field microscope can already provide images of sufficient quality.

All the HCS imagers can focus on each field and acquire fluorescence images at various magnifications and wavelengths in fully automated mode. A robotic

Table 9.1 A Partial List of High Content Imaging Instruments

Instrument	*Manufacture*	*Technology*	*Live Cell*	*Liquid Handling*	*Data Management*	*Information*
ArrayScan VTi	ThermoFisher	Widefield with Apotome	Available	Available	STORE	http://www.cellomics.com
ImageXpress Micro	Molecular Devices	Widefield	Available	Available	MDCStore	http://www.moleculardevices.com
ImageXpress Ultra	Molecular Devices	Laser Scanning Confocal	No	No	MDCStore	http://www.moleculardevices.com
IN Cell Analyzer 1000	GE Healthcare	Widefield with structured light	Yes	Yes	In Cell Miner	http://www.gelifesciences.com
IN Cell Analyzer 3000	GE Healthcare	Laser Scanning Confocal	Yes	Yes	In Cell Miner	http://www.gelifesciences.com
Opera	PerkinElmer	Laser confocal with Nipkow disk	Available	Available	File directory	http://las.perkinelmer.com
CellWoRx	Applied Precision	Widefield with oblique illumination	No	No	STORE	http://www.api.com
Pathway 435	BD Biosciences	Widefield with Nipkow disk	No	No	File directory	http://www.atto.com
iCyte	Compucyte	Widefield Laser scanning	No	No	File directory	http://www.compucyte.com
MIAS-2	Maia Scientific	Widefield	No	No	File directory	http://www.maia-scientific.com
Explorer ^{e}X3	Acumen	Widefield Laser scanning	No	No	File directory	http://www.ttplabtech.com

plate handler can usually be integrated with the imager for a large screening effort. Several instruments also provide live imaging and liquid handling capabilities. In Table 9.1 we provide a partial list summarizing their main features. Technical details can easily be obtained by contacting the specific vendors. One has to consider many factors before choosing a high content imager that fits a specific need. Obviously, price, capability, IP status, and maintenance are all issues that need to be considered carefully. The most important thing is to test several instruments with real samples and questions in mind. For example, if counting responsive cells is the only desired application, one can choose a high speed instrument such as Explorer with low magnification and totally avoid image analysis. But if detailed intracellular structures such as endosomes are the primary interest, one has to go for an instrument with a higher magnification and a more sensitive camera, perhaps with confocal capability. High content imagers generate a large volume of raw data, especially when fitted with an automated plate loader. If large scale screening is the main application, one has to consider data archiving solutions. A database built in with the instrument is good for handling a large, multiuser environment but sufficient IT support is often required. All data generated for this chapter was from an upgraded version of Cellomics ArrayScan VI or MetaXpress Ultra Confocal imagers.

9.6 Image Analysis

Image analysis is a complex topic, and comprehensive coverage of the topic is beyond the scope of this chapter. At this moment most of the HCS instrument vendors provide some level of image analysis capabilities which are developed for specific biological applications, such as nuclear translocation, object content and texture measurements, vesicle counting, and neurite length/branch scoring. Vendor software usually gets the most use, largely because the proprietary format used by different vendors makes it hard to export and analyze in other software packages. But most vendor software is sufficient in generating wellular summary of image data for simple analysis. Generally speaking, most of the HCS image analysis starts with nuclei recognition, because nuclei are spatially well separated with sharp boundaries. A nuclear counter stain such as Hoechst33342 is most often used to outline the nuclear boundary. Thus, they can be easily recognized and accurately separated with almost any boundary detection algorithm. For the same reason, bright spots such as endosomes and Golgi can be detected and segmented easily [21].

Accurate cell boundary detection is a complex task. Because cells are irregularly shaped and frequently overlapped, efforts to accurately identify cell boundaries often suffer from serious under or over-segmentation problems. In practice, a proportional ring expansion or watershed expansion from the nuclear area is often used to approximate the cellular area. More complex iterative decision making processes have been shown to be able to detect cell boundary accurately. First, over-segmentation of cell bodies and then joining at a later stage with complex, environment-sensitive criteria were employed in these exercises [22, 23]. Once boundary detection is achieved, it is rather straightforward to score a multitude of parameters for each object, such as dimension, intensity, and texture features. Cellomics and Metamorph software, for example, provide more than 30 parameters for each channel. A more comprehensive list was described by Haralick [24].

In addition to vendor software there are several attempts being made to develop generic image and data analysis software that can be used with all acquisition platforms. These include the free and open source image and data analysis software Cell Profiler, from the Broad/MIT and Whitehead Institutes (http://www.cellprofiler.org/index.htm) and Zebrafish image analysis software (ZFIQ) from the Methodist Hospital in Houston, Texas (http://www.cbi-platform.net/download.htm). More advanced image analysis software can be found in the Definiens software package (http://www.definiens.com/), which uses a novel over-segmentation and reconstruction approach to identify cellular regions of interest. Finally, another example of a free and robust (but rudimentary) software package is Image J from the NIH (http://rsb.info.nih.gov/ij). This package is extremely useful for viewing and quickly manipulating images for display and analysis. This software has a recordable macro language for analysis of images in batch and has a large user group base to freely exchange macros.

9.7 Data Analysis

In contrast to all the assays, hardware, and image analysis software developed for HCS in the past 10 years, little has been done on the data analysis aspects, one reason being that most application developers have neither experience nor the desire to handle a large amount of data, and without proper integration of data analysis pipeline, HCS has little use for high throughput screening and remains on par with a manually operated fluorescence microscope. In an effort to make HCS practical in a drug screen scenario, we have developed a series of tools for data extraction, normalization, dose response estimation, and multiparameter cell classification. We will use most of the rest of the chapter to discuss these tools, and introduce examples when necessary.

9.7.1 Data Process Pipeline

HCS generates a large amount of primary image data. For example, a typical medium scale screen of 25,000 compounds in duplicate can easily be achieved in one week with one HCS instrument and limited liquid handling automation access. Assuming four individual channels and four image fields were collected in each treatment condition, the experiment generates ~600 GB of raw image data, and a full HTS screen at 1M compounds will generate over 25 TB of raw image data (Figure 9.2). For small scale exploration such as HCS assay development, a local hard drive is sufficient. A corporate storage and archive system is necessary if screening effort runs at full scale. Some image analysis routines also generate a large amount of data, especially when every single cell analysis data is stored. We estimated that 2.5 billion parameters will be generated in the typical medium-sized experiment, as mentioned above, and a full million compound screen will produce over 100 billion data points. The dataset is big enough that it cannot be handled by

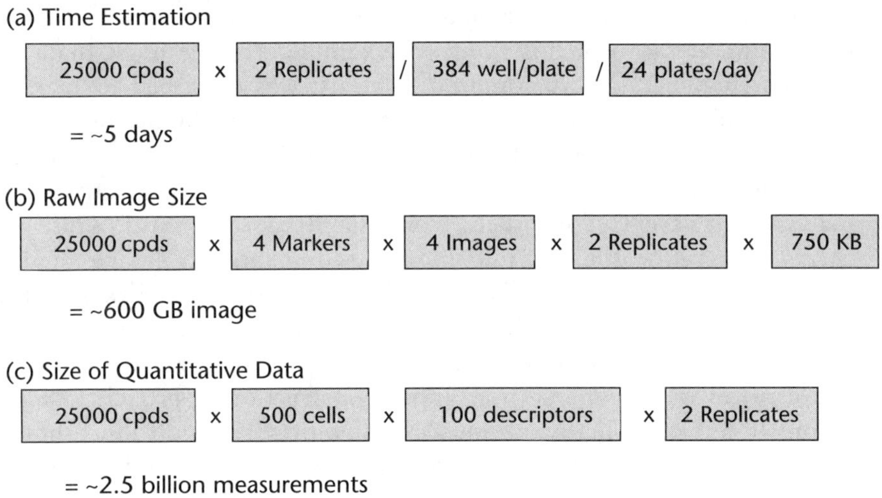

Figure 9.2 HCS data size. (a) A typical medium sized HCS screen with 25,000 compound treatments in duplicates took one week to complete, assuming imaging takes one hour per 384 well plates. (b) 600 GB of raw image data was generated, assuming four image fields for each treatment and four marker channels were taken with a 2 × 2 binned 1M pixel CCD camera and 12-bit digitizing format. (c) 2.5 billion measurements were generated, assuming 500 cells were quantified with 100 descriptors each in every treatment condition.

an Excel spreadsheet used by most bench biologists unless dramatic data reduction was achieved.

In our lab, a relational database was provided by Cellomics for storage of all raw image and image analysis data. Cellomics, like many other vendors, also provides simple data visualization software at the plate, well, and cell levels, but not at the screen level. Data treatment is limited to simple statistics such as averaging and Z function. None of the current commercial vendors provides tools for large scale data normalization and data mining, which are absolutely necessary for a successful high throughput screening campaign.

Here we developed/adopted a series of data analysis tools for a large numerical dataset resulting from image analysis algorithms. These data analysis tools are grouped into three modules, preprocessing, data mining, and data integration modules (Figure 9.3). The preprocessing module handles retrieval of cell level data from the image analysis software, data normalization, and generation of quality control plots. Most of the data reduction was achieved by a variety of data mining modules generating treatment level data. A series of data mining modules has been built in our lab. We will describe the dose response curve/confidence estimation module and the automated cytometry classification module in detail. A data driven factor model was also recently published, and we will briefly introduce the concept.

9.7.2 Preprocessing Normalization Module

Systematic errors, such as plate-to-plate and well-to-well variations, often exist in a high content screen. Data normalization is a necessary procedure to minimize the

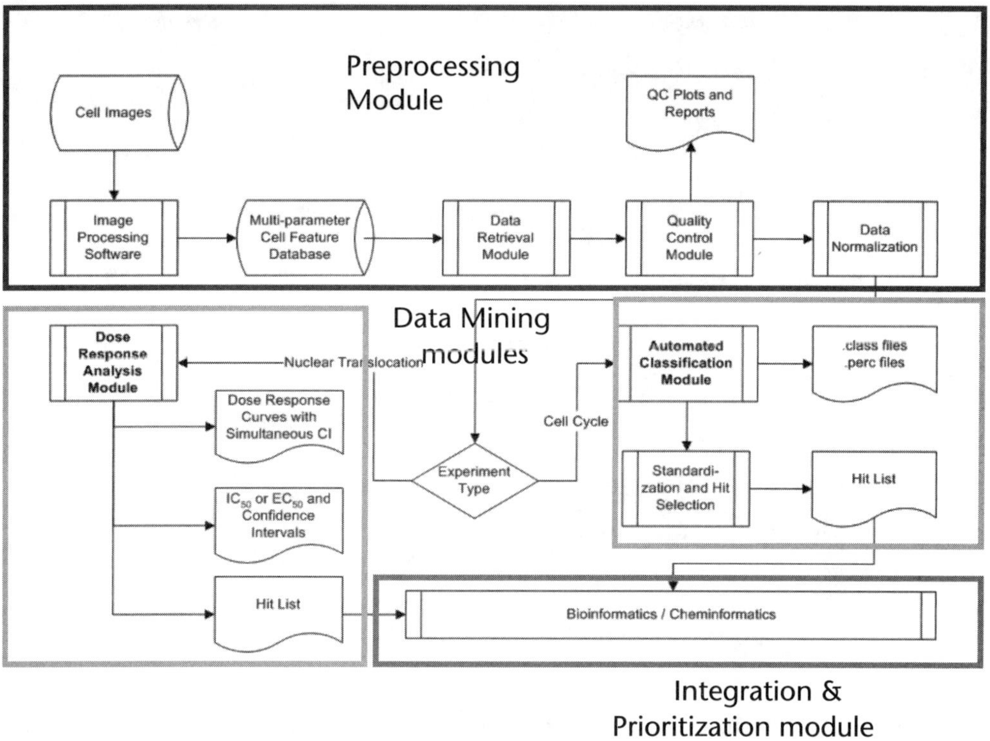

Figure 9.3 HCS data processing pipeline. Flowchart of HCS data processing pipeline consisting of preprocessing, data mining, and data integration modules. The preprocessing module extracts image analysis data and performs normalization. Data mining modules reduce the cell centric data to treatment centric data. The multiple data mining module can be tailored to specific needs such as dose response curve estimation, cytometry analysis, and morphometry analysis. The integration module links treatment centric data back to treatment annotation files.

impact of such errors. For each parameter, we first calculated the median for each well, x_{ij}, which is the median for the jth well (for a 384-well plate, $j = 1,2,\ldots,384$) on the ith plate. The following multiplicative model was then used:

$$x_{ij} = p_i \cdot w_j \cdot \varepsilon_{ij}$$

where p_i is the plate effect, w_j is the well effect, and the residue, ε_{ij} is the value after normalization.

Applying logarithmic transformation to both sides of the above equation converts it into an additive model:

$$\log x_{ij} = \log p_i + \log w_j + \log \varepsilon_{ij}$$

ε_{ij}, which was subsequently used to correct the value of each cell in well j on plate i, was then determined using the median polish method by alternatively removing row and column medians [25]. The effects of normalization on plate-to-plate variation and plate edge effect—two typical systematic errors in high throughput screening—are shown in Figure 9.4.

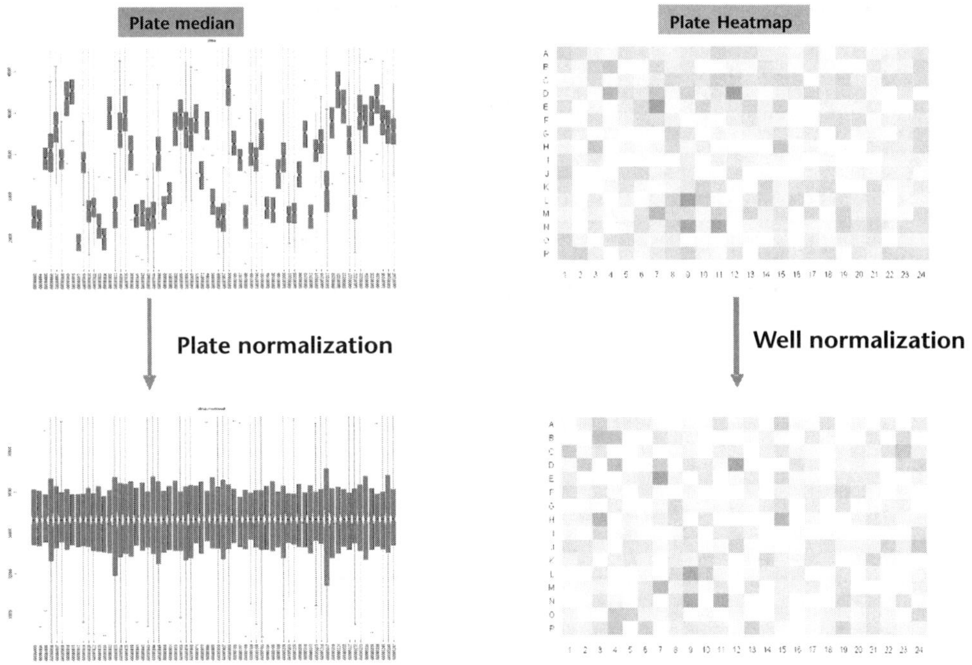

Figure 9.4 Preprocessing data normalization. Normalization removes plate-to-plate and well-to-well variations, thus allowing uniform analysis of the entire HCS dataset in other modules.

9.7.3 Dose Response and Confidence Estimation Module

EC50 estimation is the most commonly used method for determining the compound effect on cells. Here we used a four-parameter method to fit the dose response curve using all single cell measurements:

$$f(x) = \alpha + (\beta - \alpha)/(1 + exp[(x - \lambda)/\theta]$$

where α and β are the low and high responses, respectively. λ is the EC50 value, and θ is the slope of the dose response curve reflecting cooperativity.

Quantitative image analysis often results in significant data variation between cells and a rather small window of response to treatment (Figure 9.5). Fortunately we were still able to get an accurate measurement of the treatment effect because it is relatively easy to collect information for hundreds of cells in each HCS experiment, thus greatly increasing the confidence level of our estimation. But that comes with a price: acquiring more data requires more storage space and slows down the data analysis process. One important question here is what minimal cell number is necessary for a sufficiently confident measurement. The rule of thumb that was used was to measure 500 cells per treatment. To have an accurate estimation, we performed both theoretical analysis and numerical simulation. We found that the confidence interval (CI) for the cumulative distribution function is inversely proportional to the square root of the cell numbers. As shown in Figure 9.5, we can reach CI $<$ 0.1 at about 250 cells.

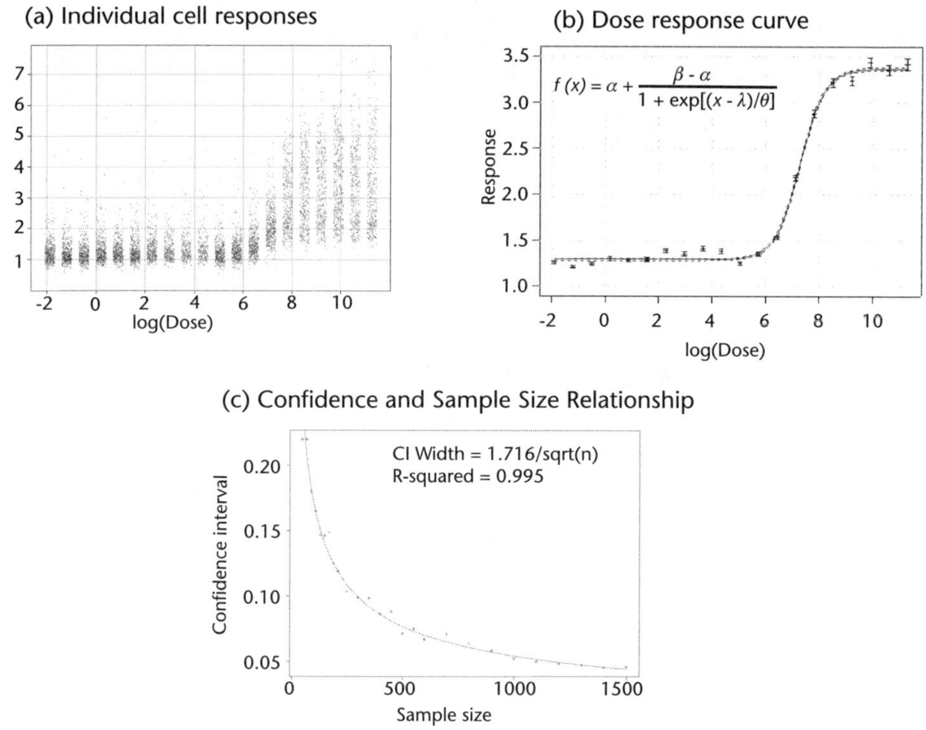

Figure 9.5 Dose response and confidence interval estimation. (b) Dose response curve was derived from (a) single cell responses using a four-parameter model. 95% confidence bands of each treatment point were shown by error bars, and curve fitting was shown by dotted lines. (c) The confidence interval of the cumulative distribution function is inversely proportional to the square root of sample size.

9.7.4 Automated Cytometry Classification Module

In a cell cycle assay, the percentage of cells in each cell cycle stage is the most crucial measurement. Cells were automatically classified into different phases of

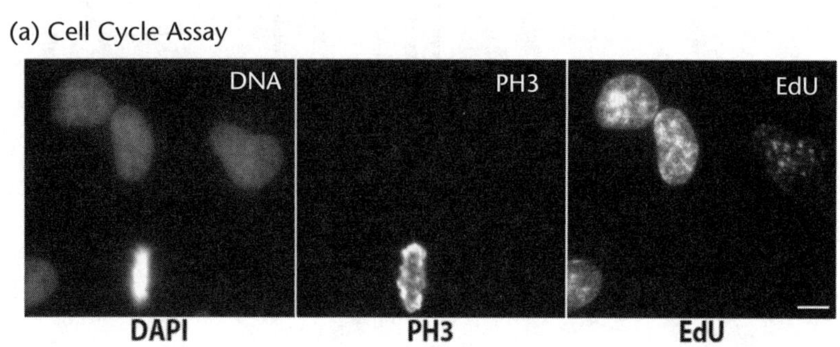

Figure 9.6 Decision tree cell cycle phase classification. (a) Cells were stained to show their levels of DNA, PH3, and EdU in each nuclei. (b) Every cell was classified as either G1 (2N DNA), G2 (4N DNA), M (PH3 positive), or S (EdU positive) phases using the automated four-parameter decision tree model. (c) The results of the classification were shown in two scatter plots where cells in different phases were shown.

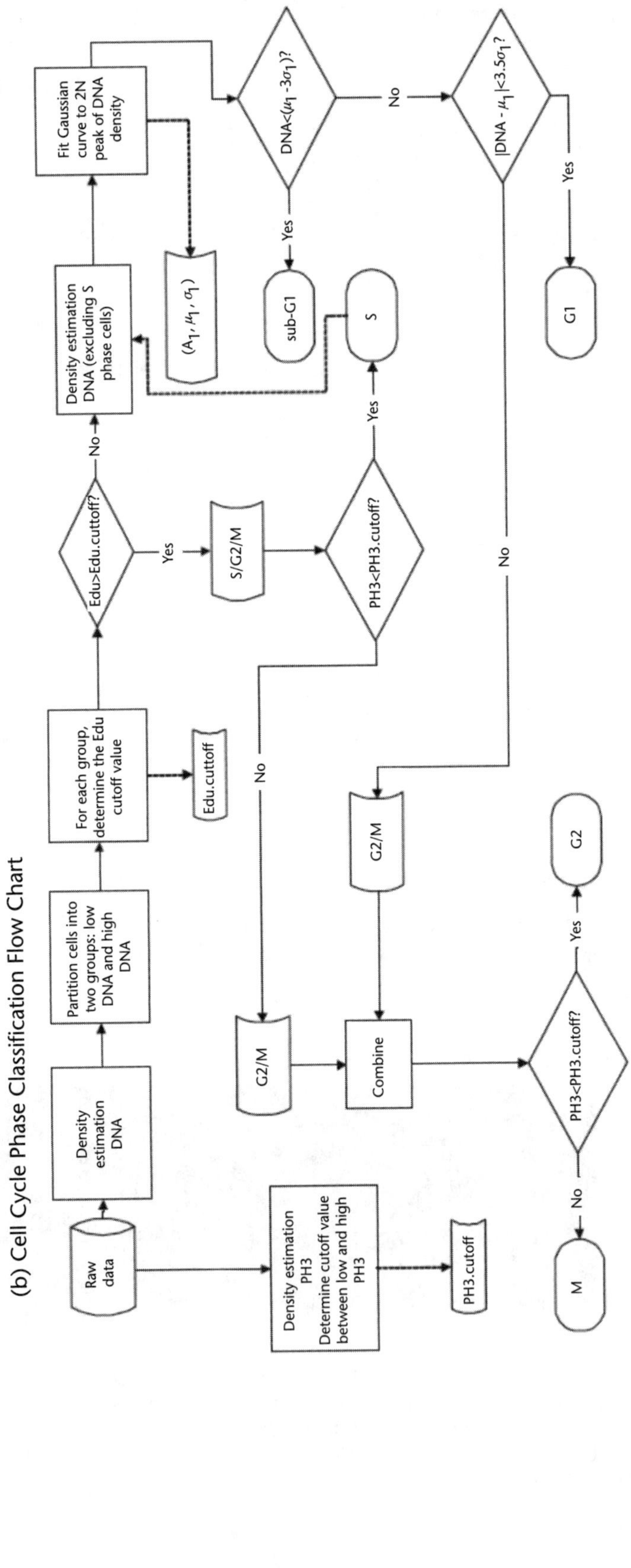

Figure 9.6 *(continued)*

(c) Cell Cycle Phase Classification Result

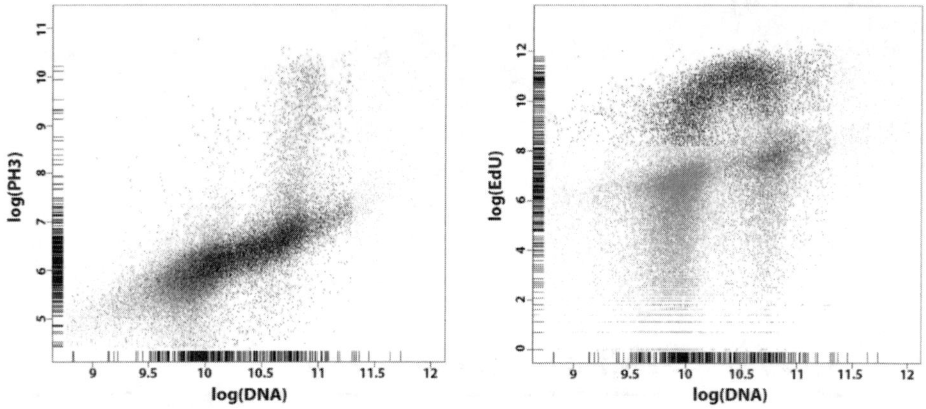

Figure 9.6 *(continued)*

the cell cycle on a plate-by-plate basis using a decision-tree based algorithm. An outline of the algorithm has the following steps (Figure 9.6):

1. For each plate, estimate the distribution of DNA content and determine a cutoff value to roughly divide the cells into two groups: low and high DNA content. This step is necessary because the EdU distribution for the low DNA group usually differs from that of the high DNA group.
2. For each group, estimate the distribution of EdU content to partition the cells into low EdU and high EdU intensity groups.
3. Cells in the high EdU group are candidates for S, G2, or M phases. Those cells with low PH3 level are classified as S-phase; otherwise, G2/M.
4. Estimate the distribution of the DNA content again; this time excluding those cells in S-phase.
5. Fit a Gaussian curve, determined by

$$f(x) = \frac{1}{\sqrt{2\pi}\sigma_1} \exp\left[\frac{-(x-\mu_1)^2}{2\sigma_1^2}\right]$$

   to the $2N$ peak of the DNA distribution.
6. Classify cells with DNA content less than $\mu_1 - 3\sigma_1$ as *sub-G1*.
7. Cells with DNA content less than $\mu_1 + 3.5\sigma_1$ and greater than $\mu_1 - 3\sigma_1$ are classified as in G1 phase.
8. Combine cells with DNA content greater than $\mu_1 + 3.5\sigma_1$ with the G2/M cells from Step 3.
9. Classify those cells from step 8 and with low PH3 content as G2, otherwise, M.

Alternatively, a supervised method, such as neural network or support vector machine, can be used (Figure 9.7). This approach requires a small, often manually curated training set [26].

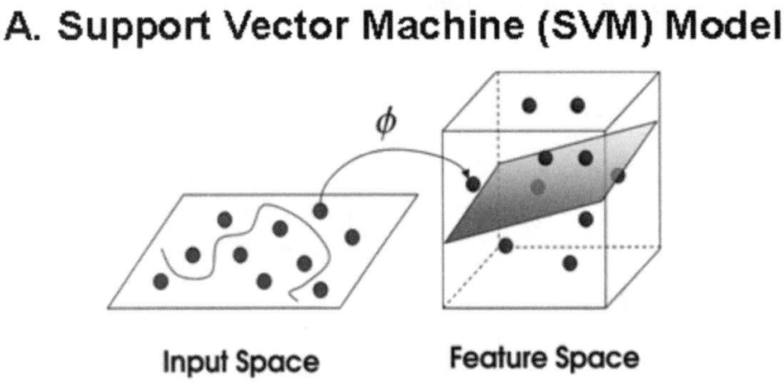

B. SVM Classification Accuracy

		Expert Annotated				
		Pro	Prometa	Meta	Ana/Telo	Lobed
SVM Predicted	Pro	**36**	1	2	0	4
	Prometa	10	**94**	7	0	1
	Meta	4	6	**95**	0	0
	Ana/Telo	2	0	0	**78**	0
	Lobed	2	0	0	1	**26**
Sensitivity (%)		66.7	93.1	91.3	98.7	83.9
Specificity (%)		97.8	93.3	96.2	99.3	99.1

Figure 9.7 Support vector machine mitotic subphase classification. (a) A supervised support vector machine (SVM) model was implemented to generate a multidimensional classification filter based on expert annotated images of cells in mitotic subphases. (b) Overall sensitivity and specificity of the method were above 90%.

The percentage of cells in each phase was then calculated for each well. Typically, most cells are in interphase (G1, S, or G2), with only a small number of cells in M-phase. Therefore, the percentages of cells in each cell cycle phase vary significantly among the phases and need to be standardized before the comparison of cell cycle profiles of different treatments can be made.

We introduced the *NZ* score for this purpose. If the percentage for well i is x_i, where $i = 1 \ldots N$ and N is the total number of wells in the reagent set, then the *NZ* score was calculated as follows:

$$NZ_i = \frac{x_i - \mu'}{1.4826 \cdot \sigma'}$$

where μ' is the median of $\{x_i; i = 1\ldots N\}$ and σ' is the median absolute deviation, defined as the median of $\{|x_i - \mu'|; i = 1\ldots N\}$. It can be shown that if x_i is normally distributed, *NZ* score is the same as the commonly used *Z* score, defined as $Z_i = (x_i - \mu)/\sigma$, where μ is the mean and σ is the standard deviation. *NZ* score was used as an alternative to the *Z* score as it is resistant to outliers which can occur frequently in high throughput screening; screen hits are by definition such outliers.

As an example, a positive S-phase *NZ* score indicates that the percentage of S-phase cells is higher than usual for the particular well and that treatment has likely resulted in a delay in S-phase progression. A negative S-phase *NZ* score indicates that the percentage of S-phase cells is low, and that treatment has likely resulted in a block at S-phase entry. Together, the four *NZ*-score numbers (one for each of G1, S, G2, M) give the cell cycle profile of a particular treatment.

9.8 Factor Analysis

A typical HCS experiment might generate gigabytes of numbers extracted from the images describing the amount and location of biomolecules on a cell-to-cell basis. Most of these numbers have no obvious biological meaning; for example, while the amount of DNA per nucleus has obvious significance, that of other nuclear measures, such as DNA texture, or nuclear ellipticity, are much less clear. This leads biologists to ignore the nonobvious measurements, even though they may report usefully on compound activities. A standard method in other fields for analyzing large, multidimensional datasets is factor analysis. It allows a large data-reduction but retains most of the information content, and quantifies phenotype using data-derived factors that are biologically interpretable in many cases. For this reason, factor analysis is highly appropriate to high content imaging, as it seeks to identify these underlying processes [27].

HCS data are contained in an $n \times m$ matrix, $\mathbf{X}$ consisting of a set of n image-based features measured on m cells. In mathematical terms, the so-called Common Factor Model posits that a set of measured random variables $\mathbf{X}$ is a linear function of common factors, $\mathbf{F}$ and unique factors, $\boldsymbol{\varepsilon}$:

$$\mathbf{X} = \mathbf{LF} + \boldsymbol{\varepsilon}$$

In HCS the common factors in $\mathbf{F}$ reflect the set of major phenotypic attributes measured in the assay. The loading matrix $\mathbf{L}$ relates the measured variables in $\mathbf{X}$ to $\mathbf{F}$. $\boldsymbol{\varepsilon}$ is a matrix of unique factors and is comprised of the reliable effects and the random error that is specific to a given variable. Rooted in this model is the concept that the total variance of $\mathbf{X}$ is partitioned into common and specific components. Therefore, after fitting the factor model and performing the rotations, we estimate the common attribute $\mathbf{F}$ on each of the k factors for each observation (i.e., cell) using a regression equation derived from the factor model (Figure 9.8) This is accomplished using the score procedure in SAS [28]. The complete factor structure and underlying phenotypic traits are outlined in Figure 9.8(d). As we can see in this case, the top six common factors have significant value and each contains a specific interpretable attribute.

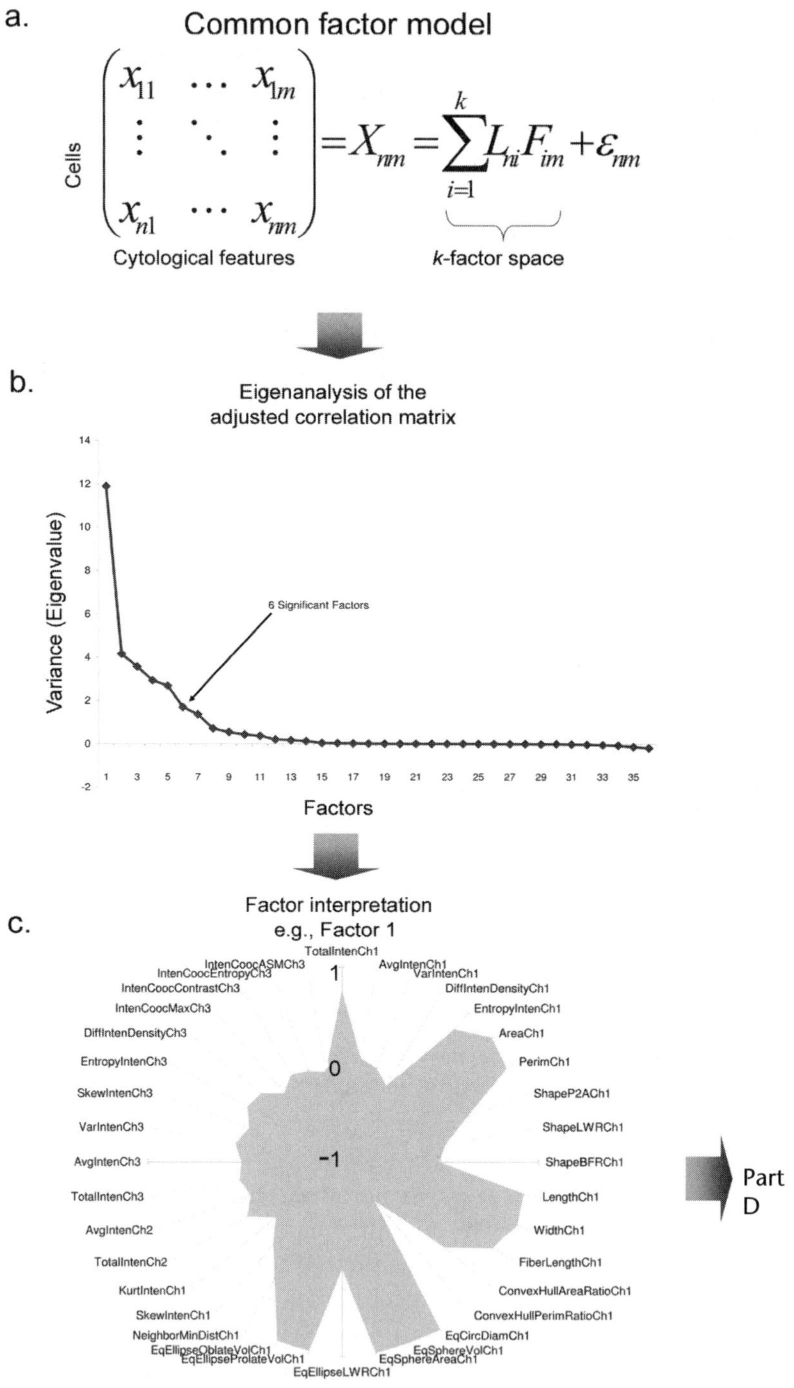

$$\begin{pmatrix} x_{11} & \cdots & x_{1m} \\ \vdots & \ddots & \vdots \\ x_{n1} & \cdots & x_{nm} \end{pmatrix} = X_{nm} = \sum_{i=1}^{k} L_{ni} F_{im} + \varepsilon_{nm}$$

Figure 9.8 Data driven factor model of cell cycle. (a) HCS data are contained in an $n \times m$ matrix, X consisting of a set of n image-based cytological features measured on m cells. The common factor model maps the n–cytological features to a reduced k–dimensional space described by a set of factors, F, that reflect the major underlying phenotypic attributes measured in the assay. (b) The dimensionality of the factor space is determined by an eigen-analysis of the correlation matrix of the data matrix, X. We determine that there are six significant factors. (c) The loading matrix for factor 1, as an example. (d) The complete factor structure of the cell cycle assay is shown in this schematic. Each of the six factors is drawn with lines connected to the cytological features with which they are most significantly correlated. Our interpretation of the phenotypic attributes characterized by each factor is shown on the right.

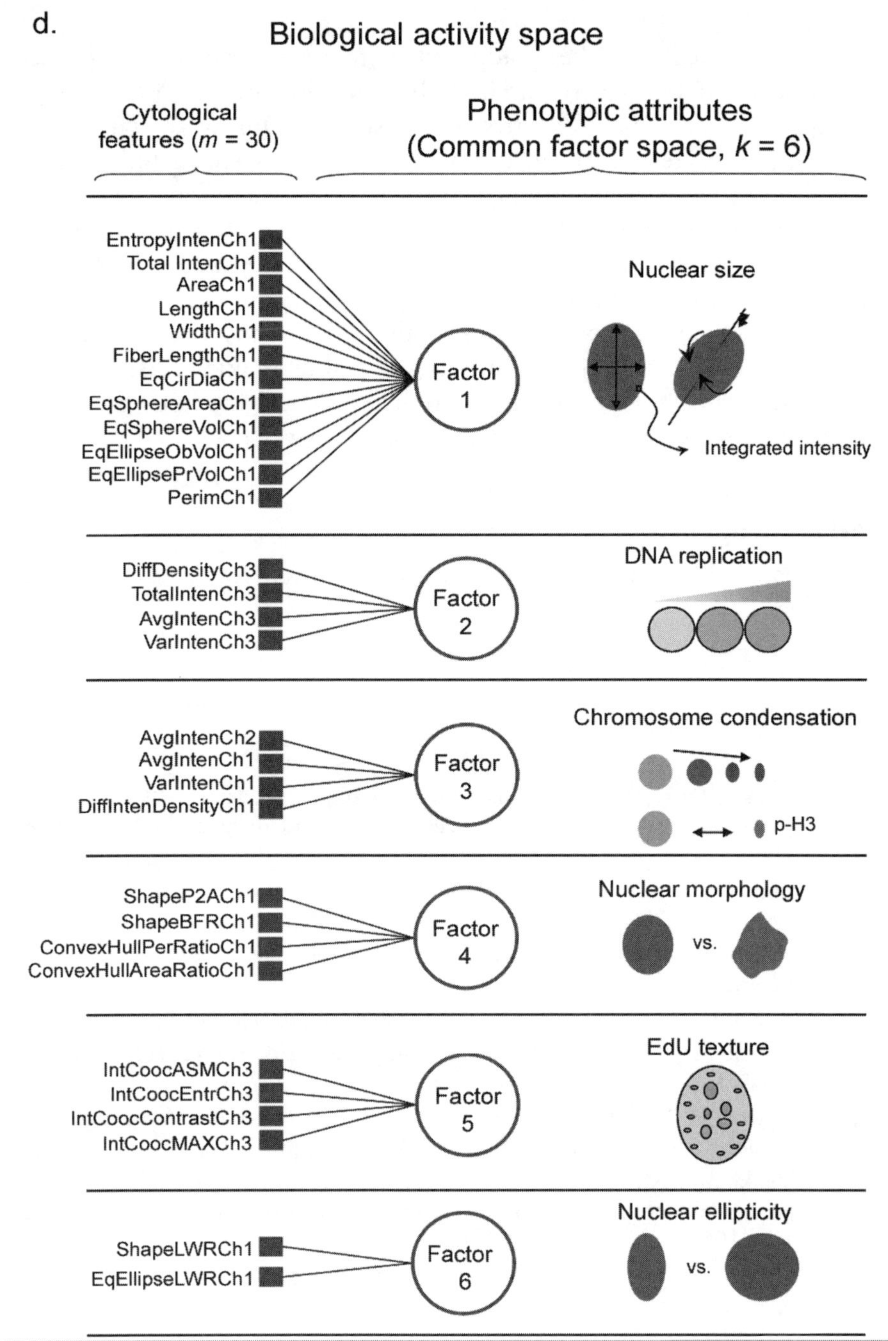

Figure 9.8 *(continued)*

9.9 Conclusion and Future Perspectives

HCS has increasingly been adopted in many aspects of drug discovery as well as system cell biology research. It has the potential to make major impacts on compound mechanistic and toxicology studies. During the past 10 years of development, many technical hurdles have been surpassed, including automated image acquisition, basic forms of image analysis, and data storage and archiving. Here we provide a practical basic framework on large scale HCS data analysis. We envision that more advanced statistical methods will be adopted in the near future to better mine the rich information contained in the HCS dataset. There are also significant efforts spent on improving image analysis, especially for morphology based HCS, and analysis of time lapsed datasets, as well as making basic image analysis tools more accessible for academic research. Statistical methods to mine and compare these rich datasets will become increasingly necessary.

Acknowledgments

We thank Jonathan Hoyt and Elizabeth McWhinnie for providing technical support.

References

[1] Taylor, D. L., J. R. Haskins, and K. A. Giuliano, (Eds.), "High Content Screening: A powerful approach to systems cell biology and drug discovery," *Methods Mol. Biol.*, 2007.

[2] Inglese, J., (Ed.), "Measuring biological responses with automated microscopy," *Methods Enzymol.*, 2006.

[3] Haney, S. A., (Ed.), *High Content Screening: Science, Techniques, and Applications*, Wiley Interscience Series, 2008.

[4] General Accounting Office report, "New drug development," 2006.

[5] Butcher, E. C., "Can cell systems biology rescue drug discovery?" *Nat. Rev. Drug Disc.*, 2005, Vol. 4, No. 6, pp. 461–467.

[6] Taylor, D. L., "Past, present, and future of high content screening and the field of cellomics," *Methods Mol. Biol.*, 2007, Vol. 356, pp. 3–18.

[7] Venkatesh, N., et al., "Chemical genetics to identify NFAT inhibitors: potential of targeting calcium mobilization in immunosuppression," *Proc. Natl. Acad. Sci. USA*, 2004, Vol. 101, No. 24, pp. 8969–74.

[8] Oakley, R. H., et al., "The cellular distribution of fluorescently labeled arrestins provides a robust, sensitive, and universal assay for screening G protein-coupled receptors," *Assay Drug Dev. Technol.*, 2002, Vol. 1, pp. 21–30.

[9] Wilson, C. J., et al., "Identification of a small molecule that induces mitotic arrest using a simplified high-content screening assay and data analysis method," *J. Screen.*, 2006, Vol. 11, No. 1, pp. 21–28.

[10] Liu, D., et al., "Screening of immunophilin ligands by quantitative analysis of neurofilament expression and neurite outgrowth in cultured neurons and cells," *J. Neurosci. Methods.*, 2007, Vol. 163, No. 2, pp. 310–320.

[11] Haney, S. A., et al., "High-content screening moves to the front of the line," *Drug Discov. Today*, 2006, Vol. 11, No. 19-20, pp. 889–894.

[12] Mitchison, T. J., "Small-molecule screening and profiling by using automated microscopy," *Chembiochem.*, 2005, Vol. 6, No. 1, pp. 33–39.

[13] Young, D. W., et al., "Integrating high-content screening and ligand-target prediction to identify mechanism of action," *Nat. Chem. Biol.*, 2008, Vol. 4, No. 1, pp. 59–68.

[14] Ding, G. J., et al., "Characterization and quantitation of NF-kappaB nuclear translocation induced by interleukin-1 and tumor necrosis factor-alpha. Development and use of a high capacity fluorescence cytometric system," *J. Biol. Chem.*, 1998, Vol. 273, No. 44, pp. 28897–28905.

[15] Knauer, S. K., "Translocation biosensors to study signal-specific nucleo-cytoplasmic transport, protease activity and protein-protein interactions," *Traffic*, 2005, Vol. 6, No. 7, pp. 594–606.

[16] Giuliano, K. A., "Fluorescent proteins in drug development," *Expert Rev. Mol. Diagn.*, 2007, Vol. 7, No. 1, pp. 9–10.

[17] Moffat, J., "A lentiviral RNAi library for human and mouse genes applied to an arrayed viral high-content screen," *Cell*, 2006, Vol. 124, No. 6, pp. 1283–1298.

[18] Srinivasan, R., et al., "Automated axon tracking of 3D confocal laser scanning microscopy images using guided probabilistic region merging," *Neuroinformatics*, 2007, Vol. 5, No. 3, pp. 189–203.

[19] Wagner, B. K., "Small-molecule fluorophores to detect cell-state switching in the context of high-throughput screening," *J. Am. Chem. Soc.*, 2008, Vol. 130, No. 13, pp. 4208–4209.

[20] Carpenter, A. E., "CellProfiler: image analysis software for identifying and quantifying cell phenotypes," *Genome Biol.*, 2006, Vol. 7, No. 10, p. R100.

[21] Barak, L. S., et al., "A beta-arrestin/green fluorescent protein biosensor for detecting G protein-coupled receptor activation," *J. Biol. Chem.*, 1997, Vol. 272, No. 44, pp. 27497–27500.

[22] Li, F., "An automated feedback system with the hybrid model of scoring and classification for solving over-segmentation problems in RNAi high content screening," *J. Microsc.*, 2007, Vol. 226, No. 2, pp. 121–132.

[23] Baatz, M., "Object-oriented image analysis for high content screening: detailed quantification of cells and sub cellular structures with the Cellenger software," *Cytometry A*, 2006, Vol. 69, No. 7, pp. 652–658.

[24] Haralick, R. M., "Statistical and structural approaches to texture," *Proceedings of the IEEE*, 1979, Vol. 67, No. 5, pp. 786–804.

[25] Tukey, J. W., *Exploratory Data Analysis*, Reading, MA: Addison-Wesley, 1977.

[26] Tao, C. Y., J. Hoyt, and Y. Feng, "A support vector machine classifier for recognizing mitotic subphases using high-content screening data," *J. Biomol. Screen.*, 2007, Vol. 12, No. 4, pp. 490–496.

[27] Spearman, C., "General Intelligence, Objectively Determined and Measured," *American Journal of Psychology*, 1904, Vol. 15, pp. 201–293.

[28] Hatcher, L. *A Step-by-Step Approach to Using SAS for Factor Analysis and Structural Equation Modeling*, Cary, NC: SAS Institute, Inc., 1994.

CHAPTER 10

Information About Color and Orientation in the Primate Visual Cortex

Youping Xiao and Ehud Kaplan

10.1 Introduction

A major goal of neuroscience is to understand how the outside world is represented in the brain. In primates, visual information plays a major role in guiding behavior, and the largest part of the sensory cortex is devoted to the analysis of visual information [1]. The visual cortex has thus attracted much research, and a great deal is now known about the representation of several important aspects of the visual world, such as orientation or size, although less is known about the functional architecture of the representation of other aspects, such as color. The availability of imaging techniques, together with the fact that the thin cortical sheet is suitable for the visualization of spatial activity patterns, have made it possible in the past two decades to gain new insights into this important issue.

Psychophysical studies have suggested that the various attributes of visual stimuli such as color, form, or motion are represented by separate channels in our brain [2]. Consistent with this hypothesis, many studies suggest that the primate cortex comprises several areas that are specialized in representing one or a few particular attributes. For instance, patients with cerebral achromatopsia [3, 4], a deficit in color perception, have damage in cortical areas that seem to correspond to the color-preferring regions as uncovered by functional imaging studies ("V4" according to [5]; "V8" according to [6]). These patients seem to be relatively normal with respect to other visual functions, suggesting that the affected color-preferring regions, or the so-called "color centers," are parts of a subpathway specialized for color perception [7]. Studies in macaque monkeys have also found areas that are specialized for processing of color information [8].

However, it had been intensely debated whether or not these attributes are represented by different compartments within early visual areas. These areas, including V1 and V2, process the visual inputs from the eyes, and pass them along to the higher visual areas that are putatively specialized for different functions. Whether the representation of different attributes is spatially separated in early visual areas has to be determined before we have a full understanding of where and how the specialized channels for various attributes are constructed.

An early influential study [9] has suggested that information about color is represented in the primary visual cortex, V1, by cells in specialized regions that are rich in the mitochondrial enzyme cytochrome oxidase (CO). These regions are referred to as the CO "blobs." Livingstone and Hubel [9] further suggested that orientation information is represented by cells in the interblob regions. However, this suggestion has been challenged by studies that found many V1 cells that were selective for both color and orientation [10, 11], although some studies did find that these two types of selectivity tended to be negatively correlated across neuronal population in V1 [11–13].

Area V2 also consists of CO-rich regions, arranged in thin and thick stripes, which are complemented by CO-poor ones (interstripes). Based on the anatomical connectivity between V1 and V2, and some electrophysiological studies, it has been proposed that color and orientation are represented by cells inside and outside the thin stripes, respectively [9]. As was the case with the studies on V1, various studies have found different degrees of segregation (or no segregation) between color and orientation selectivity among V2 neuronal populations (for a review, see [14]).

All of the electrophysiological studies have examined the type of information that was encoded by the response amplitude (or firing rate) of individual neurons. However, it is well known that the response amplitude exhibits large trial-to-trial variability [15–18], and the trial-to-trial fluctuations are correlated across large areas of cortex [19]. Therefore, a code that is based on amplitude is bound to be unreliable, even after averaging across neighboring neurons.

Studies that used optical imaging techniques have shown that the spatially distributed response pattern in some regions of V1/V2 is determined by one or more stimulus attributes, suggesting that the *spatial pattern* of responses encodes information about these attributes. Because the trial-to-trial fluctuations in response amplitude are correlated across large areas [19], the spatial pattern of responses in a small region could be relatively constant in each trial, and thus could represent stimulus attributes more reliably than could be achieved by the response amplitude of each neuron or cortical column. This chapter addresses the question of whether information about stimulus color and orientation is represented by response patterns in distinct subregions of V1/V2.

Previous imaging studies that related spatial pattern of responses to attribute values characterized each response pattern with only a few parameters, such as the peak location of a response patch [20]. Due to the large variety of the response patterns associated with various stimuli, it is impossible to model these patterns with any mathematical function that has only a few parameters. Therefore, a parameter-based approach cannot fully estimate the information encoded by the response pattern in a given region. To overcome this problem, we have estimated the amount of information about a particular attribute by determining the success rate with which the attribute can be decoded from the response pattern by a *support vector machine* (SVM), a recently developed statistical classifier [21].

10.1.1 Monitoring Activity in Neuronal Populations: Optical Imaging and Other Methods

All brain functions involve many interacting neurons, and it is therefore essential to study simultaneously the activity of many neurons, rather than record from one neuron at a time, as is often done. Imaging techniques can provide information about the spatial and temporal patterns of neuronal activity in the brain.

This information is fundamentally different from the kind of information provided by the standard electrophysiological techniques, which can report the detailed activity of single neurons with high temporal precision, but can tell us only very little about the *spatial distribution* of neural activity across neuronal populations. Those electrophysiological techniques that do report population activity, such as the electroencephalogram (EEG), event related potentials (ERP), the electroretinogram (ERG), or visually evoked potentials (VEPs) have a high temporal resolution, but very poor and unreliable spatial localization and resolution. Note, however, that these elecrophysiological techniques monitor the electrical activity of neurons *directly*. In contrast, most of the imaging techniques that are currently available (except for VSD-based imaging; see below) rely on monitoring, directly or indirectly, some aspect of the metabolic activity of neurons, which reflects *indirectly* the electrical signals that flow through the neurons.

Two-deoxy-glucose imaging (2DG) uses a radioactive glucose analog as a marker of activation of neurons [22]. Positron emission tomography (PET) uses radioactive, biologically active tracers that are consumed by active neurons and synapses [23]. Functional magnetic resonance imaging (fMRI) monitors the blood oxygenation and blood volume in activated brain regions. PET and fMRI can provide three-dimensional visualization of spatio-temporal activation patterns in the brain, while 2DG is limited to spatial maps only, since the animal must be sacrificed in order for its brain to be processed. PET and fMRI suffer from a relatively poor resolution: the smallest detectable voxel is measured in millimeters, and the briefest event is measured in seconds. 2DG has a better spatial resolution, but can provide no temporal information. Finally, PET and fMRI (but not 2DG) can be used noninvasively in a behaving human subject. Optical imaging methods that are used for functional mapping in the brain include two major variants: *intrinsic* optical imaging, and imaging that relies on *voltage sensitive dyes* (VSDs), which change their transmission or fluorescence in proportion to the voltage across neuronal membranes. Intrinsic optical imaging [24, 25] requires no external dye, and uses the intensity of light reflected from the exposed brain as a mapping signal. Like fMRI, it relies on signals that are derived primarily from blood oxygenation and blood flow (with a small contribution from light scattering), but because it uses light, and because the observed tissue is exposed, its spatial resolution is far superior to that of fMRI, and such imaging can achieve a resolution of a few tens of microns [26]. In the temporal domain, however, the resolution of intrinsic optical imaging is similar to that of fMRI, and is thus measured in seconds. Optical imaging that uses VSDs, however, has a far superior temporal resolution, similar to that of electrophysiological recordings (< 1 ms), with spatial resolution that is similar to that of intrinsic imaging. Both of these optical methods suffer from a very low signal to noise (S/N) ratio, on the order of 10^{-3} to 10^{-4}. This low S/N ratio requires the collection of entormous amounts of data over long imaging periods. The data must be averaged to improve the detection of faint activated signals embedded in the noisy images.

The noise in all imaging approaches comes from several sources: the physiological processes of breathing and heart beat, digitization noise, dark current in imaging devices, and tissue movement. This kind of noise, which is usually rather large, cannot be completely eliminated or avoided at the acquisition stage.

Therefore, all imaging methods must rely on sophisticated methods of statistical analysis (see, for example, [27]) or computational classification, which sometimes requires the use of massive, parallel computation, as illustrated by the work we describe here. Typically, the noise also makes responses to single presentation of a

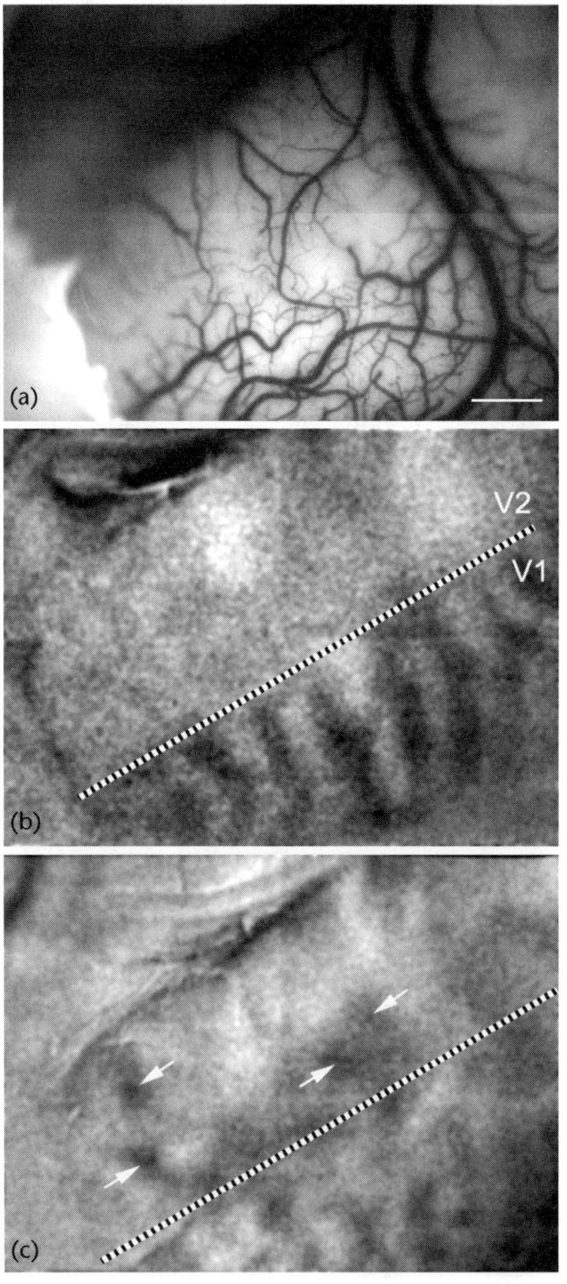

Figure 10.1 (a) A surface image taken under 560-nm light. (b) The differential image associated with left-eye stimulation versus right-eye stimulation. Note the ocular dominance columns, the end of which marks the V1/V2 border. (c) The differential image associated with color stimulation versus luminance stimulation. The dark regions indicated by arrows represent V2 regions that preferred isoluminant red/green gratings to black/white gratings. Scale bar, 1 mm.

stimulus highly unreliable in terms of determining what stimulus was presented to the investigated brain. This is true for both imaging and electrophysiology. Below we show that, with our imaging data, the application of the SVM classifier in a focused "searchlight" mode makes it possible to identify the stimulus that elicited a single response with high accuracy.

10.2 Methods and Results

A detailed description of experimental methods and the SVM algorithm can be found elsewhere [21, 26]. In each experiment, we first identified the V1/V2 border by imaging the ocular dominance (OD) columns [Figure 10.1(b)]. OD columns are present only in V1, so their termination marks the border between these two visual areas of the cortex. The OD columns were visualized by subtracting the responses elicited by stimulating the right eye from those driven by stimulation of the left eye. We then identified the CO thin stripes by imaging the V2 areas that preferred color gratings to black/white ones [regions indicated by arrows in Figure 10.1(c)] [20, 28].

To determine how the spatially distributed activity across the cortical surface represents information about stimulus color and orientation, 16 different

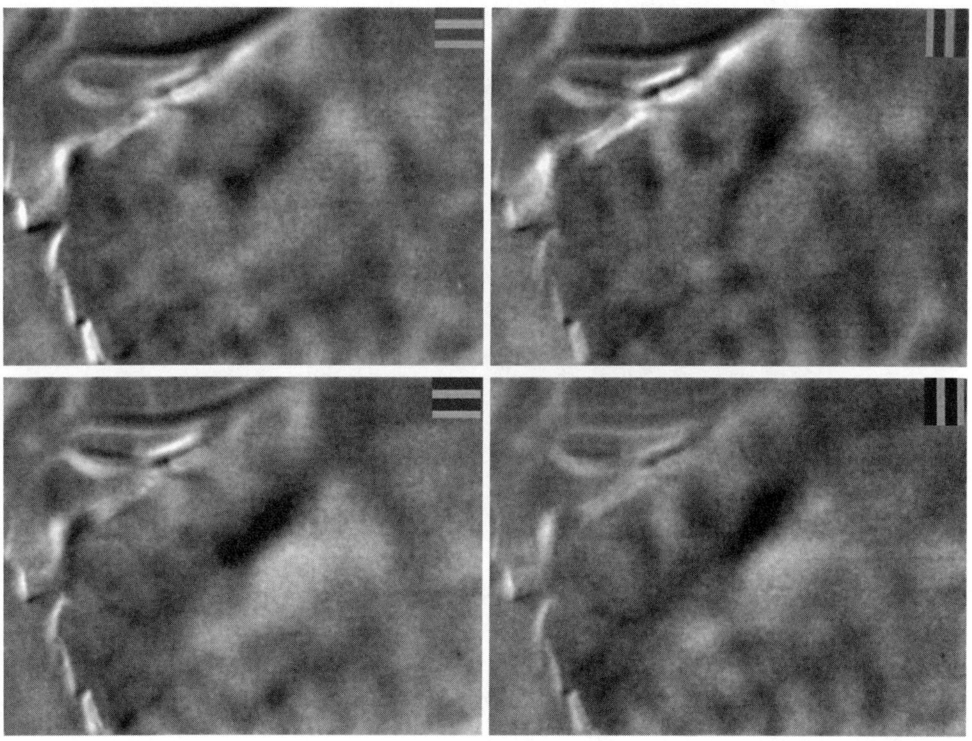

Figure 10.2 Mean responses to four different gratings. Each mean response was calculated by averaging the 32 single-trial responses to the given grating.

color/gray gratings were used as stimuli, each having one of four orientations (0, 45, 90, 135 degrees) and one of four colors (red, blue, green, yellow). To ensure that response differences were due to color and orientation only, all gratings were isoluminant at 10 cd/m^2. Each grating was presented 32 times, and was thus associated with 32 single-trial response images. Figure 10.2 shows the averaged response images that were associated with four different stimuli. This figure demonstrates that each stimulus gave rise to a unique pattern of responses across the imaged area.

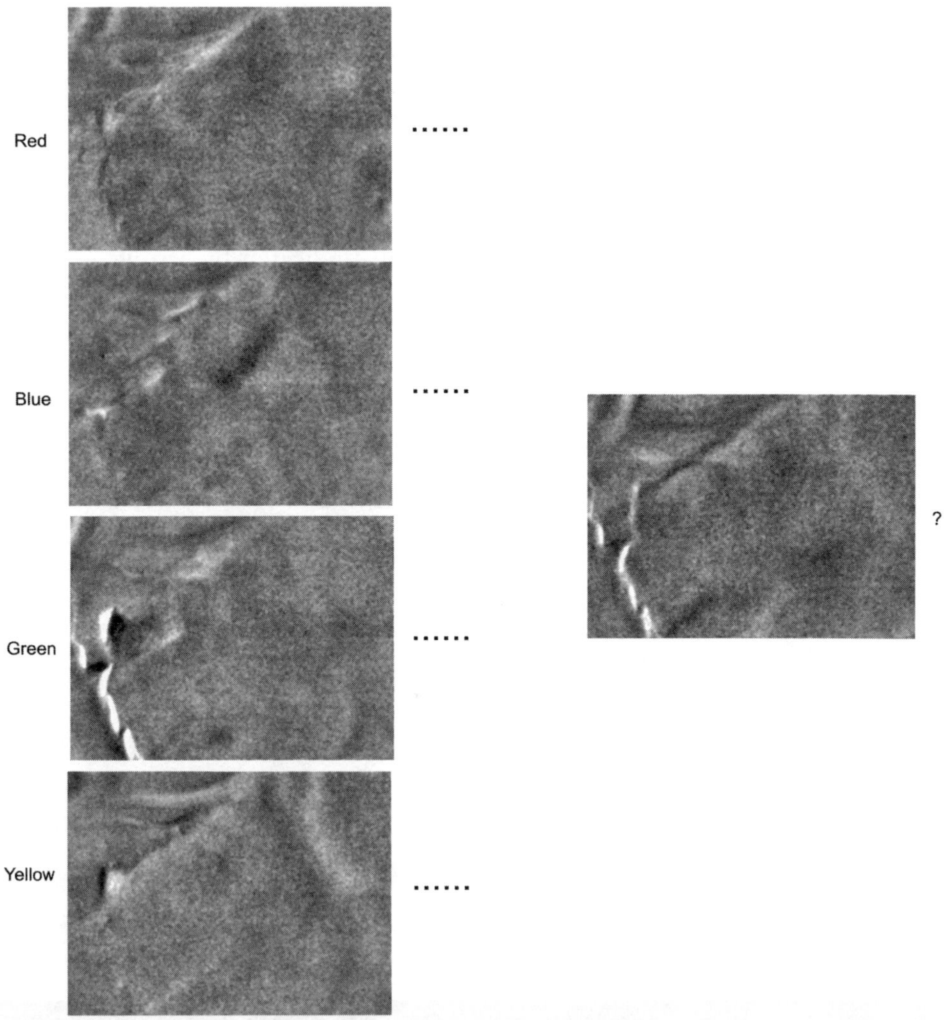

Figure 10.3 Decoding the stimulus color from a single-trial response. All single-trial responses except those from the last imaging block were grouped according to stimulus color, regardless of their orientations. These four groups of responses were used to train a set of SVM classifiers, which were then used to decode the stimulus color of each single-trial image from the last block. This procedure was repeated 32 times, each time using a different block of data, to test the SVM classifiers after they have been trained by the remaining data. The decoded stimulus color was compared with the real one for each testing response, and the percentage of correct decoding across the entire dataset represents a quantitative measure of the average amount of information about color that was encoded in a single-trial response.

Because of the large noise that is inherent in optical imaging (see above), the spatial patterns of the responses are less clear in the single-trial images than in the averaged images (Figure 10.3). To test whether the response pattern in each single-trial image contains information about stimulus color, we combined all the

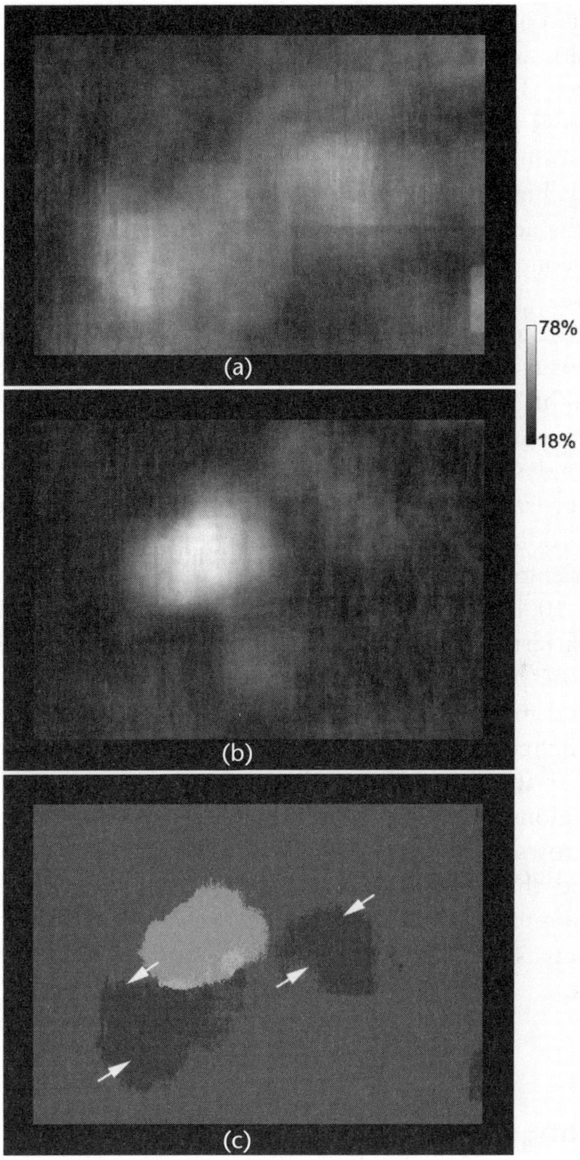

Figure 10.4 Distribution of the information about (a) stimulus color and (b) orientation. Each pixel value represents the percentage of correct decoding when the responses inside a small region (0.97 × 0.97 mm) around the given pixel were analyzed. The pixels with a value of >= 50% in either or both of these two maps were painted with pseudo-colors (black for color, light gray for orientation, and white for both), as shown in (c). The two pairs of arrows point to the same locations as those in (c), and thus mark the inferred locations of two CO thin stripes. (c) demonstrates that the information about stimulus color was largely segregated from that about stimulus orientation in areaV2, although the segregation was not complete. The former seems to be centered at the thin stripes, while the latter was centered at a region between them.

single-trial images associated with a given color into a single class, regardless of stimulus orientation. We used data from 31 of the 32 imaging blocks to train an SVM classifier and test the SVM with data from the remaining block. Each imaging block consisted of one presentation of each stimulus. In the training phase, a linear SVM [29] was derived from the collection of training pairs of single-trial image and its associated color. In the testing phase, the SVM classifier inferred the stimulus color that was associated with each test single-trial image, and the inferred color was compared with the actual one. This training-testing procedure was repeated 32 times, with each repetition using a different imaging block for testing. Across the total of 512 tests, the SVM gave 430 (84%) correct inferences about the stimulus color. So the accuracy of decoding stimulus color was 84%, significantly higher than what it would be by chance (1 in 4 colors, or 25%). Similarly, the accuracy for decoding orientation was 74%. Therefore, each single-trial image contained a significant amount of information about stimulus color and orientation.

To map the spatial distribution of information about color and orientation across the visual cortex, a "search light " approach was used [30]. In this approach, a "search light" of 494 $\times$ 494 microns (81 $\times$ 81 pixels) scanned through the entire imaged area, shifting its center location by one pixel at each stop. At each location, the SVM analysis described above was applied to the highlighted region, and the resulting accuracy was assigned to the pixel at the given center location. This approach produced two accuracy maps, one for stimulus color, and the other for stimulus orientation [Figures 10.4(a, b), respectively].

Figure 10.4(a, b) shows that information about color and orientation had different patterns of spatial distribution. To compare these patterns directly, we selected those pixels that had a value of >50% in accuracy map of color or orientation, and marked them with different colors in Figure 10.4(c) (black for color decoding, light gray for orientation decoding, and white for both). This figure suggests that the information about color and orientation was concentrated in different regions that were largely segregated from each other.

The arrows in Figure 10.4(c) point to the same locations as those in Figure 10.1(c), and thus mark the locations of two CO stripes. It is evident that the regions containing large amount of color information were centered in the thin stripes, whereas the region with orientation information were located between the thin stripes.

10.3 Discussion

Our results suggest that information about stimulus color and orientation is coded by the spatial pattern of responses in different CO stripes of V2. Previous imaging studies also suggested that the thin CO stripes and the interstripes are involved in coding color and orientation information, respectively. Cortical columns that were selective for stimulus orientation were found in the interstripes and thick stripes; whereas columns that responded preferentially to color stimuli were found in the thin stripes [20, 28]. Furthermore, the thin stripes were found to contain orderly organized color maps, in which stimulus color was coded by the location of response peaks [20]. Compared to the previous imaging studies of this topic, our current one has the following advantages.

First, in the previous studies, the coding of color and orientation was studied with *different* stimuli. The orientation maps were imaged with black/white gratings, while the color maps were obtained with color gratings. The results of these studies could have been biased by the difference in stimulus preference in different stripes. In contrast, the current study used a single set of gratings that varied in both color and orientation, and determined the distribution of information about color or orientation of the same stimuli. This paradigm is similar to our everyday experience, in which we have to extract information about several attributes of a single stimulus.

Second, previous studies have either determined the selectivity of each pixel independently, or characterized each response patch with a single parameter, such as peak location. Therefore, those studies have failed to uncover the information that was coded by *correlated* activity across many pixels. Such information can be effectively extracted by the linear SVM that was used in the current study, as long as the correlation between pixels is largely linear. The high accuracy of color or orientation decoding by the SVM at some parts of V2 suggests that a significant amount of information was coded by the linear correlation between neighboring pixels, and such information was distributed unevenly in V2. However, we cannot exclude the possibility that the regions with low accuracy rate might contain information that was coded by *non*linear interactions among pixels. To address this issue, nonlinear SVMs need to be used in future studies.

In addition, previous studies have determined which cortical locations were, on average, selective for orientation or color, and those regions were therefore implicated as the locations that the brain uses to determine those aspects of the stimulus. Our approach has measured directly how accurately a classifier could determine the color or orientation of the presented stimulus from a single response.

The SVM "searchlight" calculation presents extreme computational demands, which can be met realistically only with very powerful computers, which employ many processors working in parallel. In our work, we have used the IBM Blue-Gene parallel computer, with up to 4,096 nodes. Performing the calculation on a standard workstation would have taken approximately 18 months. On Blue Gene it took only a few hours.

One disadvantage of the "searchlight" method is that a "searchlight" behaves like a kernel of a low pass filter. Any sharp change in the amount of information across the cortical surface would be blurred by the current approach. Therefore, the current study might have underestimated the extent by which the coding of different attributes was spatially segregated in V2.

Despite this disadvantage, the information provided by the analysis we present here can be very valuable to the neuroscientist. By delineating the brain regions that process color and orientation information, respectively, investigators can now make more focused and productive measurements. For instance, multielectrode arrays can be targeted to areas that process one visual attribute or another, and this direct measurement of electrical neuronal activity could lead to new insights about the nature of the coordinated computation that the brain performs. Needless to say, analysis of the results of such studies will also require massive computation by parallel, multiprocessor machines.

Acknowledgments

This work had been supported by NIH grants EY16371, EY16224, NIGMS-GM71558, and EY12867 Core Grant.

References

[1] Hubel, D. H., "Eye, Brain and Vision," *Scientific American Library*, 1988.

[2] Magnussen, S., "Low-level memory processes in vision," *Trends Neurosci.*, Vol. 23, 2000, pp. 247–251.

[3] Meadows, J. C., "Disturbed perception of colours associated with localized cerebral lesions," *Brain*, Vol. 97, 1974, pp. 615–632.

[4] Damasio, A., et al., (1980) "Central achromatopsia: behavioral, anatomic, and physiologic aspects," *Neurology*, Vol. 30, 1980, pp. 1064–1071.

[5] McKeefry, D. J., and S. Zeki, "The position and topography of the human color center as revealed by functional magnetic resonance imaging," *Brain*, Vol. 120, 1997, pp. 2229–2242.

[6] Hadjikhani, N., A. K. Liu, et al., "Retinotopy and color sensitivity in human visual cortical area V8," *Nature Neuroscience*, Vol. 1, 1998, pp. 235–241.

[7] Zeki, S., (1990) "A century of cerebral achromatopsia," *Brain*, Vol. 113, 1990, pp. 1721–1777.

[8] Conway, B. R., and D. Y. Tsao, "Specialized color modules in macaque extrastriate cortex," *Brain*, Vol. 113, 1990, pp. 1721–573.

[9] Livingstone, M. S., and D. H. Hubel, "Anatomy and physiology of a color system in the primate visual cortex," *Journal of Neuroscience*, Vol. 4, 1984, pp. 309–356.

[10] Lennie, P., J. Krauskopf, and G. Sclar, "Chromatic mechanisms in striate cortex of macaque," *J. Neurosci.*, Vol. 10, 1990, pp. 649–669.

[11] Leventhal, A. G., et al., "Concomitant sensitivity to orientation, direction, and color of cells in layers 2, 3, and 4 of monkey striate cortex," *Journal of Neuroscience*, Vol. 15, 1995, pp. 1808–1818.

[12] Zeki, S., "Colour coding in the cerebral cortex: the reaction of cells in monkey visual cortex to wavelengths and colours," *Neuroscience*, Vol. 9, 1983, pp. 741–765.

[13] Creutzfeldt, O. D., et al., "Neuronal representation of spectral and spatial stimulus aspects in foveal and parafoveal area 17 of the awake monkey," *Experimental Brain Research*, Vol. 68, 1987, pp. 541–564.

[14] Shipp, S., and S. Zeki, "The functional organization of area V2, I: specialization across stripes and layers," *Vis. Neurosci.*, Vol. 19, 2002, pp. 187–210.

[15] Schiller, P. H., B. L. Finlay, and S. F. Volman, "Short-term response variability of monkey striate neurons," *Brain Res.*, Vol. 105, 1976, pp. 347–349.

[16] Dean, A. F., "The variability of discharge of simple cells in the cat striate cortex," *Exp. Brain Res.*, Vol. 44, 1981, pp. 437–440.

[17] Snowden, R. J., S. Treue, and R. A. Andersen, "The response of neurons in areas V1 and MT of the alert rhesus monkey to moving random dot patterns," *Exp. Brain Res.*, Vol. 88, 1992, pp. 389–400.

[18] Croner L. J., K. Purpura, and E. Kaplan, Response variability in retinal ganglion cells of primates. *Proc. Natl. Acad. Sci. USA*, Vol. 90, 1993, pp. 8128–8130.

[19] Arieli, A., et al., "Dynamics of ongoing activity: explanation of the large variability in evoked cortical responses," *Science*, Vol. 273, 1996, pp. 1868–1871.

[20] Xiao, Y., Y. Wang, and D. J. Felleman, "A spatially organized representation of colour in macaque cortical area V2," *Nature*, Vol. 421, 2003, pp. 535–539.

[21] Vapnik, V., *Statistical Learning Theory*, New York: John Wiley & Sons, Inc., 1998.

[22] Kennedy, C., et al., "Metabolic mapping of the primary visual system of the monkey by means of the autoradiographic [^{14}C]deoxygluycose technique," *Proc. Natl. Acad. Sci. USA*, Vol. 73, 1976, pp. 4230–4234.

[23] Ter-Pogossian, M. M., M. E. Phelps, and E. J. Hoffman, "A positron-emission transaxial tomograph for nuclear imaging (PETT)," *Radiology*, Vol. 114, No. 1, 1975, pp. 89–98.

[24] Blasdel, G. G., and G. Salama, "Voltage-sensitive dyes reveal a modular organization in monkey striate cortex," *Nature*, Vol. 321, 1986, pp. 579–585.

[25] Grinvald, A., et al., "Functional architecture of cortex revealed by optical imaging of intrinsic signals," *Nature*, Vol. 324, 1986, pp. 361–364.

[26] Xiao, Y., et al., "Hue maps in primate striate cortex," *Neuroimage*, Vol. 35, 2007, pp. 771–786.

[27] Sirovich, L., and E. Kaplan, "Analysis Methods for Optical Imaging," in *In Vivo Optical Imaging of Brain Function*, R. D. Frostig (Ed.), CRC Press LLC, 2002, pp. 43–76.

[28] Roe, A. W., and D. Y. Ts'o, "Visual topography in primate V2: multiple representation across functional stripes," *Journal of Neuroscience*, Vol. 15, 1995, pp. 3689–3715.

[29] Cristianini, N., and J. Shawe-Taylor, *An Introduction to Support Vector Machines and Other Kernel-Based Learning Methods*, Cambridge University Press, 2000.

[30] Haynes, J. D., et al., "Reading hidden intentions in the human brain," *Curr. Biol.*, Vol. 17, 2007, pp. 323–328.

CHAPTER 11

High-Throughput Analysis of Microdissected Tissue Samples

J. Rodriguez-Canales, J. C. Hanson, M. A. Tangrea, S. Mukherjee, H. S. Erickson, P. S. Albert, Z. K. Majumdar, R. F. Bonner, T. J. Pohida, M. R. Emmert-Buck, and R. F. Chuaqui

11.1 Introduction

Molecular studies of human diseases can be performed using various experimental models,including human cells in vitro (cell lines), laboratory animals, and human tissue (ex vivo) specimens. Each approach is valuable, but each has advantages and disadvantages [1]. For example, cell lines are generally straightforward to grow in the laboratory and a large amount of high quality biomolecules can be obtained for molecular analysis. However, cell lines do not necessarily maintain all the molecular features of the tissues from which they were derived and may differ from corresponding cells in tissues, with a wide divergence of expressed genes [2, 3]. As an example, in one study proteomic profiles showed less than 20% identity between cell lines and matched tissues [4].

The direct application of molecular techniques to tissue samples provides a unique opportunity for the study of diseases. Molecular changes identified in clinical samples in particular may most closely reflect the biological alterations as they occur in humans in vivo than research models and thus be of particular value [3]. However, molecular analysis of clinical tissue samples involves a variety of challenges. Tissue sample acquisition and handling is a complex process in its own right and oftentimes only a limited number of cases adequate for a particular study are available [5–7]. If bulk tissue samples are analyzed, then the quantity of biomolecules is generally not limiting. Conversely, if a microdissection-based approach is employed such that specific, phenotype-defined cell populations are studied, then the small DNA, mRNA, or protein amounts that are recovered can be problematic as the yield of biomolecules from dissected material is typically low, in the nanogram (ng) range [6, 8–10]. For nucleic acids, DNA or RNA amplification methods can be applied to facilitate analysis. The low yield, however, is a challenge for proteomic studies given the lack of available amplification approaches and thus only highly sensitive protein analysis methods can be applied to microdissected samples. Finally, the quality of the biomolecules from tissue specimens may not be optimal, in particular when dealing with archival material

Disclaimer: Several of the authors of this chapter are inventors on allowed NIH patents covering the technologies that are described, and receive royalty payments through the NIH Technology Transfer program.

that has been fixed and embedded. Both experimental design and subsequent data analysis must take into account this possibility [11].

In order to analyze pure cell populations, tissue microdissection techniques have become critical to the field of molecular pathology. In the present chapter we review the molecular approaches that have been applied to dissected cells in order to generate high-throughput molecular profiling data. Typically, a multidisciplinary approach is required for these efforts, as knowledge of tissue handling and processing, histopathology, imaging and signal quantitation, physics and engineering for the development of microscopy and dissection instruments, and bioinformatics need to be integrated together in order to efficiently perform the profiling experiments and to analyze the large datasets that are produced.

11.2 Microdissection Techniques and Molecular Analysis of Tissues

11.2.1 General Considerations

Tissue specimens from patients offer the possibility of analyzing both the morphological and molecular basis of diseases, which can then be matched with clinical records. An extraordinary example of the value of integrating high throughput molecular techniques with histopathological analysis of samples is the sequencing of the 1918 Spanish Influenza Virus genome. Archival material from autopsy cases almost 90 years old allowed for the characterization and reconstruction of this pandemic virus and its subsequent study in the laboratory. This remarkable feat is leading to identification of new prognostic indicators for the illness and potentially to the development of novel antiviral therapies [12, 13]. Similar molecular pathology-based investigations of disease processes such as cancer are also revealing novel etiological insights and generating new targets for clinical intervention [14–16].

11.2.2 Fixation—A Major Consideration When Working with Tissue Samples

Formalin fixation followed by paraffin embedding (FFPE) is the usual practice for processing clinical samples and is the gold standard for histopathological diagnosis. However, this method limits the feasibility of molecular analysis to some extent [7, 11, 17, 18]. The main drawback of FFPE is the damage to biomolecules that ensues, particularly to RNA and proteins, due to the cross-linking of biomolecules induced by the fixation process [7, 17, 18]. DNA and to some extent RNA can still be retrieved from the samples and analyzed, although RNA extraction from FFPE samples and subsequent analysis is challenging [7, 17, 18]. Alternative fixatives, including ethanol-based ones, are under development and assessment as a potential means to preserve biomolecules as well as morphological detail for diagnosis [19, 20].

Snap-freezing of tissues is the optimal procedure for molecular analysis in terms of the quality and quantity of biomolecules that can be recovered; however, these samples are not as readily available for study as most pathology laboratories have only recently started collecting snap-frozen tissue in biorepositories [21, 22]. Moreover, the histological detail in these specimens is inferior to that in FFPE samples, and the resource needs are increased when using frozen tissue with respect to freezer space and cryostat sectioning.

11.2.3 Why Is Microdissection Important When Using Tissue Samples?

Tissue samples have an intricate three-dimensional structure composed of heterogeneous cell populations. Epithelial cells, fibroblasts, blood and neural cells are all typically present in a given tissue. In addition, clinical samples often include both normal and pathological cell types, including a range and diversity of disease phenotypes. Since tissues are heterogeneous, "contaminating" cells can interfere with the study of a specific cell population of interest such as cancer cells or components of the tumor microenvironment. In a recent study of gene expression of metastatic breast carcinoma comparing microdissected samples versus whole tissue samples, Harrell et al. identified more than 1,000 genetic changes between the primary tumor and the metastasis using laser capture microdissection (LCM). Less than 1% of these changes were identifiable in a nonmicrodissected analysis of the same samples [23].

Furthermore, the recent focus on specific cellular components of the tumor microenvironment demonstrates the importance of microdissection for molecular studies [24–26]. As an example, in a recent study on GSTP1 methylation in prostate carcinoma, the authors were able to identify an epigenetically unique tumor-associated stromal microenvironment by using LCM [5]. Similarly, Richardson et al. used LCM to identify gene expression differences between the tumor-associated stroma and the normal (nontumor related) stroma that could not otherwise be discerned, again highlighting the unique discoveries that can be made using a microdissection-based approach [6].

11.2.4 Tissue Microdissection Techniques

11.2.4.1 Manual Dissection Under Microscopic Vision

Manual microdissection is performed on a stained histology slide using a fine needle or razor blade to procure the tissue under low microscopic magnification. This technique is effective when an investigator needs to retrieve cells from a relatively large region of a tissue section. However, the method is operator-dependent and tedious, and does not offer the precision of today's laser dissection instruments.

11.2.4.2 Infrared Laser (IR)–Based Systems

In the 1990s, investigators at the National Institutes of Health (NIH) developed a laser-based system for isolating pure cell populations from a tissue section under direct microscopic visualization [27, 28] (Figure 11.1). The technique, called laser capture microdissection (LCM) uses an infrared laser beam to activate a thermoplastic film that can bind the targeted cells and isolate them from the tissue [Figure 11.2(a)]. The low-energy infrared laser does not damage the biomolecules, leaving them amenable to subsequent molecular analysis. LCM and related laser dissection techniques have significantly advanced the biomedical research field by allowing for integration of molecular biology with anatomic pathology, in particular with highly defined and purified cell types that exhibit a specific phenotype [29]. The LCM system is operator-dependent and does require some training in histopathology, since the desired cells need to be first identified by morphology under the microscope in order to target them with the laser beam.

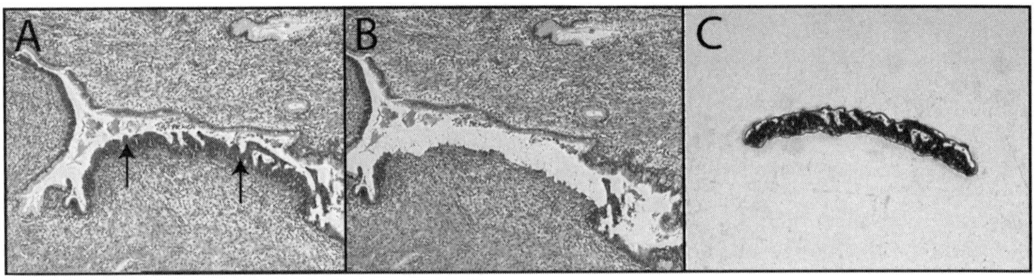

Figure 11.1 LCM of an intraepithelial lesion of the uterine endocervix (endocervical glandular dysplasia). Dysplastic epithelium (arrows) was specifically procured from a tissue section. (a) Tissue before dissection. (b) Tissue after dissection. (c) LCM cap image showing the dissected cells. 10×, Hematoxylin and Eosin stain.

11.2.4.3 Ultraviolet (UV) Laser-Based Systems

Ultraviolet-based dissections systems have also been developed, including the laser microdissection system and pressure catapulting technology (P.A.L.M. MicroLaser

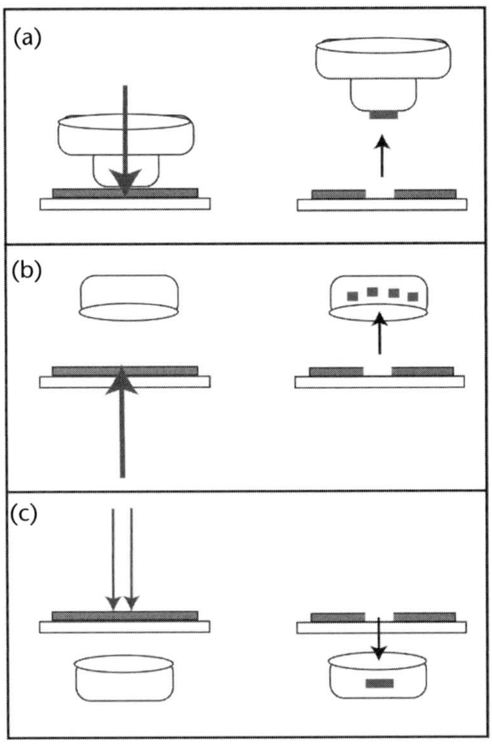

Figure 11.2 Schematic image showing laser-based microdissection methods, including LCM, PALM, and the LMD cutting system. (a) A transparent LCM cap containing a transfer film is placed on top of the tissue under microscopic visualization. The laser beam is directed through the cap to all sites in the field containing desired cells, activating the film. The cap is then lifted, procuring all targeted cells. (b) A UV laser is used to generates photonic pressure and catapult cells into a microcentrifuge tube after UV cutting. (c) Desired cell areas are cut by a UV laser and fall by gravity into a collection tube.

Technologies GmbH) based on microdissection with a UV laser that generates photonic pressure and catapults the dissected cells into a microcentrifuge tube cap [30, 31]. This system provides a noncontact method, differing from LCM in this regard [Figure 11.2(b)]. PALM is partially automated but is also operator-dependent since the researcher needs to identify and microscopically select the cells for dissection. A similar cutting laser microdissection is the Leica LMD (Leica Microsystems), which differs from the PALM in that the dissected cells fall by gravity into a tube instead of being catapulted by pressure [32] [Figure 11.2(c)]. Finally, another UV laser–based system is the MMI Cellcut system (Molecular Machines & Industries) [33] whose main advantage is avoidance of contamination via minimizing environmental contact during dissection. It has been reported that UV-induced biomolecule damage can occur in cells lying directly under the laser cutting path and the damage can be significant if the cut region is large in relation to the microdissected area [34].

11.2.4.4 Combined Laser Capture-Cutting Systems (IR and UV Lasers)

Molecular Devices, Inc., offers two new semi-automated instruments that combine infrared and UV lasers, Veritas and Arcturus XT. Veritas is a closed system with improved imaging software. More recently, the Arcturus XT was developed, adding additional contrast phase illumination and fluorescence capabilities along with the dual laser approach. Overall, the enhanced optics, dissection flexibility, and improved software all facilitate more rapid and precise dissections.

11.2.4.5 Ultrasonic-Based Microdissection

This is a recently developed approach in which microdissection is performed using ultrasonic vibration as a cutting system [35]. A piezoelectric actuator generates a high frequency and low amplitude vibration (approximately 16 to 50 kHz, and 0 to 3 μm) that cuts the tissue [35]. Preliminary data shows that the method is reliable and the biomolecules in the tissue are not damaged during the dissection process [35].

11.2.4.6 High Throughput, Operator-Independent Microdissection (Expression Laser Microdissection, xMD)

One of the major challenges in the microdissecion field is that all of the laser-based techniques, although they have contributed greatly to the field of molecular pathology, are operator-dependent and thus somewhat laborious to utilize. The rate-limiting step in most dissection studies is the laser operator who must carefully select the cells for procurement one by one, a process that can often require days or weeks of effort to procure enough cells for analysis. Therefore, new microdissection systems are needed, especially those that increase the speed of dissection and thus decrease operator time. More efficient yields of dissected cells are particularly relevant for proteomics since there is no tool available for protein amplification.

In 2004 expression microdissection (xMD) was developed by investigators at the NIH. The method is a high throughput process for procuring specific cell populations from tissues [36]. Although still in development, xMD potentially offers significant advantages over LCM and its variations, such as a significant increase in dissection speed and precision. Cells are selected based on molecular targeting, thus the method is operator-independent and the investigator does not need to individually identify cells for dissection. At present, standard antibody-based immunohistochemistry (IHC) is used to stain the target cells based on expression of a specific protein. A clear film is then placed over the tissue section (as opposed to the dye-containing film required for LCM) and the applied laser energy is then absorbed selectively by the stained cells, bonding them to the film (Figure 11.3).

In principle, any stain can be used for xMD targeting, including colorimetric in situ hybridization for gene expression, or even intrinsic tissue pigments like melanin. At a typical dissection scan rate of 50,000 cells/second, a large number of

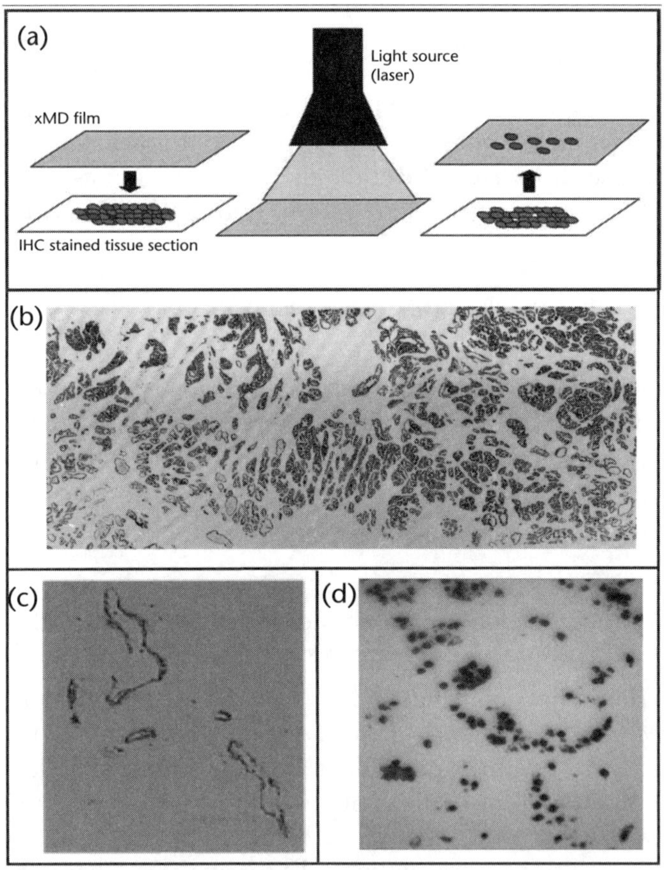

Figure 11.3 Expression microdissection (xMD). (a) Schematic representation of the xMD process. After staining the desired cell population with a specific antibody (immunohistochemistry), a clear film is placed over the tissue section and the entire tissue is irradiated by a laser. Only the stained cells absorb enough laser energy to generate heat and activate the film. (b) Large dissections of prostate epithelial cells. (c, d) Dissection of fine targets using xMD: (c) endothelial cells (*Source:* [38]); and (d) nuclei.

specific cells from histological regions can be rapidly procured [Figure 11.3(b)]. In addition, serial sections from the same tissue block can be generated and dissected thereby producing high yields of biomolecules from dissected cells.

Using xMD, fine targets can also be stained and procured, such as small individual cells (lymphocytes, endothelial cells, and so forth) or even subcellular structures like nuclei [37, 38] [Figure 11.3(c, d)]. Also, selective dissections of cells based on molecular status rather than morphological structure can facilitate unique dissection-based analysis. For example, cells in different phases of the cell cycle, or expressing specifically mutated proteins such as p53 positive can be selectively retrieved and analyzed.

11.3 DNA Analysis of Microdissected Samples

11.3.1 General Considerations

DNA was the first biomolecule to be analyzed via LCM as it is more stable than RNA or proteins and therefore easier to handle in the laboratory. Spectrophotometry, Pico-Green assay, and real-time PCR with control genes are used to assess the amount and quality of DNA recovered. The average size of fragments is typically assessed using lab on-a-chip techniques such as the Bioanalyzer system (Agilent Technologies). In general one can expect a DNA yield from microdissected tissues of approximately 6 picograms (pg) per cell and successful PCR can be performed with as few as 100 cells [34]. However, for consistent PCR results, especially in quantitative assays or the more complex global genomic amplification reactions, it is advisable to work with higher DNA inputs, on the order of at least a few nanograms per reaction. In a standard 3,000 LCM shot dissection of frozen tissue, one is able to procure at least 50 ng of DNA, providing ample amounts of nucleic acid for repeated and/or multiple reactions per dissection.

Microdissected samples have been analyzed by quantitative PCR [10], direct sequencing for mutation detection [39, 40], and microsatellite marker assessment for identification of genomic deletions [27, 41–45]. Whole genome amplification techniques have also been applied to increase the amount of DNA for subsequent global genomic assays such as single nucleotide polymorphism (SNP) chips [46–50].

11.3.2 Loss of Heterozygosity (LOH)

One of the first applications of LCM-based DNA samples was analysis of genomic deletions in cancers (loss of heterozygosity or LOH) using amplification of polymorphic markers [27]. The process of carcinogenesis is thought to occur as a step-wise accumulation of genetic alterations, including activation of oncogenes and inactivation of tumor suppressor genes (TSGs) [41]. LOH corresponds to a DNA deletion that occurs in tumors and is a common mechanism of inactivation of TSGs [43]. Identification of deletions at a genetic locus at a level of more than 20% suggests the presence of a TSG [43].

A major difficulty in the identification and characterization of deletions is the need to analyze pure tumor cell populations, since contaminating nontumor cells will mask their detection. Microdissection of pure tumor cells via LCM, coupled with the ability to amplify polymorphic microsatellite markers, overcomes this problem and facilitates identification of TSGs. Cloning of the gene responsible for Multiple Endocrine Neoplasia Type I (MEN1) is one example of the use of microdissection in a successful gene hunt [44].

11.3.3 Global Genomic Amplification

DNA analysis of dissected cells is generally performed as a single microsatellite reaction, even though up to eight markers can be amplified using a multiplex PCR approach [45].

However, more complex reactions that involve DNA amplification of the entire genome are also important for high throughput DNA analysis and several amplification schemes have been developed along these lines. Methods include degenerated oligonucleotide primed PCR (DOP-PCR) [46], the use of short primers (random hexamers) [47], and the restriction of DNA with subsequent ligation of adapters that allow for downstream amplification [48]. Thousands of single nucleotide polymorphic loci (SNP) can be analyzed on a chip or via comparative genomic hybridization (CGH) after global amplification, even from LCM samples [49, 50].

11.3.4 Epigenetic Analysis

Epigenetics refers to genomic changes that affect gene expression without involving modification of the primary DNA sequence. One of the most important types of epigenetic alteration is DNA methylation. In human DNA cytosine-guanosine rich sites (CpG islands) are concentrated near transcription start sites in the promoter region of genes and can be chemically modified by the addition of a methyl group to carbon number 5 of cytosine. Hypermethylation of CpG islands in gene promoter interferes with the attachment of the cell's transcription machinery, thus reducing the expression of the corresponding gene. It is thought that methylation of gene promoters plays a particularly key role in the inactivation of specific genes in tumors [51].

The methylation status of human DNA can be determined by a chemical treatment method The exposure of unmethylated DNA to sodium bisulfite leads to the conversion of cytosines to uracils; however, methylated cytosines are protected from this sequence modification. By applying PCR with sequence specific primers after bisulfite treatment, it is possible to identify the presence or absence of methylated CpG islands in gene promoters (methylation specific PCR, MSP) [52].

More recently, Pyrosequencing has been developed as an approach to quantify the methylation status of individual CpGs [39] and the technique has been successfully applied to LCM samples [5].

11.3.5 Mitochondrial DNA Analysis

Mitochondria contain DNA (mitochondrial DNA, mtDNA) that encode for proteins important in cellular energy production. Mitochondrial DNA has several characteristics that differ from nuclear DNA; it is relatively small (containing approximately 16.5 kilobases), circular, and several copies are present per cell (200 to 2,000 mitochondria are contained in each cell; 5 to 10 copies of mtDNA are present in each mitochondria). Recently, mtDNA mutations have been shown to occur frequently in cancers, representing a potentially useful tumor marker [39, 40]. Since multiple copies of mtDNA are contained in each cell, sufficient yields of mtDNA for amplification can be obtained even with minute microdissections. Therefore, LCM samples are particularly applicable to mtDNA mutation analysis [53]. Furthermore, if additional template is needed, the entire mtDNA genome can be amplified in a single step reaction using long distance polymerases [54].

11.4 mRNA Analysis of Microdissected Samples

11.4.1 General Considerations

Analysis of RNA from microdissected tissues presents numerous challenges, with bias being cited as the main "threat to validity" [55–58]. Therefore, appropriate experimental design must be used in order to generate reliable data. In the case of mRNA studies, candidate genes are typically discovered using array-based techniques such as expression microarrays, and subsequently validated by quantitative RT-PCR (qRT-PCR).

Microdissected tissues yield low quantities of mRNA. Using epithelial cells as an example, one cell contains approximately 6 pg of RNA, thereby producing approximately 160 ng of RNA from a 10,000 cell dissection [6, 9, 59]. When the starting material is so low, one must utilize nucleic acid amplification prior to microarray analysis. Gene expression profiling techniques include differential display [60], cDNA Library construction [61], serial analysis of gene expression [62, 63], and expression microarrays [61, 63, 64].

11.4.2 Expression Microarrays

Microarrays were originally developed for differential mRNA analysis [60, 65, 66] and are now a prevalent high throughput expression method. The method is a robust analysis tool, straightforward to use, and able to perform parallel and quantitative analysis of thousands of genes from many samples [67]. To date, the most commonly used format is the GeneChip (Affymetrix, Inc., Santa Clara, California) [65].

Expression arrays have been used to generate expression profiles of diverse cancer types and phenotypic and histologic subtypes [6, 8, 59, 68–77]. In addition to mRNAs, microarrays are now being used to analyze miRNAs and RNAis [78–83].

11.4.3 Quantitative RT-PCR

Quantitative RT-PCR (qRT-PCR) can measure small quantities of mRNA from microdissected tissue samples and is useful for validation of expression array data. For example, qRT-PCR has been used regularly in the validation of prostate cancer biomarkers [6, 74, 84, 85]. There are some drawbacks, however, as qRT-PCR probes are expensive, frozen tissue specimens are required, and the number of samples and genes that can be analyzed by qRT-PCR is limited, being 100-fold to 1,000-fold less than gene expression microarrays.

Two novel approaches have been developed in an attempt to increase the number of samples and number of transcripts that can be analyzed by qRT-PCR. The first is the OpenArray (BioTrove, Inc., Woburn, Massachusetts) that combines a microtiter plate and a fluidics format to produce a 24-fold higher analytical throughput with a 64-fold smaller reaction volume. The OpenArray technique has particularly good application to microdissected tissue analysis due to its minimal starting template volumes [86]. For example, a typical microdissection of 10,000 cells yields enough RNA to analyze, in triplicate, nine genes of interest and one endogenous control housekeeping gene. The improvements to qRT-PCR made by the OpenArray allows for the analysis of ten times more transcripts from microdissected tissues.

The second novel quantitative gene expression technique is the QuantiGene Reagent System (Panomics, Inc., Fremont, California) [57]. It does not use PCR, but applies a branched DNA detection approach [57]. QuantiGene allows for the use of FFPE tissue samples as starting template because cross-linking does not affect the assay and the RNA degradation only reduces the detection signal by two to three-fold [87]. Also, no data normalization using an endogenous control needs to be conducted because RNA is measured directly using a limit of detection call method [88]. These improvements make the QuantiGene technique applicable to microdissected tissue RNA analysis for both frozen and archival samples.

11.5 Protein Analysis of Microdissected Samples

11.5.1 General Considerations

Proteomic studies differ significantly from nucleic acid–based studies. Typically, snap-frozen tissue is utilized to preserve protein integrity and larger amounts need to be collected in order to have enough material for study. In an effort to overcome this barrier, more sensitive protein assays are currently being developed. Alternatively, replicate dissections from serial tissue recuts can be utilized to obtain enough material for analysis, especially in the case where the target is present in low abundance. As an example, a typical western blot requires $\sim$60,000 cells to obtain the 15 μg of protein needed for analysis. This is in contrast to the 10,000 cells needed for DNA or mRNA analysis where amplification methods can be applied.

There are several methods of protein analysis that have been established in the last 30 years; western blot, two-dimensional polyacrylamide gel electrophoresis (2D-PAGE) protein arrays, and mass spectrometry. Each of the techniques can be used to study microdissected samples.

11.5.2 Western Blot

Western blots are based on electro-transfer of proteins from a polyacrylamide gel to a membrane. A heterogeneous protein lysate, from a tissue source or cell line, is first separated by size via sodium dodecyl sulfate–polyacrylamide gel electrophoresis

(SDS-PAGE). The proteins separate based on size because SDS imposes a negative charge to them. The separated proteins from the gel are then electro-transferred to the membrane which can be probed for a specific protein of interest using a targeting antibody.

Western blotting offers several benefits, including the ability to evaluate low abundance proteins (due to the initial separation of the complex protein source by size), to determine the molecular size of the protein/antigen of interest, and the quantification of expression [89]. The main disadvantage is that only one protein is evaluated at a time. However, more recently, there has been a movement towards multiplex western blots by incorporating multiple, fluorescently labeled antibodies [90] or quantum dots [91] to detect several different antigens on the same blot [91]. Another disadvantage of the technique is the requirement for significant sample input, usually in the range of 10 to 20 μg. This protein yield is challenging to obtain from dissected samples. Therefore, a high throughput microdissection approach such as xMD may become particularly useful for western blot analysis, especially if moderate or lower abundant proteins need to be studied from dissected cells.

11.5.3 Two-Dimensional Polyacrylamide Gel Electrophoresis (2D-PAGE)

Originally described in 1975 by O'Farrell, 2D-PAGE remains a prominent technique in the field of proteomics [92]. In this method, proteins are first separated according to their isoelectric point or charge (first dimension), and then by their molecular weight (second dimension) creating a two-dimensional "fingerprint" of the protein content of a sample. Proteins can subsequently be visualized by a variety of staining methods, including Coumassie Blue or silver staining, although there are also staining alternatives that are more amenable to downstream mass spectrometry analysis.

One limitation to the use of LCM with 2D-PAGE is the problem of sensitivity, as the detection of less abundant proteins by 2D-PAGE is difficult [93]. A drawback to 2D-PAGE generally (for both dissected and nondissected samples) is the inability to detect insoluble proteins, as well as those with extremes of charge or size (less than 10 kiloDaltons or greater than 200 kDa) [94]. This class includes a large group of biologically important membrane proteins that require strong detergents for solubilization. Strong detergents cause protein unfolding and a uniformity of charge that prevents the isoelectric focusing step from effectively separating the proteins [94].

Despite these shortcomings, 2D-PAGE studies can produce important proteomic information. Such "fingerprints" can then be further probed with analyzed by western blotting or mass spectrometry to identify particular proteins of interest or characterize post-translational modifications [94]. Thus, 2D-PAGE remains a vital analysis tool for the field of proteomics.

A relatively new twist on the standard method of 2D-PAGE evaluates two samples simultaneously by differential fluorescence labeling. This technique has been termed 2D-DIGE or two-dimensional difference gel electrophoresis [95]. The two samples, labeled with distinct fluorophores, are distinguished on the same gel, enabling direct proteomic comparison and eliminating the problem of intergel variability [96].

11.5.4 Mass Spectrometry

Mass spectrometry (MS) is a sensitive albeit complex technique that enables analysis of samples either on the individual protein/peptide level or on the scale of the entire proteome. While many instrument types and techniques for MS exist [97], the central concept is to identify proteins and peptides based upon their mass-to-charge ratio. Typically, a sample is ionized from either the solid or liquid state into charged particles. This ionization can be accomplished in one of two ways, either from the solid phase by embedding a sample in an energy absorbing matrix for laser desorption (matrix-associated laser desorption/ionization, MALDI) [98] or from the liquid phase through a tiny, charged capillary tube (electrospray ionization, ESI) [99, 100]. These charged particles (ions) then travel through the air into a mass spectrometer. The ions are measured by an electrical detector and the mass-to-charge ratios of the corresponding proteins determined based on time of flight.

In comparison to 2D-PAGE, mass spectrometry is more sensitive (on the order of amino acids up to 10^5 Daltons/charge) [97] and is particularly useful for proteins of larger size (via enzymatic digestion) or strong charge (via ionization) [101] that are not amenable to a gel format. For the field of proteomics, mass spectrometry has been used in two main applications: to identify and describe single protein structure [102], and to characterize proteome-level peptide fingerprints of various tissue types [103].

Microdissected samples and prepurification separation techniques can be incorporated to further focus the mass spectrometry analysis [104]. For example, Li et al. used LCM, isotope-coded affinity tags (ICAT), and two-dimensional liquid chromatography followed by MS/MS to identify and quantify 261 differentially expressed proteins in hepatocellular carcinoma [104]. ICAT works by the same principle as DIGE in gel electrophoresis in that each sample is labeled with a different unique marker to facilitate differentiation of proteins amongst two samples. In the case of ICAT, heavy or light isotopes are used instead of the fluorescent markers of DIGE [105].

11.5.5 Protein Arrays

A recently developed proteomics technique is the protein microarray. These chips enable high throughput sample evaluation of a specific set of proteins [106]. Protein microarrays have been designed in two main formats, known as forward and reverse. Forward phase or antibody arrays spot known antibodies onto a matrix, typically a nitrocellulose-coated slide, in a predetermined array pattern [107]. A protein lysate is then applied to the antibody array and bound analytes are detected either directly or via a secondary antibody [108]. Reverse phase protein arrays operate by spotting protein lysates onto the matrix surface [106]. Arrays are then probed with known antibodies. In this method, sample lysates are spotted as serial dilutions to allow for protein quantification when compared to a set of internal standards. Both forward- and reverse-phase methods have been used to assess protein-protein interactions, post-translational modifications, and small molecule binding assessment [109].

Protein arrays can be used in the analysis of tissue-derived specimens. For example, Grubb et al. successfully used reverse phase arrays to study signaling differences in cell populations isolated from prostate cancer specimens [110]. This was one of the first studies completed with the reverse phase array technology and demonstrates the feasibility of high throughput array analysis of multiple patient samples using LCM-quality material. However, because of the complexity of a multiplexed reaction, initial results from these array platforms needs to be independently validated in follow-up studies.

An alternative protein array approach for the analysis of microdissected samples is the surface enhanced laser desorption/ionization (SELDI) mass spectrometer system to generate protein profiles [111]. SELDI is based on the use of ProteinChip arrays containing 8 to 24 spots onto which tissue lysates are applied. This system can measure molecules less than 20 kDa in size, and its sensitivity makes it ideally suited for analyzing microdissected samples [112].

11.6 Statistical Analysis of Microdissected Samples

11.6.1 General Considerations

In the following section qRT-PCR is used as an example of the challenges that are faced when applying statistical analysis to microdissected samples. These difficulties include: variability in sample selection, the variability in dissection, and variability inherent in the downstream molecular analysis protocols.

11.6.2 Quantification of Gene Expression

qRT-PCR is based on determining the cycle threshold (C_T) at which the RT-PCR product is detected in a sample. C_T values are usually normalized in order to make meaningful comparisons between groups of biological samples. From the C_T value of the gene of interest, for each sample the C_T value of a housekeeping gene (or average of a few genes) is subtracted. The resulting quantity is called a ΔC_T. Analysis is done by either: (1) comparing the ΔC_T in sample A (for example, tumor cells) or B (normal tissue) across groups or conditions, or (2) comparing changes in ΔC_T between tumor and adjacent normal tissue across groups or conditions. For example, one may be interested in analyzing the effect of two therapeutic agents on gene expression in prostate cancer. Specifically, this would involve comparing the mean normalized C_T values (ΔC_T) in tissue across the two treatment groups, indicating whether a particular gene expression level is increased in tumors treated with one agent versus the other. Alternatively, interest may be on comparing the difference in normalized C_T values between tumor and adjacent normal tissue ($\Delta\Delta C_T$) across the two treatment groups. This addresses whether there are differential changes in gene expression in the tumor relative to in adjacent normal tissue across the two treatment arms.

It is important to decide on the type of normalization strategy when planning a study, especially with respect to the number of housekeeping genes that are necessary [9]. Unfortunately, there is no general agreement as to which single gene is a stable control for all cell types [113], and averaging the expression level of two housekeeping genes is a better normalization strategy than using just a single gene [9].

11.6.3 Sources of Variation When Studying Microdissected Material

The sources of variation in the qRT-PCR process include variation in the dissection process and technical variation in the downstream molecular analysis. Variation due to microdissection is an important factor to consider since only a few thousand cells out of the entire samples are analyzed. Our experience, however, is that dissection variance is relatively small if the process is carefully performed. In general, for established protocols such as qPCR or qRT-PCR, the variation of the downstream analysis itself (technical variation) is negligible. However, as a quality control measure, three qPCR replicates are usually performed on each sample and the average of the replicates is used for analysis.

11.6.4 Comparisons of Gene Expression Between Two Groups

Comparisons of the ΔC_T or $\Delta\Delta C_T$ between groups or conditions can usually be done with a two-sample t-test, whether the researcher is dealing with microdissected material or not. This test assumes that the outcome variables are normally distributed with the same variance in the two groups. Although the t-test is not sensitive to departures from these assumptions, it is wise for investigators to check whether these assumptions hold. Alternative statistical tests should be considered if there are marked departures from normality or there is evidence for different variances in the two groups. For example, the nonparametric Wilcoxion rank sum test does not assume that the data in the two groups follow a normal distribution. Further, the Welch t-test does not assume equal variance in the two groups.

In many analyses, the expression levels for many genes will be compared across groups. Thus, it is not obvious what P-value is required to be confident that a particular gene expression comparison is statistically significant. In most studies, whether they are based on LCM samples or nonmicrodissected material, a P-value smaller than 0.05 is used as a benchmark for statistical significance. However, when there are a large number of genes being compared, such a criterion will produce a large number of false-positive results. For example, if an investigator is comparing gene expression values of 100 genes across two groups, they should expect five genes to be classified as "significantly" different across the two groups just by chance alone. This is an inherent problem with performing multiple statistical tests in the same analysis. Various approaches have been proposed to control for these false positive results. One strategy, called a Bonferroni correction, is to adjust the significance level by dividing by the total number of comparisons made. For example, if one is comparing 50 genes across groups, a comparison is considered statistically significant if the P-value is less than 0.05/50=0.001. Such a comparison guarantees that the chance of generating a false positive result is below 5%. Although such an approach adequately controls the overall false-positive rate, in many situations, it may have very little power (i.e., have a low probability of detecting a meaningful difference). This may be a particular problem when the number of genes being examined is large (say above 10 genes). An alternative approach is to control the type of false discovery rate (FDR) rather than the probability of any false positive result. Specifically, this approach controls the proportion of false positive results at a particular probability. For example, controlling the FDR at 10% would allow for 10% of declared differentially expressed genes to be falsely identified.

11.6.5 Microarray Analysis

One of the major areas of focus among the bioinformatics community is the analysis of microarray (chip) data, where thousands of data points are generated in a single experiment either for gene mRNA (oligonucleotide expression array as an example) or DNA deletion studies (SNP chips as an example) [114]. Although a thorough discussion of this topic is beyond the scope of this chapter, it is important to raise some important issues. In general, there are two types of analytic problems: (1) class comparison and (2) class prediction.

Class comparison involves comparing the high dimensional gene expression profiles across groups or conditions (for example, high versus low grade tumors, responsive versus resistant diseases, normal versus tumor cells, stroma versus epithelium cells). One approach is to compare genes one by one in order to identify overexpressed mRNAs. Here, the problems of multiple comparisons that were discussed above for qRT-PCR are magnified since thousands of genes are compared across groups rather than only dozens of genes. Therefore, it becomes especially essential to control the probability of a false positive result using multiple comparisons procedures such as controlling the false discovery rate (FDR). One approach that is commonly used is to do all testing at a 0.001 significance level (i.e., a gene is differentially expressed when the p-value <0.001 using a t-test). A gene list is then constructed based on the statistically significant differences between genes, and a false discovery rate can be computed. An overall test of whether the patterns in gene expression are different between groups can be constructed using a permutation test. Specifically, we can scramble the class labels (group identifiers) and redo the analysis many times (say, 5,000). The p-value for a test of whether the gene expression profiles are different across groups can be computed as the proportion of times the number of significant genes is above the number of "significant" genes in the scrambled datasets. Visually, these gene expression patterns can be compared across groups by multidimensional scaling (MDS). MDS compresses differences between sample expression profiles into three eigenvectors for plotting in three-dimensional space.

Class prediction involves developing a predictor of disease outcome from high-dimensional microarray data. There are many methods for developing a class predictor including discriminate analysis, logistic regression, neural network methodology, and classification trees (see [114] for a comparison of approaches). It is particularly important to emphasize that any predictive model needs to be validated on a completely independent dataset. Validating a predictive model on the same dataset for which the model was developed can result in over-fitting and an overoptimistic assessment of the quality of the predictive model. One approach that is commonly used is to split the dataset into a training set in which the predictive model is developed and a test-set in which the predictive model is validated. This can be done by splitting the data in two and fitting the predictive models in the first half of the data and evaluating the accuracy of the predictions in the second half.

Software is commonly available to do microarray analysis. For example, two publicly available software packages from the NIH are BRB-array tools (http://linus.nci.nih.gov/BRB-ArrayTools.html) and the National Cancer Institutes microarray analysis program, mAdb (http://nciarray.nci.nih.gov, NIH, Bethesda, Maryland).

11.7 Conclusions

One of the main hurdles in biomedicine is to integrate experimental approaches utilized in basic science laboratories with clinical information. Microdissection of patient tissue samples may be valuable in this regard since these specimens closely reflect molecular status in vivo, and dissection provides a means for procuring specific, purified populations of cells for study. As an example of this strategy, candidate clinical markers have been identified by direct comparison of normal and tumor cells. Pancreatic cancer is a neoplasia with a poor response to therapies and no available method for early diagnosis [115]. However, recently, HSP27 was identified as a potential marker of early disease in a study based on LCM [116]. In a different study, comparative genomic hybridization (CGH) performed on microdissected archival specimens identified chromosome 8q gain as a potential prognostic marker in resectable pancreatic adenocarcinoma [117]. And KIAA0101, a candidate derived from cDNA array analysis of microdissected samples is a promising novel target for the development of anti-cancer therapeutic drugs [118]. In prostate tissues, an LCM-based study using differential display identified POV1 (PB39), a gene over-expressed in aggressive tumors [61]. More recently, POV1 has been confirmed to play a role in the transition of premalignant lesions to cancer and is a potential target for early treatment [117]. All of these examples highlight the use of tissue microdissection to reveal insights into the basic biology and clinical behavior of disease processes. The list of new clinical markers with potential diagnostic, therapeutic, or prognostic implications is continuously growing as researchers are now routinely applying new high throughput molecular techniques to specific cell populations dissected from human tissues.

References

[1] Emmert-Buck, M.R., et al., "Molecular profiling of clinical tissue specimens. Feasibility and applications," *Am. J. Pathol.*, Vol. 156, 2000, pp. 1109–1115.

[2] Bright, R.K., et al., "Generation and genetic characterization of immortal human prostate epithelial cell lines derived from primary cancer specimens," *Cancer Res.*, Vol. 57, 1997, pp. 995–1002.

[3] http://cgapmf.nih.gov/ProstateExample/ProstateTissueProcessing/CellLines.html.

[4] Ornstein, D.K., et al., "Proteomic analysis of laser capture microdissected human prostate cancer and in vitro prostate cell lines," *Electrophoresis*, Vol. 21, 2000, pp. 2235–2242.

[5] Rodriguez-Canales, J., et al., "Identification of a Unique Epigenetic Sub-Microenvironment in Prostate Cancer," *J. Pathol.*, Vol. 211, 2007, pp. 410–419.

[6] Richardson, A., et al., "Global expression analysis of prostate cancer-associated stroma and epithelia," *Diagn. Mol. Pathol.*, Vol. 16, 2007, pp. 189–197.

[7] Gannot, G., and J.W. Gillespie, "Tissue Processing," Ch. 3, pp. 27–42, in Emmert-Buck, M.R., Gillespie, J.W., Chuaqui, R.F., (Eds.), *Dissecting the Molecular Anatomy of Tissue*, New York: Springer Verlag, 2005.

[8] Best, C.J.M., et al., "Molecular Alterations in Primary Prostate Cancer After Androgen Ablation Therapy," *Clin. Cancer Res.*, Vol. 11, 2005, pp. 6823–6834.

[9] Erickson, H.S., et al., "Assessment of Normalization Strategies for Quantitative RT-PCR Using Microdissected Tissue Samples," *Lab. Invest.*, Vol. 87, 2007, pp. 951–962.

[10] Hanson, J., et al., "Identification of Gene Promoter Methylation in Tumor-Associated Stromal Cells," *J. Natl. Cancer Inst.*, Vol. 98, 2006, pp. 255–261.

[11] Perlmutter, M.A., "Comparison of Snap Freezing versus Ethanol Fixation for Gene Expression Profiling of Tissue Specimens," *J Mol. Diagn.*, Vol. 6, 2004, 6, pp. 371–377.

[12] Tumpey T.M., et al., "Characterization of the reconstructed 1918 Spanish influenza pandemic virus," *Science*, Vol. 310, 2005, pp. 77–80.

[13] Kash J.C., et al, "Genomic analysis of increased host immune and cell death responses induced by 1918 influenza virus," *Nature*, Vol. 443, 2006, pp. 578–581.

[14] Pen, A., Moreno, M. J., Martin, J., and Stanimirovic, D. B., "Molecular markers of extracellular matrix remodeling in glioblastoma vessels: microarray study of laser-captured glioblastoma vessels," *Glia*, Vol. 55, 2007, pp. 559–572.

[15] Turashyli, G., et al., "Novel markers for differentiation of lobular and ductal invasive breast carcinomas by laser microdissection and microarray analysis," *B. M. C. Cancer*, Vol. 7, 2007, p. 55.

[16] Zeng, Z., et al., "Analysis of gene expression identifies candidate molecular markers in nasopharyngeal carcinoma using microdissection and cDNA microarray," *J. Cancer Res. Clin. Oncol.*, Vol. 133, 2007, pp. 71–81.

[17] Leiva, I.M., et al., "Handling of clinical tissue specimens for molecular profiling studies," *Curr. Issues Mol. Biol.*, Vol. 5, 2003, pp. 27–35.

[18] Gillespie, J.W., et al., "Evaluation of non-formalin tissue fixation for molecular profiling studies," *Am. J. Pathol.*, Vol. 160, 2002, pp. 449–457.

[19] Stanta, G., et al., "A novel fixative improves opportunities of nucleic acids and proteomic analysis in human archive's tissues," *Diagn. Mol. Pathol.*, Vol. 15, 2006, pp. 115–123.

[20] Olert, J., et al., "HOPE fixation: a novel fixing method and paraffin-embedding technique for human soft tissues," *Pathol. Res. Pract.*, Vol. 197, 2001, pp. 823–826.

[21] Naber, S.P., et al., "Role of the frozen tissue bank in molecular pathology," *Diag. Mol. Pathol.*, Vol. 1, 1992, pp. 73–79.

[22] Naber, S.P., et al., "Continuing role of a frozen-tissue bank in molecular pathology," *Diag. Mol. Pathol.*, Vol. 5, 1996, pp. 253–259.

[23] Harrell, J.C., et al., "Contaminating cells alter gene signatures in whole organ versus laser capture microdissected tumors: a comparison of experimental breast cancers and their lymph node metastases," *Clin. Exp. Metastasis*, Vol. 25, 2008, pp. 81–88.

[24] Buckanovic, R. J., et al., "Use of immuno-LCM to identify the in situ expression profile of cellular constituents of the tumor microenvironment," *Cancer Biol. Ther.*, Vol. 5, 2006, pp. 635–642.

[25] Silasi, D. A., et al., "Detection of cancer-related proteins in fresh-frozen ovarian cancer samples using laser capture microdissection," *Methods Mol. Biol.*, Vol. 414, 2008, pp. 35–45.

[26] Verma, M., Wright, G. L. Jr., Hanash, S. M., Gopal-Srivastava, R., and Srivastava, S., "Proteomic approaches within the NCI early detection research network for the discovery and identification of cancer biomarkers," *Ann. N. Y. Acad. Sci.*, Vol. 945, 2001, pp. 103–115.

[27] Emmert-Buck, M.R., et al., "Laser Capture Microdissection," *Science*, Vol. 274, 1996, pp. 998–1001.

[28] Bonner, R.F., et al., "Laser Capture Microdissection: Molecular Analysis of Tissue," *Science*, Vol. 278, 1997, pp. 1481–1482.

[29] Fend, F., et al., "Laser capture microdissection in pathology," *J. Clin. Pathol.*, Vol. 53, 2000, pp. 666–672.

[30] Schermelleh, L., et al., "Laser microdissection and laser pressure catapulting for the generation of chromosome-specific paint probes," *Biotechniques*, Vol. 27, 1999, pp. 362–367.

[31] Micke, P., et al., "Laser-assisted cell microdissection using the PALM system," *Methods Mol. Biol.*, Vol. 293, 2005, pp. 151–166.

[32] Kolble, K., "The LEICA microdissection system: design and applications," *J. Mol. Med.*, Vol. 78, 2000, pp. B24–B25.

[33] http://www.molecular-machines.com/products/lasermicrodissection.html.

[34] Espina, V., et al., "Laser-capture microdissection," *Nat. Protoc.*, Vol. 1, 2006, pp. 586–603.

[35] Sun, L., et al., "A novel ultrasonic micro-dissection technique for biomedicine," *Ultrasonic.*, Vol. 44, Suppl. 1, 2006, pp. e255–e260.

[36] Tangrea, M.A., et al., "Expression microdissection: operator-independent retrieval of cells for molecular profiling," *Diagn. Mol. Pathol.*, Vol. 13, 2004, pp. 207–212.

[37] Rodriguez-Canales, J., et al., "Expression microdissection: A new immuno-based dissection technology for cellular and nuclear procurement," *Lab. Invest.*, Vol. 88, 2008, p. 371A.

[38] Grover, A.C., et al., "Tumor-associated endothelial cells display GSTP1 and RARbeta2 promoter methylation in human prostate cancer," *J. Transl. Med.*, Vol. 4, 2006, article number 13.

[39] Parr, R. L., Dakubo, G. D., Thayer, R. E., McKenney, K., and Birch-Machin, M. A, "Mitochondrial DNA as a potential tool for early cancer detection," *Hum. Genom.*, Vol. 2, 2006, pp. 252–257.

[40] Baysal, B. E, "Role of mitochondrial mutations in cancer," *Endocr. Pathol.*, Vol. 17, 2006, pp. 203–212.

[41] Fearon, E. R., and Vogelstein, B, "A genetic model for colorectal tumorigenesis," *Cell*, Vol. 61, 1990, pp. 759–767.

[42] Negrini, M., et al., "Definition and refinement of chromosome 11 regions of loss of heterozygosity in breast cancer: identification of a new region at 11q23.3," *Cancer Res.*, Vol. 55, 1995, pp. 3003–3007.

[43] Hampton, G.M., et al., "Loss of heterozygosity in sporadic human breast carcinoma: a common region between 11q22 and 11q23.3," *Cancer Res.*, Vol. 54, 1994, pp. 4586–4589.

[44] Chandrasekharappa, S.C., et al., "Positional cloning of the gene for multiple endocrine neoplasia-type 1," *Science*, Vol. 276, 1997, pp. 404–407.

[45] Wistuba, I.I., et al., "Genome-wide Allelotyping Analysis Reveals Multiple Sites of Allelic Loss in Gallbladder Carcinoma," *Cancer Res.*, Vol. 61, 2001, pp. 3795–3800.

[46] Telenius, H., et al., "Degenerate oligonucleotide-primed PCR: general amplification of target DNA by a single degenerate primer," *Genomics*, Vol. 13, 1992, pp. 718–725.

[47] Barker, D.L., et al., "Two Methods of Whole-Genome Amplification Enable Accurate Genotyping Across a 2320-SNP Linkage Panel," *Genome Res.*, Vol. 14, 2004, pp. 901–907.

[48] Hu, N., et al., "Genome-Wide Association Study in Esophageal Cancer Using GeneChip Mapping 10K Array," *Cancer Res.*, Vol. 65, 2005, pp. 2542–2546.

[49] Lechpammer, D., and Sgroi, C., "Laser Capture Microdissection: a rising tool in genetic profiling of cancer," *M. Exp. Rev. Mol. Diagn.*, Vol. 4, 2004, pp. 429–430.

[50] Verhagen, P.C., et al., "Microdissection, DOP-PCR, and comparative hybridization of paraffin-embedded familial prostate cancers," *Cancer Genet. Cytogenet.*, Vol. 122, 2000, pp. 43–48.

[51] Grønbaek, K., Hother, C., and Jones, P.A, "Epigenetic changes in cancer," *APMIS*, Vol. 115, 2007, pp. 1039–1059.

[52] Shames, D.S., Minna, J.D., and Gazdar, A.F, "Methods for detecting DNA methylation in tumors: from bench to bedside," *Cancer Lett.*, Vol. 251, 2007, pp. 187–198.

[53] Chen, J. Z., Gokden, N., Greene, G. F., Mukunyadzi, P., and Kadlubar, F. F, "Extensive Somatic Mitochondrial Mutations in Primary Prostate Cancer Using Laser Capture Microdissection," *Cancer Res.*, Vol. 62, 2002, pp. 6470–6474.

[54] Kreuder, J., Repp, R., Borkhardt, A., and Lampert, F, "Rapid detection of mitochondrial deletions by long-distance polymerase chain reaction," *Eur. J. Ped.*, Vol. 154, 1995, p. 996.

[55] Ransohoff, D.F., McNaughton Collins, M., and Fowler, F. J., "Why is prostate cancer screening so common when the evidence is so uncertain? A system without negative feedback," *Am. J. Med.*, Vol. 113, 2002, pp. 663–667.

[56] Ransohoff, D.F., "Rules of evidence for cancer molecular-marker discovery and validation," *Nat. Rev. Cancer*, Vol. 4, 2004, pp. 309–14.

[57] Zhang, A., et al., "Small interfering RNA and gene expression analysis using a multiplex branched DNA assay without RNA purification," *J. Biomol. Screen.*, Vol. 10, 2005, pp. 549–556.

[58] Twombly, R., "Identity Crisis: Finding, Defining, and Integrating Biomarkers Still a Challenge," *J. Natl. Cancer Inst.*, Vol. 98, 2006, pp. 11–12.

[59] Best, C. J., et al., "Molecular differentiation of high- and moderate-grade human prostate cancer by cDNA microarray analysis," *Diagn. Mol. Pathol.*, Vol. 12, 2003, pp. 63–70.

[60] Krizman, D.B., et al., "Construction of a representative cDNA library from prostatic intraepithelial neoplasia," *Cancer Res.*, Vol. 56, 1996, pp. 5380–5383.

[61] Chuaqui, R.F., et al., "Identification of a novel gene up-regulated in clinically aggressive human prostate cancer," *Urology*, Vol. 50, 1997, pp. 301–307.

[62] Khattra, J., et al., "Large-scale production of SAGE libraries from microdissected tissues, flow-sorted cells, and cell lines," *Genome Res.*, Vol. 17, 2007, pp. 108–116.

[63] Velculescu, V.E., Zhang, L., Vogelstein, B., and Kinzler, K.W, "Serial analysis of gene expression," *Science*, Vol. 270, 1995, pp. 484–7.

[64] Granjeaud, S., Bertucci, F., and Jordan, B.R, "Expression profiling: DNA arrays in many guises," *Bioessays*, Vol. 21, 1999, pp. 781–90.

[65] Lockhart, D.J., et al., "Expression monitoring by hybridization to high-density oligonucleotide arrays," *Nat. Biotechnol.*, Vol. 14, 1996, pp. 1675–1680.

[66] Cheng, Q., et al., "Identification and characterization of genes involved in the carcinogenesis of human squamous cell cervical carcinoma," *Int. J. Cancer*, Vol. 98, 2002, pp. 419–426.

[67] Russo, G., Zegar, C., and Giordano, A., "Advantages and limitations of microarray technology in human. Cancer," *Oncogene*, Vol. 22, 2003, pp. 6497–6507.

[68] Lipshutz, R.J., et al., "Expression Monitoring by Hybridization to High-Density Oligonucleotide Arrays," *Nat. Biotechnol.*, Vol. 14, 1996, pp. 1675–1680.

[69] Dhanasekaran, S. M., et al., "Delineation of prognostic biomarkers in prostate cancer," *Nature*, Vol. 412, 2001, pp. 822–826.

[70] Luo, J., et al., "Human prostate cancer and benign prostatic hyperplasia: molecular dissection by gene expression profiling," *Cancer Res.*, Vol. 61, 2001, pp. 4683–4688.

[71] Luo, J.H., et al., "Expression Analysis of Prostate Cancers," *Mol. Carcinog.*, Vol. 33, 2002, pp. 25–35.

[72] Welsh, J.B., et al., "Analysis of gene expression identifies candidate markers and pharmacological targets in prostate cancer," *Cancer Res.*, Vol. 61, 2001, pp. 5974–5978.

[73] Singh, D., et al., "Gene expression correlates of clinical prostate cancer behavior," *Cancer Cell*, Vol. 1, 2002, pp. 203–209.

[74] Ashida, S., et al., "Expression of Novel Molecules, MICAL2-PV (MICAL2 Prostate Cancer Variants), Increases with High Gleason Score and Prostate Cancer Progression," *Clin. Cancer Res.*, Vol. 12, 2006, pp. 2767–2773.

[75] Mattie, M.D., et al., "Optimized high-throughput microRNA expression profiling provides novel biomarker assessment of clinical prostate and breast cancer biopsies," *Mol. Cancer*, Vol. 5, 2006, p. 24.

[76] Ma, S., et al., "The significance of LMO2 expression in the progression of prostate cancer," *J. Pathol.*, Vol. 211, 2007, pp. 278–285.

[77] True, L., et al., "A Molecular Correlate to the Gleason Grading System for Prostate Cancer," *Proc. Natl. Acad. Sci. U. S. A.*, Vol. 103, 2006, pp. 10991–10996.

[78] Williams, N.S., et al., "Identification and validation of genes involved in the pathogenesis of colorectal cancer using cDNA microarrays and RNA interference," *Clin. Cancer Res.*, Vol. 3, 2003, pp. 931–946.

[79] Huang, Z.G., Ran, Z.H., Lu, W., and Xiao, S.D, "Analysis of gene expression profile in colon cancer using the Cancer Genome Anatomy Project and RNA interference," *Chin. J. Dig. Dis.*, Vol. 7, 2006, pp. 97–102.

[80] Pallante, P., et al., "MicroRNA deregulation in human thyroid papillary carcinomas," *Endocr. Relat. Cancer*, Vol. 13, 2006, pp. 497–508.

[81] Silveri, L., Tilly, G., Vilotte, J.L., and Le Provost, F., "MicroRNA involvement in mammary gland development and breast cancer," *Reprod. Nutr. Dev.*, Vol. 46, 2006, pp. 549–556.

[82] Visone, R., et al., "Specific microRNAs are downregulated in human thyroid anaplastic carcinomas," *Oncogene*, Vol. 26, 2007, pp. 7590–7595.

[83] Hsu, S.D., et al., "miRNAMap 2.0: genomic maps of microRNAs in metazoan genomes," *Nuc. Ac. Res.*, Nov. 19, 2007.

[84] Micke, P., et al., "In situ identification of genes regulated specifically in fibroblasts of human basal cell carcinoma," *J. Invest. Dermatol.*, Vol. 127, 2007, pp. 1516–1523.

[85] Zhao, H., Ramos, C. F., Brooks, J. D., and Peehl, D. M., "Distinctive gene expression of prostatic stromal cells cultured from diseased versus normal tissues," *J. Cell. Physiol.*, Vol. 210, 2007, pp. 111–121.

[86] Morrison, T., James H., Garcia J., Yoder K., Katz A., Roberts D., Cho J., Kanigan T., Ilyin S.E., Horowitz D., Dixon J.M., and Brenan C.J.H., "Nanoliter high throughput quantitative PCR," *Nuc. Ac. Res.*, Vol. 34, 2006, p. e123.

[87] Knudsen, B.S., Allen, A.N., McLerran, D.F., Vessella, R.L., Karademos, J., Davies, J.E., Maqsodi, B., McMaster, G.K., and Kristal, A.R., "Evaluation of the branched-chain DNA assay for measurement of RNA in formalin-fixed tissues," *J. Mol. Diagn.*, Vol. 10, 2008, pp. 169–176.

[88] Canales, R.D., et al., "Evaluation of DNA microarray results with quantitative gene expression platforms," *Nat. Biotechnol.*, Vol. 2, 2006, pp. 1115–1122.

[89] Dennis-Sykes, C.A., Miller, W.J., and McAleer, W.J., "A quantitative Western Blot method for protein measurement," *J. Biol. Stand.*, Vol. 13, 1985, pp. 309–314.

[90] Gingrich, J.C., Davis, D.R., and Nguyen, Q., "Multiplex detection and quantitation of proteins on western blots using fluorescent probes," *Biotechniques*, Vol. 29, 2000, pp. 636–642.

[91] Bakalova, R., Zhelev, Z., Ohba, H., and Baba, Y., "Quantum dot-based western blot technology for ultrasensitive detection of tracer proteins," *J. Am. Chem. Soc.*, Vol. 127, 2005, pp. 9328–9329.

[92] O'Farrell, P.H., "High resolution two-dimensional electrophoresis of proteins," *J. Biol. Chem.*, Vol. 250, 1975, pp. 4007–4021.

[93] Kondo, T., et al., "Application of sensitive fluorescent dyes in linkage of laser microdissection and two-dimensional gel electrophoresis as a cancer proteomic study tool," *Proteomics*, Vol. 3, 2003, pp. 1758–1766.

[94] Ong, S.E. and, Pandey, A., "An evaluation of the use of two-dimensional gel electrophoresis in proteomics," *Biomol. Eng.*, Vol. 18, 2001, pp. 195–205.

[95] Unlu, M., Morgan, M.E., and Minden, J.S., "Difference gel electrophoresis: a single gel method for detecting changes in protein extracts," *Electrophoresis*, Vol. 18, 1997, pp. 2071–2077.

[96] Molloy, M.P., Brzezinski, E.E., Hang, J., McDowell, M.T., and VanBogelen, R.A., "Overcoming technical variation and biological variation in quantitative proteomics," *Proteomics*, Vol. 3, 2003, pp. 1912–1919.

[97] Domon, B., and Aebersold, R., "Mass spectrometry and protein analysis," *Science*, Vol. 312, 2006, pp. 212–217.

[98] Karas, M.B.D., Bahr, U., and Hillenkamp, F., "Matrix-Assisted Ultraviolet Laser Desorption of Non-Volatile Compounds," *Int. J. Mass. Spectrom. Ion Proc.*, Vol. 78, pp. 53–68.

[99] Fenn, J.B., Mann, M., Meng, C.K., Wong, S.F., and Whitehouse, C.M., "Electrospray ionization for mass spectrometry of large biomolecules," *Science*, Vol. 246, 1989, pp. 64–71.

[100] Costello, C.E., "Time, life ... and mass spectrometry. New techniques to address biological questions," *Biophys. Chem.*, Vol. 68, 1997, pp. 173–188.

[101] Payne, A.H., and Glish, G.L, "Tandem mass spectrometry in quadrupole ion trap and ion cyclotron resonance mass spectrometers," *Methods Enzymol.*, Vol. 402, 2005, pp. 109–148.

[102] Srebalus Barnes, C.A. L.A, "Applications of mass spectrometry for the structural characterization of recombinant protein pharmaceuticals," *Mass Spectrom. Rev.*, Vol. 26, 2007, pp. 370–388.

[103] Rodland, K.D., "Proteomics and cancer diagnosis: the potential of mass spectrometry," *Clinical Biochem.*, Vol. 37, 2004, pp. 579–583.

[104] Li, C., et al., "Accurate qualitative and quantitative proteomic analysis of clinical hepatocellular carcinoma using laser capture microdissection coupled with isotope-coded affinity tag and two-dimensional liquid chromatography mass spectrometry," *Mol. Cell. Proteomics*, Vol. 3, 2004, pp. 399–409.

[105] Gygi, S.P., et al., "Quantitative analysis of complex protein mixtures using isotope-coded affinity tags," *Nature Biotechnol.*, Vol. 17, 1999, pp. 994–999.

[106] MacBeath, G., and Schreiber, S.L, "Printing proteins as microarrays for high-throughput function determination," *Science*, Vol. 289, 2000, pp. 1760–1763.

[107] Miller, J.C., et al., "Antibody microarray profiling of human prostate cancer sera: antibody screening and identification of potential biomarkers," *Proteomics*, Vol. 3, 2003, pp. 56–63.

[108] Lal, S.P., Christopherson, R.I., and dos Remedios, C.G., "Antibody arrays: an embryonic but rapidly growing technology," *Drug Discovery Today*, Vol. 15, Suppl. 18, 2002, pp. 143–149.

[109] Liotta, L.A., et al., "Protein microarrays: meeting analytical challenges for clinical applications," *Cancer Cell*, Vol. 3, 2003, pp. 317–325.

[110] Grubb, R.L., et al., "Signal pathway profiling of prostate cancer using reverse phase protein arrays," *Proteomics*, Vol. 3, 2003, pp. 2142–2146.

[111] Wright, Jr., G.L., et al., " Proteinchip surface enhanced laser desorption/ionization (SELDI) lass spectrometry: a novel protein biochip technology for detection of prostate cancer biomarkers in complex protein mixtures," *Prostate Cancer Prostat. Dis.*, Vol. 2, 2000, pp. 264–276.

[112] Paweletz, C.P., et al., "Rapid protein display profiling of cancer progression directly from human tissue using a protein biochip," *Drug. Devel. Res.*, Vol. 49, 2000, pp. 34–42.

[113] Vandesompele, J., et al., "Accurate normalization of real-time quantitative RT-PCR data by geometric averaging of multiple internal control genes," *Genome Biol.*, Vol. 3, 2002, RESEARCH0034.

[114] Simon, R., et al., *Design and Analysis of DNA Microarray Investigations*, New York: Springer, 2003.

[115] Schleicher, C., et al., "Gain of Chromosome 8q: A Potential Prognostic Marker in Resectable Adenocarcinoma of the Pancreas? Cancer Progression," *Clin. Cancer Res.*, Vol. 12, 2006, pp. 2767–2773.

[116] Melle, C., et al., "Protein Profiling of Microdissected Pancreas Carcinoma and Identification of HSP27 as a Potential Serum Marker," *Clin. Chem.*, Vol. 53, 2007, pp. 629–635.

[117] Ashida, S., et al., "Molecular Features of the Transition from Prostatic Intraepithelial Neoplasia (PIN) to Prostate Cancer: Genome-wide Gene-expression Profiles of Prostate Cancers and PINs," *Cancer Res.*, Vol. 64, 2004, pp. 5963–5972.

[118] Hosokawa, M., et al., "Oncogenic Role of KIAA0101 Interacting with Proliferating Cell Nuclear Antigen in Pancreatic Cancer," *Cancer Res.*, Vol. 67, 2007, pp. 2568–2576.

CHAPTER 12

Applications of High-Performance Computing to Functional Magnetic Resonance Imaging (fMRI) Data

Rahul Garg, A. Ravishankar Rao, and Guillermo A. Cecchi

12.1 Introduction

Magnetic resonance imaging can be used to view the internal activity in a subject's brain [1–21]. This is accomplished by measuring changes in the blood oxygenation level in different parts of the brain [15]. It is the underlying neural activity that determines blood flow to different regions of the brain. Such measurements are performed while the subject is performing specific functions or mental tasks, giving rise to the term functional magnetic resonance imaging.

There are many techniques to measure activity in a subject's brain, including direct electrical recordings, and electroencephalography (EEG) [13]. Each technique has its advantages and disadvantages. Obtaining electrical recordings is an invasive procedure, but gives high-resolution spatial and temporal information. EEG recordings are noninvasive and provide high-temporal resolution but poor spatial resolution. Amongst the noninvasive techniques, fMRI recordings provide high spatial resolution. The temporal resolution is about 2 seconds, which is reasonable for capturing a variety of responses in a subject's brain during different tasks.

In a typical fMRI experiment, subjects are presented with some external stimulus while they are being scanned inside an fMRI machine. Some examples of external stimulus consist of instructions to perform certain tasks (such as finger tapping), flashing of images of certain types (tools, buildings, pleasant or unpleasent images, familiar faces, and so forth), and different types of sounds. In such experiments, the subjects may also be expected to respond (say, by pressing a button) to the external stimulus during the course of fMRI scan. The timings as well as details of the external stimuli and responses are recorded along with the fMRI data.

12.1.1 fMRI Image Analysis Using the General Linear Model (GLM)

The fMRI data is analyzed to test a specific hypothesis about the brain function, such as which areas in the brain become active during a finger tapping task. The analysis of fMRI data has traditionally been done using the general linear model method [9]. In this method, a suitable design matrix is constructed to answer questions about the hypothesis. The design matrix consists of the time series of different external stimuli and responses. The voxel time series of the fMRI data are correlated with the design matrix. The brain areas that respond to specific patterns of activity are obtained by multiplying the pseudoinverse of the design matrix with the fMRI data.

The success of the general linear model can be attributed to its simplicity, accompanying statistical inferencing methods, and computational tractability. Typical fMRI experiments collect data for 10 to 100 sessions. Each session consists of 200 to 400 brain volumes, where each volume may be collected at a resolution of 3-mm voxels ($64 \times 64 \times 32$ voxels per volume). Such experiments may easily generate from 1 to 10 GB of data. Since the GLM requires just a linear scan of the data and simple matrix operations, it can implemented in a simple high-level interpretted language such as MATLAB.

12.1.2 fMRI Image Analysis Based on Connectivity

It is well known that the brain is a highly interconnected structure, with dense local connections and sparse long-range connections. An example of a long-range connection is the projection from the prefrontal cortex areas responsible for planning and behavior generation to the visual areas that are responsible for representing detailed information about a visual scene. One of the limitations of the GLM method is that it does not proide information about the connectivity between different brain regions that are involved in a given task.

Hence, there is a growing interest in studies involving functional connectivity in the brain [5]. In addition to activation patterns, researchers are now studying connectivity patterns in the brain and how these are modulated by different tasks. Connectivity involves studying the response of every pair of voxels in the brain. If there are N active voxels in the scan, there could be $N(N-1)/2$ possible connections. In case of a scan with 30,000 active voxels, there are approximately 500 million potential connections. It is impossible to carry out such an analysis on traditional desktop systems without making simplifying assumptions.

In this chapter, we examine both a novel mathematical model applicable to this problem, and a parallel implementation of the model to improve its performance. We show that by using high-performance computing, it is possible to start answering questions about functional connectivity of brain. Moreover, it is also possible to employ complex nonlinear models and algorithms to understand brain dynamics. In this chapter we demonstrate this using a scalable method to compute the structure of causal links over large scale dynamical systems that achieves high efficiency in discovering actual functional connections. The method is based on the Granger causality analysis of multivariate linear models, solved by means of a sparse regression approach, and can deal with autoregressive models of more than 10,000 variables.

12.2 The Theory of Granger Causality

Granger has defined causality in terms of predictability of stochastic processes. A process X_i is said to have a causal influence on another process X_j if *predictability* of X_i at a given time instant is improved by using the past values of X_j.

Formally, let X represent a vector of N stationary stochastic processes. Let $X(t)$ represent the N dimensional vector of random variables at time t with $X_i(t)$ as its i^{tj} component. Let $\overline{X}(t)$ represent the set of past random variables in X (i.e., $\overline{X}(t) = \{X(t-j) : j = 1,2,\ldots\infty\}$). Let $P(A|\overline{B})$ represent the optimal unbiased least-square predictor of the random variable A using only the random variables in the set $\overline{B}$. Thus $P(X_i(t)|\overline{X}_i(t))$ represent the optimal unbiased least-square predictor of $X_i(t)$ using only the past values of $X_i(t)$. Let $\sigma^2(A|B)$ be the variance of $A - P(A|B)$.

We say that process X_j has a causal influence on the process X_i in the context of processes X, if

$$\sigma^2(X_i(t)|\overline{X}(t)) < \sigma^2(X_i(t)|\overline{X}(t)\backslash\overline{X}_j(t))$$

In other words, if the error variance of an optimal unbiased predictor of X_i using the past values of X, increases by excluding past values of X_j, then X_j is said to have causal influence on X_i.

The above definition of causality is very general. It is defined in an existential form, for arbitrary stationary stochastic processes, using the abstract notion of "optimal unbiased least variance predictor." The definition is not constructive, that is, it does not specify how this predictor should be computed. This makes it extremely difficult to determine Granger causality (in its purest form) because computing the optimal unbiased least-square predictor for arbitrary stochastic processes is a nontrivial task. In order to apply this definition in practice, people (including Granger himself) have resorted to causality in the context of *linear models*.

Note that the above definition of causality is critically dependent on the context of stochastic processes being studied. As Granger pointed out himself, the apparent causality relationship between two stochastic processes may disappear by inclusion of a confounding process in the context (X) of processes being considered. Also note that the above is a rigorous mathematical definition of the word "causality" in the present context. This may or may not reflect the metaphysical concept of cause and effect. Moreover, the causal relationships discovered by invoking the above definition may be erroneous due to the following factors:

- Weakness in the mathematical concept defining causality;
- Inaccuracies arising from the inability to account for all the confounding processes while discovering causal relationships;
- Inability to sample the underlying stochastic processes at a suitable temporal resolution;
- The (linear) simplifications used for tractability of computation.

12.2.1 The Linear Simplification

Let X be represented as an N dimensional row vector. We model the multivariate stochastic process as a linear combination of its past values and independent, identically distributed (iid) noise. Such a representation is also known as a multivariate autoregressive model. Formally,

$$X(t) = \sum_{\tau=1}^{k} X(t-\tau)A(\tau) + E(t) \quad (12.1)$$

where k is called the *model order*, $A(\tau)_{\tau=1\ldots k}$ are the model parameters in the form of k matrices of size $N \times N$ (with coefficients $a_{ij}(t)$), and $E(t)$ is an N-dimensional row vector of noise with zero mean and a covariance of R. For any $t_1 \neq t_2$, $E(t_1)$ and $E(t_2)$ are identically distributed and uncorrelated.

In this model, if $a_{ij}(t) > 0$ for some t, then past values of X_i improve the predictability of X_j, and therefore, X_i is said to have causal influence on X_j. The parameter t is called the the causality lag between X_i and X_j.

To infer the causal relationships in the linear simplification, we need to know the model parameters $\{a_{ij}(t)\}$. These may be estimated from a realization of the process X. Let $\{x(t)\}_{t=1\ldots T}$ be a realization of the stochastic process X and $\{e(t)\}_{t=1\ldots T}$ be a realization of the iid noise E. This realization must satisfy

$$x(t) = [x(t-1)x(t-2)\ldots x(t-k)]\left[A'(1)A'(2)\ldots A'(k)\right]' + e(t) \quad (12.2)$$

for all $t \in [k+1, \ldots T]$. The above set of equations can be written in a compact matrix form as follows. Let Y be a matrix of size $(T-k) \times N$, Z be a matrix of size $(T-k) \times Nk$, W be a matrix of size $(T-k) \times N$, and $\mathcal{N}$ be a matrix of size $(T-k) \times N$ given by stacking the rows in (12.2) for different values of T.

Now, (12.2) may equivalently be written as

$$Y = ZW + \mathcal{N} \quad (12.3)$$

where Y and Z are derived from a realization x of the process X, W contains all the model parameters ($a_{ij}(t)$), and $\mathcal{N}$ is derived from realization e of the noise E.

The maximum likelihood estimate (W_{MLE}) of model parameters is given by the standard least square solution of (12.3); that is,

$$W^{MLE} = \arg\min_{W} \sum_{j=1}^{N} ||Y_j - ZW_j||_2^2 \quad (12.4)$$

$$= \arg\min_{a_{ij}(\tau)} \sum_{j=1}^{N} \sum_{t=k+1}^{T} \left[x_j(t) - \sum_{\tau=1}^{k}\sum_{l=1}^{N} x_l(t-\tau)a_{lj}(\tau)\right]^2 \quad (12.5)$$

where Y_j represents jth column of Y and W_j represents the jth column of W. Equation (12.4) has a unique solution only if (12.3) is not underdetermined; that is,

$$(T-k)N \geq N^2 K$$

$$\Rightarrow T \geq (N+1)k$$

In general, for reliable estimates of the model parameters, (12.3) must be sufficiently overdetermined; that is the number of observations of the process X must be significantly larger than the number of model parameters ($(T-k)N >> N^2 k$).

If the model is sparse (i.e., the number of nonzero coefficients in $\{a_{ij}(\tau)\}$ is significantly smaller than the total number of coefficients (N^k)), then one can use techniques of sparse regression to find a reliable solution to (12.3).

12.2.2 Sparse Regression

Consider a multivariate linear regression model of the form

$$Y = ZW$$

where Y is a known $n_1 \times 1$ response vector, Z is a known $n_1 \times n_2$ regressor matrix, and W is the unknown model vector of size $n_2 \times 1$ to be determined using the response Y and regressor Z. Such a problem is typically solved with techniques such as ordinary least square regression, ridge regression, or subset selection [21]. For these techniques to work, a general requirement is that $n_1 >> n_2$. However, recent research shows that if W is sparse then it may be recovered even if $n_2 > n_1$ using the lasso regression [6, 18].

The lasso regression [18] solves the problem

$$\min_W ||Y - ZW||_2^2 \tag{12.6}$$

$$\text{s.t.} ||W||_1 \leq t \tag{12.7}$$

where $||X||_2^2$ represents the sum of squares of the coefficients of X (square of L2 norm of X) and $||X||_1$ represents the sum of absolute values of the coefficients of X (its L1 norm). The parameter t is the regression parameter usually chosen after cross-validation. In the extreme case, when t is infinitesimally small, the optimal solution W corresponds to the regressor (column of Z) which is "closest to" Y. In the other extreme when t is sufficiently large, (12.7) ceases to be a constraint and the optimal solution W corresponds to classical the least square solution of (12.3).

It can be verified that for any t, there exists a λ such that the program (12.6), (12.7) is equivalent to the following optimization problem:

$$\min_W ||Y - ZW||_2^2 + \lambda |W|_1 \tag{12.8}$$

The programs (12.6), (12.7), and (12.8) can be solved efficiently using a technique called least angle regression [6].

12.2.3 Solving Multivariate Autoregressive Model Using Lasso

The estimation of multivariate autoregressive coefficients in (12.3) may be viewed as a regression problem where Y is the response variable, Z is the matrix containing the regressors, and W is the model to be determined. In this case, the maximum likelihood estimate of (12.4) becomes the least square solution to the regression problem. The lasso formulation thus becomes

$$W^{sparse} = \arg\min_{W} \sum_{j=1}^{N} \left[||Y_j - ZW_j||_2^2 + \lambda ||W_j||_1 \right] \tag{12.9}$$

Note that the coefficients of W_j only appear in the jth term of the above sum. So, this problem may be decomposed into N independent lasso regression problems of size $(T - k) \times Nk$, as

$$W_i^{sparse} = \arg\min_{W_j} ||Y_j - ZW_j||_2^2 + \lambda ||W_j||_1$$

Equivalently, for all j the coefficients of multivariate autoregressive model are given by

$$\{a_{lj}(\tau)\} = \arg\min_{\{u_l(\tau)\}} \sum_{t=k+1}^{T} \left[x_j(t) - \sum_{\tau=1}^{k} \sum_{l=1}^{N} x_l(t-\tau) u_l(\tau) \right]^2 + \lambda \left[\sum_{\tau=1}^{k} \sum_{l=1}^{N} |u_l(\tau)| \right] \tag{12.10}$$

for all j. Equation (12.10) essentially fits a sparse regression model on the time course of a voxel j using the past k time shifted values of all the other voxels. Before the regression is started, the data is normalized:

$$\sum_{t=1}^{T} x_j(t) = 0$$

$$\sum_{t=1}^{T} [x_j(t)]^2 = 1$$

The goodness of fit of the regression is captured using the notion of *predictability* (p_j), which is defined as

$$p_j = 1 - \frac{\sum_{t=k+1}^{T} \left[x_j(t) - \sum_{\tau=1}^{k} \sum_{l=1}^{N} x_l(t-\tau) a_{lj}(\tau) \right]^2}{\sum_{t=k+1}^{T} (x_j(t))^2} \tag{12.11}$$

It may be verified using the properties of the lasso regression that the predictability varies from 0 to 1. If the predictability of a voxel is 1, then its time course can be predicted exactly using the past k values of other voxels. On the other hand, if a voxel has zero predictability, then its time course is orthogonal to (independent of) the shifted time course of all the other voxels.

12.3 Implementing Granger Causality Analysis on the Blue Gene/L Supercomputer

Solving a lasso regression problem for every voxel in the brain is computationally demanding. However, using high-performance computing it is possible to get a solution in a reasonable amount of time. The Granger causality analysis was implemented on the Blue Gene/L [8] supercomputer. We first describe the Blue Gene/L (BG/L) hardware and software environment and then some key elements in the implementation of this analysis.

12.3.1 A Brief Overview of the Blue Gene/L Supercomputer

The Blue Gene/L supercomputer has a modular design which can be seamlessly scaled to meet the computational requirements of its users. The smallest available unit is called a "rack," which occupies space comparable to a refrigerator and packs 2,048 processors. One of the key elements of its design was to optimize for power efficiency. The Blue Gene/L rack has a peak power consumption of 27.5 kilowatts which is 5 to 10 times less than any other system with similar computational capabilities built using contemporary server/desktop processors [8]. As a result, more processors can be packed in a single rack and many racks can be assembled to form a large supercomputer. The largest current installation of the Blue Gene/L system is at the Lawrence Livermore National Labs, in Livermore, CA, which has 104 racks (i.e., 212,992 processors) with a peak performance of 596 Tflops (1 Tflop $\equiv 10^{12}$ floating-point operations per second).

The basic-building block of the Blue Gene/L supercomputer is the BG/L compute chip [16]. It is based on a system-on-a-chip (SoC) design and contains most of the elements needed for the system including two embedded PowerPC 440 processors with floating-point enhancements, complete circuit for all of its communication networks, and embedded DRAM based L3 cache among other elements. It has been integrated on an application specific integrated circuit (ASIC) which connects to external DRAM memory, typically between 512 MB and 1 GB. The two embedded processors in the chip share this memory. Each of the processors has a dual floating point unit, which can perform up to four floating point operation per cycle [20]. The Blue Gene/L processors operate at 700 MHz and thus have a computational capacity of 2.4 giga floating-point operations per second (GFLOPS). Two compute chips are assembled together in a BG/L "node-card." Thirty two node-cards are packed in a "node-board" [4]. Eight node-boards constitute a "mid-plane" and two such mid-planes constitute a Blue Gene/L "rack," which has 2,048 processors and a peak computational capacity of 5.6 TFLOPS (1 TFLOP = 10^{12} floating point operations per second) and up to 1 TB of main memory. These racks can be assembled together to make larger Blue Gene/L systems. The 104-rack based Blue Gene/L system installed at the Lawrence Livermore National Laboratory has a peak capacity of 596.378 TFLOPS and is presently the world's fastest supercomputer.

Blue Gene/L has five interconnection networks for different types of communications. The most important network is organized in the form of a three-dimensional torus [1] with bidirectional links of capacity 1.44 Gbps (gigabits per second). This network is primarily used for point-to-point communication across different processors in the system. There is another network for the "collective communication" operations such as broadcast or global reductions. In addition, there is a network for fast hardware-based barrier operation, another for monitoring the health of the system along with a gigabit Ethernet network for input-output (I/O) to the external world.

Blue Gene/L is a diskless system which is connected to an array of file servers through gigabit Ethernet links. The Blue Gene/L system runs a minimal operating system called the compute-node kernel (CNK) [11] which provides very basic functionality of job creating, termination, and I/O. To avoid performance slowdown due to operating system jitter and daemons [2, 17], the compute-node kernel only allows one process at every processor. These processes are a part of a large MPI job that runs of partitions of the Blue Gene/L system, which may range from 32 nodes (64 processors) to 104K nodes (2.13×10^5 processors). It is also possible to run a full Linux operating system on the Blue Gene/L processors, which is not supported due to performance reasons.

Blue Gene/L provides C, C++, and FORTRAN programming environment [11] with MPI [3] and POSIX support. Most of the MPI programs written for other parallel systems can be compiled and run on Blue Gene/L. The BG/L software environment provides several optimized libraries such as ESSL, FFT, massv, and debugging and performance monitoring tools such as totalview and HPM [10].

12.3.2 MATLAB on Blue Gene/L

The first step in parallelizing the Granger causality analysis is to have an efficient implementation of Lasso regression on the BG/L compute-node kernel. A variant of an algorithm called least-angle regression (LARS) can efficiently solve (in some cases) the Lasso regression problem [6]. Some MATLAB-based implementations of LARS and its Lasso variant are also publicly available. The Lasso implementation we used solves the constraint formulation [i.e., (12.6) and (12.7)] of the problem, whereas the Granger causality analysis needs a solution to the penalty formulation (12.8). It was simple to make changes to the original Lasso implementation to solve the penalty formulation.

In order to run this code on Blue Gene/L, one needs to run MATLAB on Blue Gene/L which is next to impossible. An open source alternative to MATLAB called Octave [14] is freely available for download, along with its C/C++ source code. Moreover, a wrapper called MPI Toolbox for Octave (MPITB) [12] is freely available that exports most of MPI functions to Octave. Thus, using MPITB and Octave, it is possible to write MPI-based parallel programs using the MATLAB scripting language. With little effort, it was possible to compile Octave and MPITB on the Blue Gene/L supercomputer As a result the MATLAB implementation of Lasso could be directly used on Blue Gene/L.

12.3.3 Parallelizing Granger Causality Analysis

As noted earlier, the Granger causality analysis can be decomposed into N independent Lasso regression problems, where N is the number of active voxels in the brain. Since N is typically large (30,000 or so), each of the regression problems could be run independently in parallel on different BG/L processors. Although this decomposition was simple to implement, we encountered some typical parallel programming issues, namely *load imbalance*, *input/output latency*, and *output consistency*. We discuss these issues in more detail and describe our solution.

- *Load imbalance.* The progress of a parallel program is determined by the progress of its slowest process. Therefore, one needs to ensure that every process of the program takes almost the same time to carry out its allocated work. Although one could almost equally divide the voxels among the processes, it does not ensure that all the processes take almost the same amount of time. The time taken to run the Lasso regression on a dataset of fixed size is not constant. In addition to the size of the data, it also depends on the actual values stored in the data, primarily because the number of iterations of the LARS algorithm needed to find a solution is dependent on how quickly the residual error decreases. If N voxels are divided equally among P processes, each process gets N/P voxels. If N/P is large, then one might expect the variability in the average time taken to run the Lasso regression to decrease following the law of large numbers. However, when the number of processes was large, the variability was significant and slowed down the entire computation.
- *Input/output latency.* Only one node needs to read the input data. After the data is read by an "input node," it can be broadcast efficiently to other nodes. This turns out to be faster than every node reading the same data. To avoid duplication of computation, the node reading the data can also carry out all the preprocessing needed for the analysis, before broadcasting the data to rest of the nodes. In a naive implementation, while the input node reads and processes the data, the other nodes keep waiting for the data, wasting the precious compute cycles. This leads to the loss of efficiency.
- *Output consistency.* Once the regression is completed for a voxel, the results need to be written to a file. If all the nodes write their results to different files, then the number of files become very large and difficult to manage (at least 2,048 for a single rack Blue Gene/L system). Moreover all files need to be merged into a single (or a few) files, which itself ends up taking a long time. On the other hand, if all the nodes write to the same output file, the data can become garbled, because MPI does not provide any guarantee about the order in which the output from multiple MPI processes will be written to the file.

The above problems have been solved by designating two MPI processes (or nodes) as special nodes—the *master node* and the *input node* as shown in Figure 12.1. The rest of the MPI processes are called *worker nodes*. The input node reads the data, carries out the required preprocessing, and broadcasts the data to the worker nodes. The master node allocates work to the worker nodes, collects results from worker nodes, and writes them to files. The worker nodes carry out the Lasso regressions on the voxels assigned to them by the master node.

The worker nodes first send a "request-for-work" message to the master node, which responds with the identity of voxel that has to be regressed. After the regression is done, the worker node sends another request-for-work message to the master. This ensures that as long as there is significant work to be done (i.e., regression needs to be done for at least P voxels), none of the workers stay idle, thereby solving the problem of load-balancing.

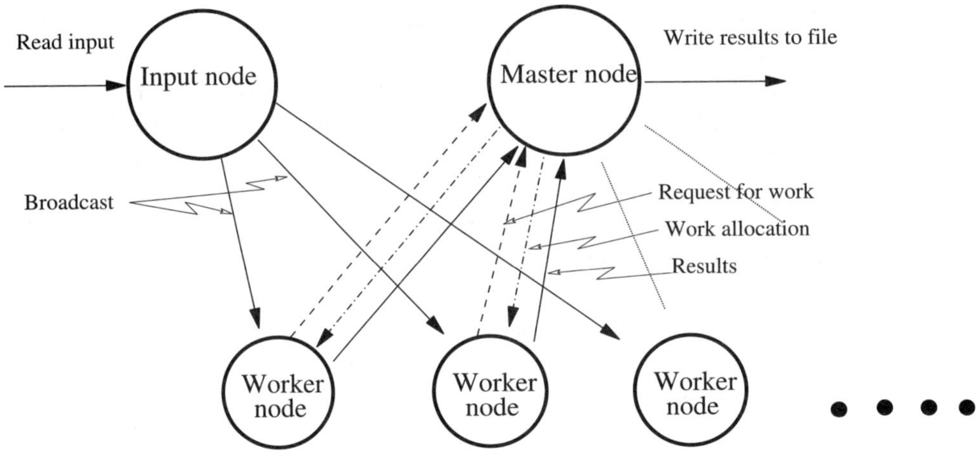

Figure 12.1 Parallel implementation of the Granger analysis–communication between input node, master node and worked nodes.

For output consistency, the master node was also assigned the task of writing all the results to the output files. The worker nodes send the results to the master node before requesting more work. The master node collects these results and writes to the output files.

For handling the input latency, we used the standard pipelining technique to overlap the input and preprocessing with computation. In a typical analysis, there are multiple sessions of data to be processed. Once the input node has finished reading, preprocessing, and broadcasting the data, it does not wait for the worker nodes to finish their computations. Instead, it moves to reading the next session, while the worker nodes carry out the computations. This sequence of events is depicted in Figure 12.2. Once the next session is read and preprocessed, the input node waits for the worker nodes to finish the computations and gets the preprocessed data for the next session. The compute time is typically larger than the time

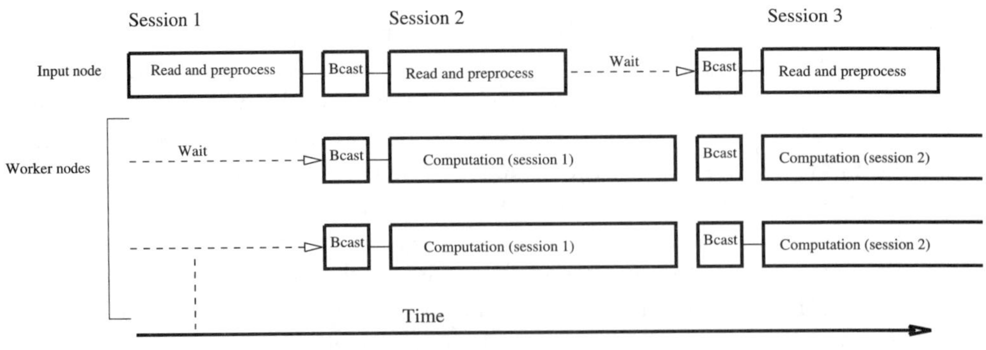

Figure 12.2 Timing relationship between input node and worked nodes.

to read and preprocess the data. Therefore, the worker nodes do not need to spend any time waiting for the input data (except in the first session).

The time taken to run Granger analysis depend on various parameters such as the spatial resolution of the fMRI data, number of volumes, the parameter λ of Lasso regression, and the number of processors. For fMRI data with spatial resolution of $32 \times 64 \times 64$ voxels, and 500 volumes, the 1,024 processor Blue Gene/L system typically takes between 30 and 60 minutes to run the Granger causality analysis on a single session. A typical fMRI experiment with 100 sessions thus takes 1 to 2 rack-days of Blue Gene/L compute time.

The implementation outlined above demonstrates how high-performance computing can be used to carry out Granger causality analysis on the fMRI data. This implementation, though far from optimal, provides a practical way to quickly prototype the analysis technique and get results in a reasonable time. This implementation distributes voxels to processors which independently carry out the regression. As a result, parallelization of the Lasso regression code was not required. (We took an off-the-shelf implementation of Lasso based on LARS).

There are several limitations of this implementation which may need to be addressed if the Granger causality was to be used routinely. The entire code is based on a MATLAB implementation of the LARS algorithm. MATLAB scripts are interpreted and inherently much slower than similar codes in C. Rewriting this code is C is expected to give significant performance improvements.

The code scales well up to a thousand processors, but the gains of adding more processors begin to taper off beyond 1,000 processors. There are two main reasons for this. Firstly, the master node becomes the bottleneck as the number of processors is increased. There is a single master node in the present implementation, which acts as a single point of contact for distributing the work and writing the results to files. As the number of worker nodes increases, the master needs to work faster distributing work to more processors and writing their results. This can scale only to a certain point after which the master node becomes the bottleneck and slows down all the worker nodes. This problem is not fundamental and can be overcome by using MPI-IO [19] for writing the results and by having more than one master node when the number of processors is large.

The second problem, which is more fundamental, is due to the fact that regression computations for different voxels end up taking different amounts of time. In the extreme case, when the number of worker nodes is equal to the number of voxels, every worker node carries out exactly one regression. In this case, the time taken by the program will be almost equal to the maximum time taken in all the regressions. This can be significantly more than the average time taken to run the regressions. The only way to overcome this limit is by using a parallel implementation of the LARS algorithm. The algorithm is based on matrix multiplications which can be parallelized. On fMRI size datasets, such a parallelization is expected to yield significant performance improvements. However, there will still be limits to such scaling.

An optimal implementation should employ both the strategies; that is, it should partition the processors in groups and assign groups of voxels to the partitions. Each processor group should then run a parallel version of the LARS algorithm on all the voxels assigned to the group. Such a parallelization strategy, along with an optimized C/C++ implementation, is expected to lead to drastic performance and scalability improvements. In the meantime, our Octave-based implementation provides a quick and effective way to evaluate the usefulness of the technique.

12.4 Experimental Results

12.4.1 Simulations

We carry out simulations of sparse multivariate autoregressive models (MAR) to generate its realizations. The parameters of the MAR model and the size of its realization were chosen to reflect the sizes of typical fMRI datasets.

We then use the Lasso regression to estimate model parameters (as discussed in Section 12.2.3). We compare the estimated parameters with the actual parameters of the MAR model.

Let $\{a_{ij}(t)\}$ be the parameters of the MAR model and $\{\hat{a}_{ij}(t)\}$ be the model parameters estimated by the regression. Let $S(\alpha) = \{(i,j,t) : |a_{ij}(t)| > \alpha\}$ and $\hat{S}(\alpha) = \{(i,j,t) : |\hat{a}_{ij}(t)| > \alpha\}$. We use the following metric for our evaluation:

- *Precision.* It is defined as the ratio of numbers of true nonzero coefficients estimated to the total number of nonzero coefficients estimated. Formally precision $p = |S(0) \cap \hat{S}(0)|/|\hat{S}(0)|$.
- *Recall.* It is defined as the ratio of the number of true nonzero coefficients estimated to the total number of nonzero coefficients present in the model. Formally recall $r = |S(0) \cap \hat{S}(0)|/|S(0)|$.
- *Thresholding.* Generally, it is more important to discover the causal relationships that are strong. This can be done by considering only the coefficients that are above a given threshold. The precision and recall may respectively be defined with respect to a threshold α as $p(\alpha) = |S(\alpha) \cap \hat{S}(\alpha)|/|\hat{S}(\alpha)|, r(\alpha) = |S(\alpha) \cap \hat{S}(\alpha)|/|S(\alpha)|$.
- *Correlations.* We use two measures of Pearson correlations between estimated and actual model parameters. The first measure c_{true} measures the correlations between $a_{ij}(t)$ and $\hat{a}_{ij}(t)$ over the true nonzero coefficient estimates (i.e., over *true positives* only). The second measure $c_{nonzero}$ corresponds to the correlations between actual and estimated parameters over *true positives*, *false positives*, and *false negatives*. Formally, $c_{true} = < a_{ij}(t), \hat{a}_{ij}(t) >_{(i,j,t) \in S(0) \cap \hat{S}(0)}$ and $c_{nonzero} =< a_{ij}(t), \hat{a}_{ij}(t) >_{(i,j,t) \in S(0) \cup \hat{S}(0)}$.

12.4.2 Simulation Setup

We generated k random sparse graphs with 10,000 vertices and 50,000 edges. Edges in these graphs were assigned random weights normally distributed with zero mean and unit variance. A MAR(k) model was constructed using these graphs. The edge weights were used to define the coefficients $\{a_{ij}(t)\}$ of the MAR(k) model. A realization of the MAR process was obtained using iid noise with zero mean and identity covariance matrix. If the MAR process was not convergent, weights were scaled down uniformly until the process became convergent. We obtained 500 time points of the realization. This realization was used to estimate the model parameters using the Lasso regression. We report results only for $k = 1$ in this paper.

12.4.3 Results

Figure 12.3 shows the precision and recall curves as a function of the Lasso regularization penalty λ [see (12.8)]. As the penalty is increased, the Lasso regression estimates fewer nonzero coefficients. This improves the precision (i.e., the coefficients estimated to be nonzero are more likely to be nonzero) at the cost of recall (i.e., many nonzero coefficients are estimated to be zero). Figure 12.4 shows the precision-recall trade-off for different thresholds. It is evident from these figures that as the threshold increases the precision and recall improves. Thus, the regression consistently makes less errors in estimating larger MAR coefficients.

The precision and recall measures only indicate the accuracy in estimating whether a coefficient is nonzero or not. These measures do not convey any information about the accuracy in the values of the MAR coefficients. The correlation measures indicate the accuracy in estimating MAR coefficient values. Figure 12.5 shows the c_{true} as a function of regularization penalty λ for different thresholds. For all thresholds, c_{true} increases as λ is increased. This indicates that higher regularization penalty (λ) results in fewer nonzero estimates, but with better accuracy. This measure might be a bit misleading because it is only based on true positives and does not consider false positives and false negatives. The measure $c_{non\text{-}zero}$

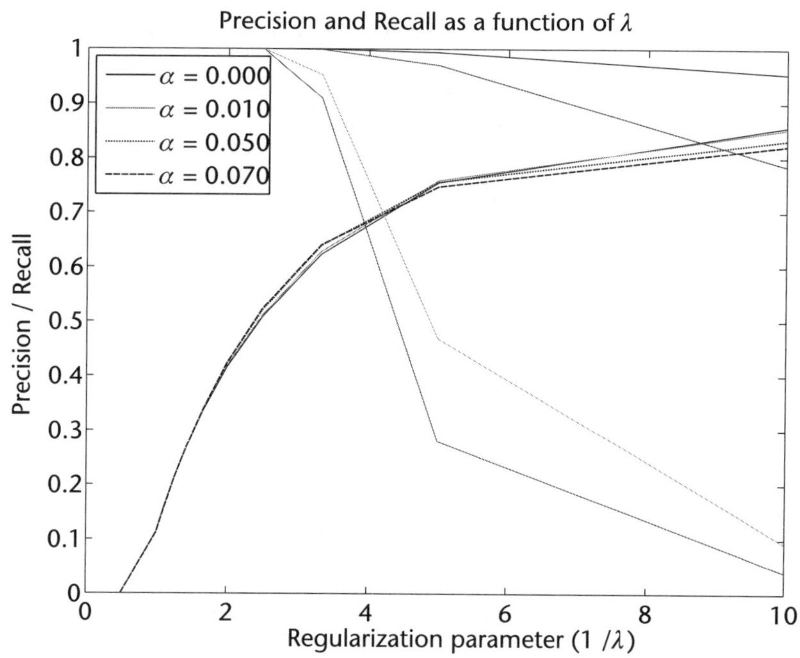

Figure 12.3 Precision and recall as a function of λ.

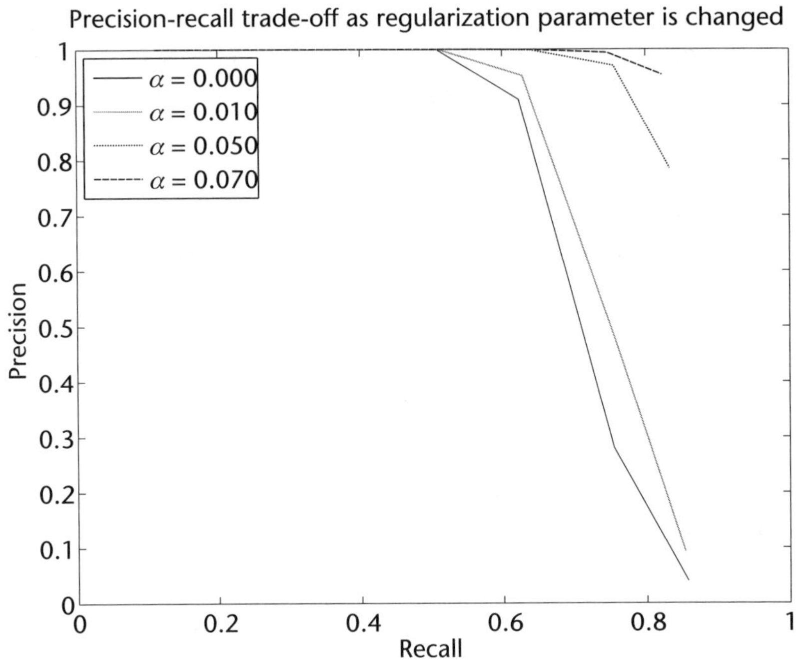

Figure 12.4 Precision-recall trade-off.

considers all of these and is plotted in Figure 12.6 as a function of λ. Note that with a suitable choice of λ, the correlations can be made as high as 0.9. This demonstrates that MAR coefficients (and hence the causality relationships) may be inferred with a reasonable degree of accuracy using the Lasso regression.

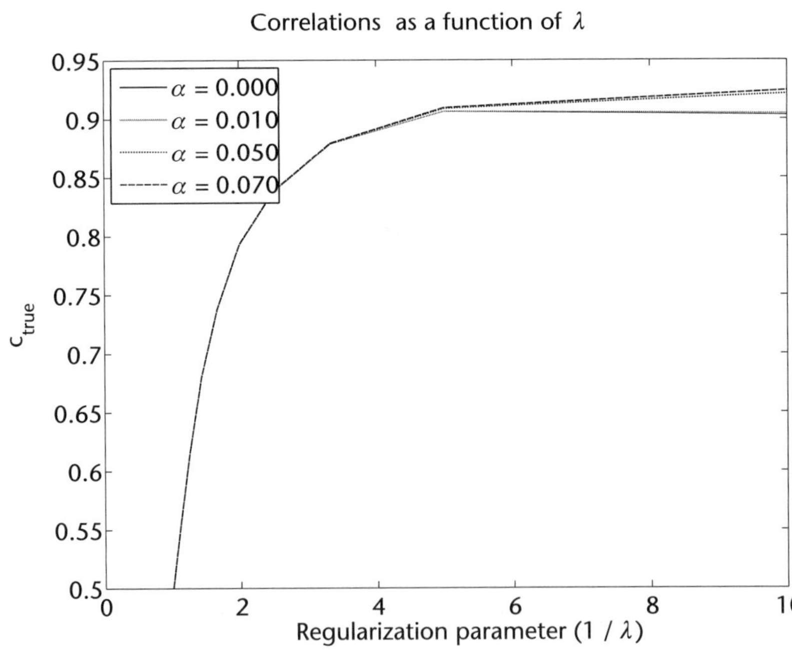

Figure 12.5 Correlations (c_{true}) as a function of λ.

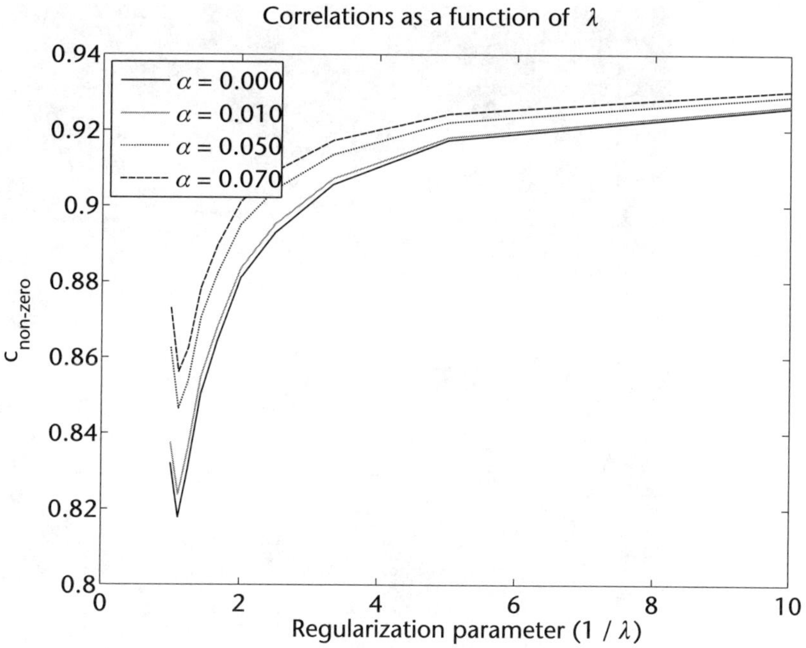

Figure 12.6 Correlations ($c_{nonzero}$) as a function of λ.

12.4.4 Analysis of fMRI Data

The Granger causality analysis gives k connectivity matrices $A(\tau)$ and a map $[1 \ldots N] \rightarrow [0, 1]$, indicating the predictability of voxels in the brain using the past values of other voxels. The map generated using predictability of voxels for fMRI data generated by a simple experiment involving a self-paced finger-tapping task is shown in Figure 12.7. The corresponding Z-statistics map obtained using the GLM analysis is shown in Figure 12.8. Clearly the Z-maps show statistically significant activations in the motor area corresponding to finger movement. It is interesting to note that the same region shows significantly higher predictability as compared to the rest of the brain. This observation has been found consistent in multiple fMRI sessions of multiple subjects carrying out the same finger-tapping task. Initial analysis of fMRI data points to the hypothesis that predictability is a good indicator of brain activations.

However, there are areas found active by the predictability maps that are not found by the GLM maps. It is very difficult to ascertain the "ground truth" relating to brain activations by using noninvasive measures such as fMRI. The most common approach to analyze fMRI data uses the general linear model (GLM) [9] which identifies activations in brain regions as a response to external stimuli represented in the form of a design matrix. By definition, GLM leaves out activations that are not linearly correlated with the design matrix. If the predictability is an indicator of brain activity, the regions found active by GLM analysis must have high predictability. Moreover, one may also find regions of high predictability that are not found to be active by GLM analysis. Our initial analysis suggests that this is indeed the case.

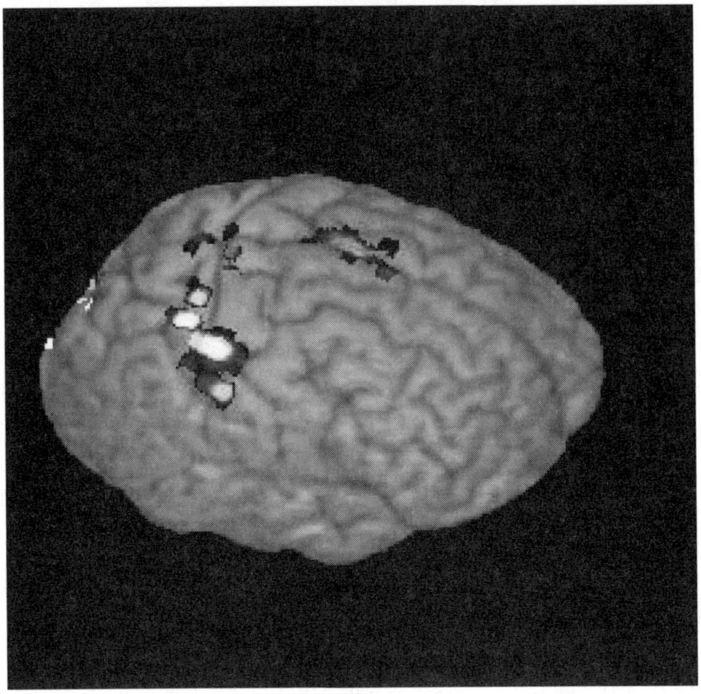

Figure 12.7 Predictability map desived from the Granger analysis.

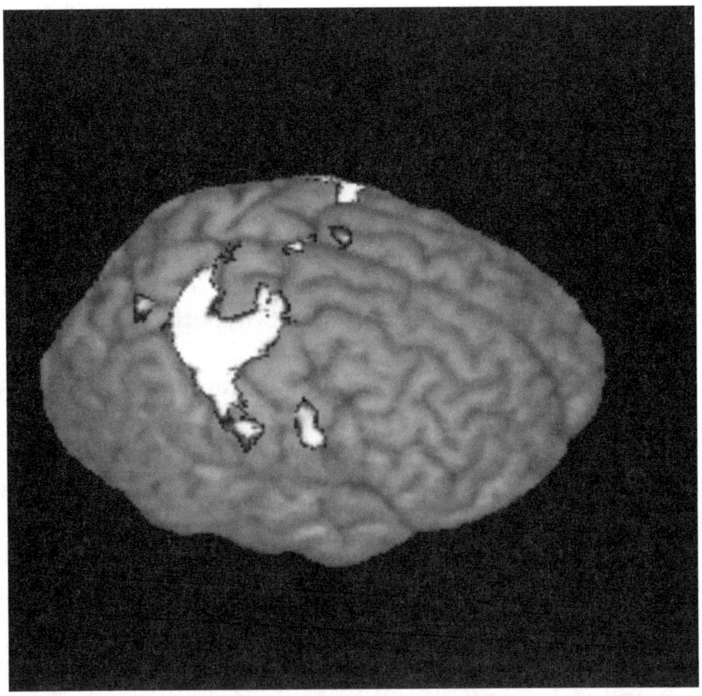

Figure 12.8 The Z-statistics map derived from the GLM analysis.

To formalize the above observation, we define a measure called *weighted coverage* of a set S_1 of voxels in space by another set S_2. First, the voxels in the two sets are clustered into groups of spatially connected 3D components (two voxels are said to be connected if they share a common face, i.e., we are using 6-connectedness in three dimensions). The weighted coverage of S_1 using S_2 is defined as ratio of the weight of clusters of S_1 that intersect with clusters of S_2 to the total weight of clusters of S_1. Clearly, the weighted coverage ranges from 0 to 1. A weighted coverage of zero implies that S_1 and S_2 have no voxels in common. Large coverage indicates that many connected components of S_1 intersect with connected components in S_2. A coverage of one implies that every connected component in S_1 intersects with a component in S_2.

In order to select the voxels that form set S_1 or S_2, we select the top p-percent of active voxels given by the predictability maps on the finger-tapping dataset considered in [7]. We also use the voxels given by the GLM maps using the same top p-percent of active voxels. We used $p = 1.25\%$ to compute S_1 and S_2, and performed 3D connected component analysis on these voxels. The coverage of GLM maps by the predictability maps has a mean of 82% and standard deviation (across subjects and conditions) of 16%. On the other hand, the coverage of predictability maps by the GLM maps has a mean of 41% and standard deviation of 30%. This confirms our hypothesis that the predictability maps discover most of the active regions found by the GLM analysis along with many other regions that are not found by the GLM analysis.

12.5 Discussion

The work presented in this chapter demonstrates how high-performance computing can be gainfully utilized to analyze fMRI data. The functional connectivity analysis of the brain using fMRI data is challenging. For a dataset with N active voxels, the number of voxel pairs grows very rapidly as $O(N^2)$. For a typical fMRI experiment with $N = 33{,}000$ active voxels, there are rougly one billion voxel pairs. With these sizes, one can only carry out very simplistic operations on a typical desktop system. Researchers usually make simplifications to get over the computational bottlenecks. Such simplifications (such as grouping the voxels into regions of interest may actually destroy the information contained in the data and thereby restrict the usefulness of these techniques.

Due to this, there are not too many techniques available to study brain connectivity. Among the techniques that are available, many make simplifications that lose information, thus limiting their utility. Moreover, these techniques are computationally demanding and have been developed over uniprocessor systems and take a long time to complete. As a result, these are not as widespread as the traditional GLM-based analysis of fMRI data.

The work presented here demonstrates that it is possible to carry out more sophisticated analysis of fMRI data. One can start analyzing brain connectivity and dynamics using the concept of Granger causality. Since fMRI data has low temporal resolution but relatively high spatial resolution, it is not possible to employ classical multivariate auto-regressive modeling techniques described in the literature. One needs to solve the problem using sparse regression techniques adopted from machine learning literature. These techniques, though computationally expensive, are tractable using modern supercomputers. One can solve a sparse regression problem for every voxel, on a high-performance machine such as the Blue Gene/L supercomputer. Furthermore, the output of such an analysis produces meaningful results that have been found to be consistent with traditional GLM analysis results. The graphs generated by this technique contain much more information which may be analyzed and visualized in many different ways. A lot of work still needs to be done to translate these results into neurologically significant findings. Research is underway towards this goal.

References

[1] Narasimha R. Adiga, Matthias A. Blumrich, Dong Chen, Paul Coteus, Alan Gara, Mark Giampapa, Philip Heidelberger, Sarabjeet Singh, Burkhard D. Steinmacher-Burow, Todd Takken, Mickey Tsao, and Pavlos Vranas, "Blue gene/l torus interconnection network," *IBM Journal of Research and Development*, 49(2-3):265–276, 2005.

[2] Saurabh Agarwal, Rahul Garg, and Nisheeth K. Vishnoi, "The impact of noise on the scaling of collectives: A theoretical approach," in *HiPC*, pp. 280–289, 2005.

[3] G. Almasi, C. Archer, J. G. Castanos, C. C. Erway, P. Heidelberger, X. Martorell, J. E. Moreira, K. Pinnow, J. Ratterman, N. Smeds, B. Steinmacher-burrow, W. Gropp, and B. Toonen, "Implementing mpi on the bluegene/l supercomputer," in *Lecture Notes in Computer Science*, Vol. 3149, pp. 833–845, 2004.

[4] Paul Coteus, H. Randall Bickford, Thomas M. Cipolla, Paul Crumley, Alan Gara, Shawn Hall, Gerard V. Kopcsay, Alphonso P. Lanzetta, Lawrence S. Mok, Rick A. Rand, Richard A. Swetz, Todd Takken, Paul La Rocca, Christopher Marroquin, Philip R. Germann, and Mark J. Jeanson, "Packaging the blue gene/l supercomputer," *IBM Journal of Research and Development*, 49(2-3):213–248, 2005.

[5] V. M. Eguluz, D. R. Chialvo, G. A. Cecchi, M. Baliki, and A. V. Apkarian, "Scale-free brain functional networks," *Physical Review Letters*, 018102:1–4, 2005.

[6] B. Efron, et al., "Least angle regression," *Ann. Statist.*, 32(1):407–499, 2004.

[7] G.A. Cecchi, et al., "Identifying directed links in large scale functional networks: application to brain fMRI." *BMC Cell Biology*, 8(Suppl 1:S5), 2007.

[8] Alan Gara, Matthias A. Blumrich, Dong Chen, George L.-T. Chiu, Paul Coteus, Mark Giampapa, Ruud A. Haring, Philip Heidelberger, Dirk Hoenicke, Gerard V. Kopcsay, Thomas A. Liebsch, Martin Ohmacht, Burkhard D. Steinmacher-Burow, Todd Takken, and Pavlos Vranas, "Overview of the blue gene/l system architecture," *IBM Journal of Research and Development*, 49(2-3):195–212, 2005.

[9] K. J. Friston, et al., "Statistical parametric maps in functional imaging - a general linear approach," *Human Brain Mapping*, 2:189–210, 1995.

[10] Xavier Martorell, Nils Smeds, Robert Walkup, José R. Brunheroto, George Almási, John A. Gunnels, Luiz De Rose, Jesús Labarta, Francesc Escalé, Judit Gimenez, Harald Servat, and José E. Moreira, "Blue gene/l performance tools," *IBM Journal of Research and Development*, 49(2-3):407–424, 2005.

[11] José E. Moreira, George Almási, Charles Archer, Ralph Bellofatto, Peter Bergner, José R. Brunheroto, Michael Brutman, José G. Castaños, Paul Crumley, Manish Gupta, Todd Inglett, Derek Lieber, David Limpert, Patrick McCarthy, Mark Megerian, Mark P. Mendell, Michael Mundy, Don Reed, Ramendra K. Sahoo, Alda Sanomiya, Richard Shok, Brian Smith, and Greg G. Stewart. "Blue gene/l programming and operating environment," *IBM Journal of Research and Development*, 49(2-3):367–376, 2005.

[12] "Mpi toolbox for octave (mpitb)," http://atc.ugr.es/javier-bin/mpitb.

[13] M.A.L. Nicolelis, "Actions from thoughts," *Nature*, 409:403–407, 2001.

[14] "The GNU octave," http://www.octave.org.

[15] S. Ogawa, T. M. Lee, A. R. Kay, and D. W. Tank, "Brain magnetic-resonance-imaging with contrast dependent on blood oxygenation," *Proc. Natl Acad. Sci. USA*, 87:9868–9872, 1990.

[16] Martin Ohmacht, Reinaldo A. Bergamaschi, Subhrajit Bhattacharya, Alan Gara, Mark Giampapa, Balaji Gopalsamy, Ruud A. Haring, Dirk Hoenicke, David J. Krolak, James A. Marcella, Ben J. Nathanson, Valentina Salapura, and Michael E. Wazlowski, "Blue gene/l compute chip: Memory and ethernet subsystem," *IBM Journal of Research and Development*, 49(2-3):255–264, 2005.

[17] Fabrizio Petrini, Darren J. Kerbyson, and Scott Pakin. "The case of the missing supercomputer performance: Achieving optimal performance on the 8, 192 processors of asci q," in *SC*, p. 55, 2003.

[18] R. Tibshirani. "Regression shrinkage and selection via the Lasso," *Journal of the Royal Statistical Society, Series B*, 58(1):267–288, 1996.

[19] Marc Snir and Steve Otto, *MPI-The Complete Reference: The MPI Core*, Cambridge, MA: MIT Press, 1998.

[20] Charles D. Wait, "IBM powerpc 440 fpu with complex-arithmetic extensions," *IBM Journal of Research and Development*, 49(2-3):249–254, 2005.

[21] S. Weisberg, *Applied Linear Regression*, New York: Wiley, 1980.

PART IV

Postprocessing

CHAPTER 13

Bisque: A Scalable Biological Image Database and Analysis Framework

Ambuj Singh and Kristian Kvilekval

13.1 Introduction

The role of microscopy is changing in that the micrograph is becoming a quantitative measurement device subject to formal analysis. The combination of high-throughput data acquisition techniques and new search technology has transformed much of biology into data-driven fields. Querying the wealth of existing data has become both a source of new hypotheses and a means of testing them. This is also true for biological images. Several large projects are underway to make data-driven research a reality in biomedical fields [26, 34], and this pursuit has been a rich source of new challenges for the information processing community. Microscopy is a fundamental tool of biomedical research, and microscopy images and measurements from images are at the very core of biomedical databases.

13.1.1 Datasets and Domain Needs

With the development of advanced in situ and in vivo imaging techniques, the number of biological images (e.g., cellular and molecular images, as well as medical images) acquired in digital form is growing rapidly. Large-scale bioimage databases are becoming available. Analysis of images has been proven to be critical for scientists to seek answers to many biological problems. For example, analysis of the spatial distribution of proteins in molecular images can differentiate cancer cell phenotypes. Comparison of in situ gene expression pattern images during embryogenesis helps to delineate the underlying gene networks. EM tomography has been used to determine the architecture and arrangement of molecular machines within cells and tissues. Novel techniques that enable millimeter-, micrometer-, and nanometer-scale observations of the same specimen are also emerging. There is enormous potential for mining information in bioimages, especially at different scales of resolution and complexity, to provide a deeper understanding of physiology and pathogenesis. This is relevant to basic sciences as well as for applied sciences and bioengineering. To cope with the great complexity and demand for high-throughput analysis, advanced techniques of bioimage informatics will be the key for any significant progress.

13.1.2 Large-Scale Image Analysis

There is a clear need for novel image informatics tools to continue the advances in our understanding of biological processes using sophisticated information processing. Methods to process unstructured images, extract feature descriptions that can represent the content effectively, and organize these descriptions into a searchable database are needed and are of immense value to the scientific community. These new methods enable biologists to look for patterns in ways that are not currently possible, and use emerging data mining tools to discover new patterns that enhance our understanding of the various biological events at the cellular and subcellular level. Scientists are able to search these image databases by image keywords like texture patterns or time-dynamics information. They can browse, navigate, and explore such databases in a manner that is fast, intuitive, and most likely to reveal connections among images.

There has been significant progress in computer vision and database research dealing with large amounts of images and video. In computer vision, content based image retrieval (CBIR) has been an active research area for over a decade. The CBIR paradigm is to represent the visual content in images using low level image descriptors such as shape, texture, color, and motion, as applicable to a particular application domain.

13.1.3 State of the Art: PSLID, OME, and OMERO

Several large projects have paved the way for large scale image analysis. One of the earliest systems developed was the Protein Subcellular Localization Image Database (PSLID) [16] which provides both management of datasets and analysis. Its features include searching for images by context (annotations) or content, ranking images by typicality within a set (e.g., to choose an image for presentation or publication), ranking images by similarity to one or more query images "searching by image content" or "relevance feedback," comparing two sets of images (hypothesis testing) (e.g., to determine whether a drug alters the distribution of a tagged protein), training a classifier to recognize subcellular patterns, and using a trained classifier to assign images to pattern classes (e.g., assigning images to "positive" or "negative"). The project provides datasets and Java-based software for analysis. Its functionality is available through a Web browser.

The Open Microscopy Environment (OME) [14] has had considerable success. Its long feature list includes support for large datasets, integrated analysis system, and support for data provenance. The original OME server was written in Perl and supports MATLAB integration. It provides access to clients through a Web browser or custom programs (VisBio).

The OMERO [15], also part of the umbrella OME project, is a Java-based system for managing, annotating, and visualizing images and metadata. It is formed out of two main subprojects: OMERO.server, and OMERO.insight. The OMERO.server is a Java-based database server, while OMERO.insight is a client application for organizing, viewing, and annotating images.

13.2 Rationale for Bisque

We review the motivation for Bisque, an image management and analysis system being developed at the Center for BioImage Informatics, University of California at Santa Barbara. We discuss the needed support tools including the digital notebook (an image and metadata annotation tool), and the types of image analysis tools developed. We highlight the flexible data models, new image processing, pattern recognition, and database methods that have been developed in order to facilitate a better understanding of the biological processes from images (which can depict the distribution of biological molecules within cells and tissues as pixels). In the following, we discuss issues with traditional database systems which have hampered the availability and impact of such systems for image analysis.

- *Data modeling:* While image data is the primary concern of an image database and analysis system, experimental data associated with the sample preparation and imaging process is vital to the interpretation of an image. Biologists and image researchers often need to append information about the process or experiment that produced the image. In current systems, experimental data and post-imaging annotations are stored either in free form textual data fields or in fixed relational database tables. Each of these solutions can be problematic as text fields provide no structure at all, while modifying relational tables can be very time consuming and can only be accomplished by database administrators. This has emerged as a significant stumbling block to the rapid ingest of image data. Experimental protocols are often dynamic, and we have found that integrating new labs and the associated datasets can be challenging.
- *Data integration:* Data integration between different groups and labs is a key challenge. While researchers may be doing similar experiments, images and metadata are usually not available to other groups for validation or reuse of data. Terminology can vary greatly between groups collecting similar data, causing unnecessary barriers to sharing of data and results.
- *Scalability:* While computer storage capacity is growing, our appetite for storage grows even faster. Automated microscopy techniques have compounded the issue. Single server solutions can severely limit the scalability of the system, through increased costs and complexity. The Bisque system scales by allowing servers to aggregate data from several sites.
- *Extensibility:* Analysis systems must be easy to use from the point of view of both users and developers. Developers are often discouraged by a steep learning curve. It should be relatively simple to integrate newly designed analyses within the analysis and database framework.

Thus, Bisque extends typical image databases with flexible data organization, database aggregation and distribution, ontology support, and an extended analysis architecture. Flexible data management has emerged as critical to supporting experimental metadata for image collections and integrating analyses and their results. Database aggregation and distribution allow users a broader view of available data seamlessly. Ontology support permits sharing of results between diverse users and laboratories.

13.2.1 Image Analysis

Basic image data contains much information for interpretation. This information is available from the actual sensed values (i.e., the pixels). Modern sensors can scan over time, space, and spectrum. Imaging techniques such as Atomic Force Microscopy (AFM) provide new view points on the real world. Image processing allows these images to be interpreted generally by humans, but also, by machines.

A fundamental action for image processing is object recognition, and the process begins with segmentation of the image. Bisque integrates several general and biologically specific segmentation methods which store derived information as object shapes or statistics. Many analysis types operate over multidimensional images looking for changes over time or space. For example, Bisque offers microtubule body tracking and is able to measure changes to microtubule lengths. This derived information can be summarized as microtubule growth analysis revealing fundamental properties.

13.2.2 Indexing Large Image Collections

Any image repository, be it a photo sharing site, a biomedical image database, an aerial photo archive, or a surveillance system, must support retrieval, analysis, comparison, and mining of images. Over the years, image datasets have gained prominence in scientific explorations due to their ability to reveal spatial information and relationships not immediately available from other data sources. Decreasing cost of infrastructure and the emergence of new tools has brought multimedia-based social networking sites and image-based online shopping sites into prominence. Existing query tools are being used in aerial photography and face recognition for automatic image annotations [24, 30]; in astronomical satellite images to accurately detect point sources and their intensities [12]; and in biology for mining interesting patterns of cell and tissue behavior [5, 9]. The development of functionally enhanced, efficient, and scalable techniques for image retrieval and analysis has the potential to accelerate research in these domains and open new frontiers for commercial and scientific endeavors.

Two problems have plagued the utility of content-based querying of multimedia databases: (1) the definition of similarity between images, and (2) the restriction of distance measurement over semantically meaningful pairs of objects and subimages. With regard to the former, a number of distance measures have been proposed such as Mahalanobis distance [25], Earth Mover's Distance [30], and even learning the appropriate distance measure [31]. The eventual goal is to identify a significant match for a pattern of interest in a query image. With regard to the latter problem, segmentation has been used to define the meaning and context of objects and segmentation-based image retrieval has been proposed. However, questions about the scalability of these techniques remain. We need techniques that can find a meaningful subregion without much user input or extensive image analysis.

Querying based on EMD: Existing distance measures for querying and mining complex images, such as the Lp norms, are insufficient as they do not consider the spatial relationship among the objects. For biological images, the spatial location is important for whether two images should be considered similar. The Earth Mover's Distance (EMD) [30] captures the spatial aspect of the different features extracted from the images. The distance between two images measures both the distance in the feature space and the spatial distance. The EMD has strong theoretical foundations [38], and is robust to small translations and rotations in an image. It is general and flexible, and can be tuned to behave like any Lp-norm with appropriate parameters. The success of using EMD in an image retrieval context over other distance functions has been shown by [30]. Computing the EMD requires solving a linear programming (LP) problem, and is therefore computationally expensive. However, it is possible to use multiresolution lower bounds to accelerate the computation of EMD. The representation of an image in feature space is condensed into progressively coarser summaries. Lower bounds for the EMD in the original resolution are computed from the coarser summaries at various resolutions. Different levels of lower bounds can be computed from different low-resolution summaries: higher-level bounds are less tight, but less expensive to compute. The bounds are reasonably tight and much faster to compute than the actual distance [21].

Localized searching: Querying specific patterns within a region helps domain scientists discover useful information and trends otherwise masked in the whole image. For example, biologists may be interested in examining only a single kind of cell in an image and finding similar patterns of just that cell from a bioimage database. Recent research in content-based retrieval has been either on whole-image retrieval or the retrieval of similar regions where regions have been predefined by segmentation. Whole-image matching fails if the pattern is contained in a specific region of an image, whereas segmentation based region retrieval fails if the pattern is spread across many segments.

An alternative is to consider a new technique in which a user marks the extent of an interesting pattern by a rectangular query region in an image and the database system retrieves top-k images having similar patterns. Both the database images and the given region are tiled and each tile is represented by a feature vector. The match of a pattern tile to a database tile is measured by a score based on a distance measure (such as the EMD) and the amount of background information. A tile having more background noise gets a low score for a match because not only does it contain little or no information but also produces false matches. The given query region is overlapped with each candidate database region. Each overlap generates a score matrix of pair-wise tile matchings. A dynamic programming heuristic can now be applied to the score matrix to choose the largest scoring connected subregion. Finally, top-k regions are returned as the best matching patterns. Results from the above retrieval technique should be semantically meaningful and statistically significant. It is also possible to compute the p-value (a measure of statistical significance) of the result in order to understand whether the retrieved pattern is rare or frequent in the database.

13.3 Design of Bisque

13.3.1 DoughDB: A Tag-Oriented Database

Often database systems allow metadata to be associated with images by defining SQL-like schema extensions. This practice has turned out to be unscalable as different datasets are added or the needs of the individual researchers change due to the strict data modeling requirement and the requirement that data be migrated during schema modifications. Bisque offers a flexible data management system which extends normal schema elements with tag-value pairs. This "soft-schema" allows local schema changes to be added to the system quickly and without the intervention of the database operators.

DoughDB is a flexible tag-value database with SQL-like query language, fine grained permissions, and user accessible provenance (data history). DoughDB is based on an extended object model where a database object may have extensible set of fields (i.e., tags) and values. At the database level, we place no restrictions on tagging and duplicate tags are allowed. A visualization of several database objects is shown in Figure 13.1. Values in the database may be numeric, string, other objects, or a list of the previous types. Each tag-value pair has an owner and permission set to be used for fine grain access to the pairs. Each pair is also time-stamped, and marked with the module execution that created it. This allows users to track the data history of values in the database.

It should be noted that other schema systems (notably SQL DB schemas) are easily mapped to DoughDB objects. Each row of a relational table can be viewed as an instance of a single DoughDB object with a fixed set of fields.

A prototype DoughDB can be implemented on top of a relational database (Postgres). An example relational organization is given below, however many organizations are possible. The core portion of the schema contains just four tables (`taggable`, `tag`, `values`, `names`) shown in Figure 13.2. The full schema contains approximately 25 tables. All DoughDB objects are represented as in-

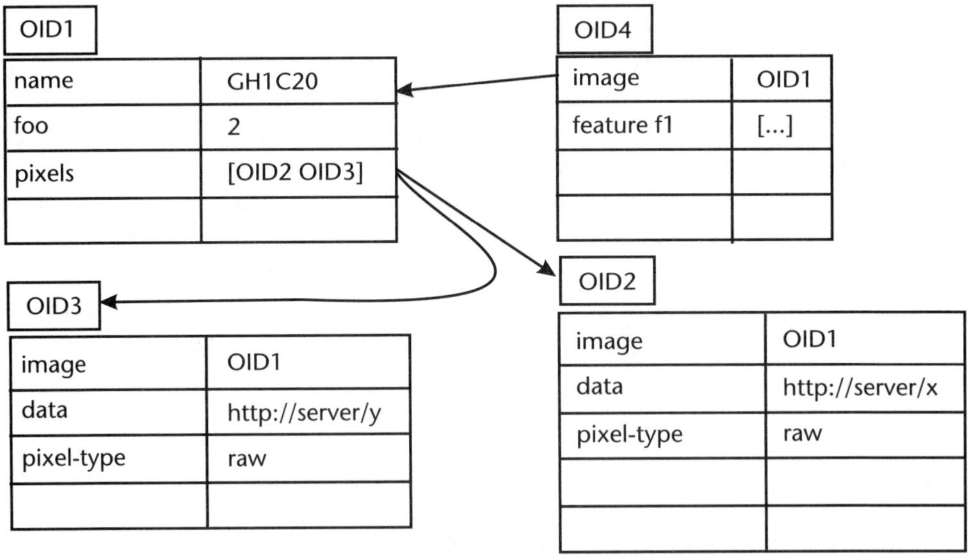

Figure 13.1 DoughDB object model: each object has unlimited fields. Fields can contain a string, number, an object reference, or a list.

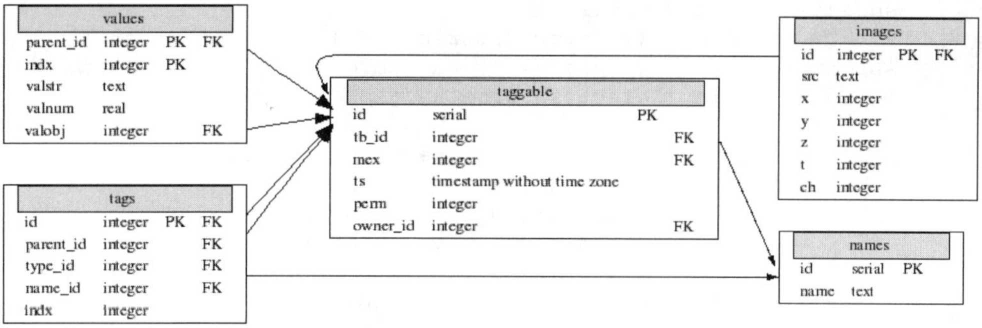

Figure 13.2 Core SQL Schema (`taggable/tag/value`) for DoughDB objects generated from postgres. DoughDB objects are derived from `taggable`, which defines a unique primary key (PK) for each object. Derived objects (`images`, `tags`) use a combined primary key and foreign key (FK) to refer to the base `taggable` object. A `taggable` object may have any number of `tags` where each tag may have a value of string, double, taggable reference, or list of the previous types as denoted by table `values`. The `names` table acts as unique string list for tag names.

stances of `taggable`. For example, the `image` object is derived from `taggable`. A DoughDB object can have any number of tags, while each `tag` can have a list of `value` associated with it. The `image` object is an example of a DoughDB object which has certain tags predefined for ease of programming or for performance reasons. Several other basic objects are defined similarly to `image` including `module`: a module definition; `mex`: a record of module execution; and `template` : a template of predefined `tags`. Other cores types can be defined at the SQL level as needed or desired. User level types are defined by instantiating an instance of `taggable` and adding tags, an operation which does not require low level access to the database schema. Each instance of `taggable` contains several fields common to all DoughDB objects. These fields include ownership, permission, and creation information, including provenance information.

Queries in DoughDB are translated into SQL and then executed by the underlying query relational engine. Efficiency is maintained through breaking complex expressions into subqueries. An example of DoughDB query language is in Figure 13.3. Queries can select object field values and relationships between objects. The query language can be extended so that users can create functional extension that will run as part of the query. This would allow extensions such as specialized string handlers or novel distance functions.

In summary, Bisque supports:

- Development of flexible data model methods for expressiveness and performance;
- A full implementation of DoughDB, including a query processor;
- Building of permission and ownership support at the object level;
- Support for data provenance based on execution histories.

```
select A where B.maskimage = A
           and L2(B.color_descriptor, [0.11, ..]) < 100;
select id from taggable as A, taggable as B1, taggable as B2
where B1.id in (select id from taggable as B1,
                               tags as b1tag,
                               values as b1val
                  where B1.id = b1tag.parent
                    and b1tag.id = b1val.parent
                    and b1tag.name = 'maskimage'
                    and b1val = A.id)
  and B2.id in (select id from taggable as B2,
                               tags as b2tag,
                               values as b2val
                  where B2.id = b2tag.parent
                    and b2tag.id = b2val.parent
                    and b2tag.name = 'color_descriptor'
                    and L2(b2val.list, [0.11, ..]) < 100));
  and B1.id = B2.id;
```

Figure 13.3 Example DoughDB query and possible translation: Find an object A such that there is an object B with tag maskimage pointing to A and the Euclidean distance between its color descriptor and the fixed vector is less than 100.

We have gained experience using our DoughDB architecture by building a test system with several thousands of 5D images (100s of thousands of 2D planes) and several million tags. DoughDB queries are translated to SQL queries, which avoid loading data from the database until explicitly needed. We have also stressed the system by creating an image database of several hundred thousand (nonbiological) images, and we have found the system to have a good performance.

13.3.2 Integration of Information Resources

The "soft-schema" previously outlined may not provide enough structure to correctly interpret data. We focus on two mechanisms to assist researchers in data modeling tasks: template and ontology management. Templates are simple predefined data models that are modifiable by individual users. Ontology integration can be used to help form better models and integrate existing models.

Templates

A "template" is a predefined group of tags that can be applied to objects in the database. In this way, a template represents a basic entity of a data model. These templates can be used as a starting point for users to define their own "soft" schemas. We currently have defined several templates based on our experience with retinal and microtubule images. Designing a good template requires expert knowledge in the problem domain, but not of database schema definitions nor database internals. Creating an initial template is a relatively low-cost operation, which can be extended at any time while the data model evolves. Templates are implemented as special DoughDB objects and supported through a template editor using a design similar to our tagging editor.

Ontology Management

Ontologies are a key tool for sharing, combining, and reasoning about data. While the soft schema described above will permit data models to grow and change naturally as the needs of researchers change, each change complicates sharing and integrating data from other sources.

The ontology system is built on the flexible data management system and thus works seamlessly with data stored there. We have built a stand-alone ontology editor and reasoning system which is being extended and integrated into our Web service environment. Ontology elements can be exported through standard ontology description languages such as OWL (Web Ontology Language).

The key components of the ontology system are:

- *Builder/editor:* The ontology builder is responsible for creating, editing, and import/export of ontologies. Bisque uses several stand-alone editors and will eventually integrate an existing Web-based editor for direct use within the system.
- *Search assistant:* The basic Bisque search system uses the ontology search system to allow searches of related terms and concepts. A special index is built to allow efficient searches of elements based on ontology structure.
- *Matcher and integration assistant:* The ontology matcher assists in integrating different data models based on related terms and concepts. An example of this is when a researcher would like to apply an analysis technique to a foreign dataset. In this case, the metadata available in the foreign data must be matched with local concepts and formats.

13.3.3 Distributed Architecture for Scalable Computing

Bisque is based on the services oriented architecture (SOA) (see Figure 13.4) to ensure that our resources can be used as a platform for many interesting applications. The core services are accessible through simple Web protocols and manipulated as XML documents. Thus, the services are available to any application framework that can access the Web and parse XML. This architecture allows the system to easily scale, supporting both remote and local databases equally. Since Bisque services are Web-based, we can utilize previously developed Web caching products and provide a language-independent suite of tools.

The Bisque architecture (Figure 13.4) defines the following components:

- *Rich Web client:* A rich Web client is based on currently available Web technologies. The key to the rich client is a set of browser executable (Javascript and Java) widgets designed to be used with our data services. Bisque has custom widgets to perform image browsing (Figure 13.5) and presentation with textual and graphical markup (Figure 13.6), analysis presentation (overlay masks, SVG objects, histograms), and analysis functionality. Other specialized capabilities include a graphical markup and image viewer that work on

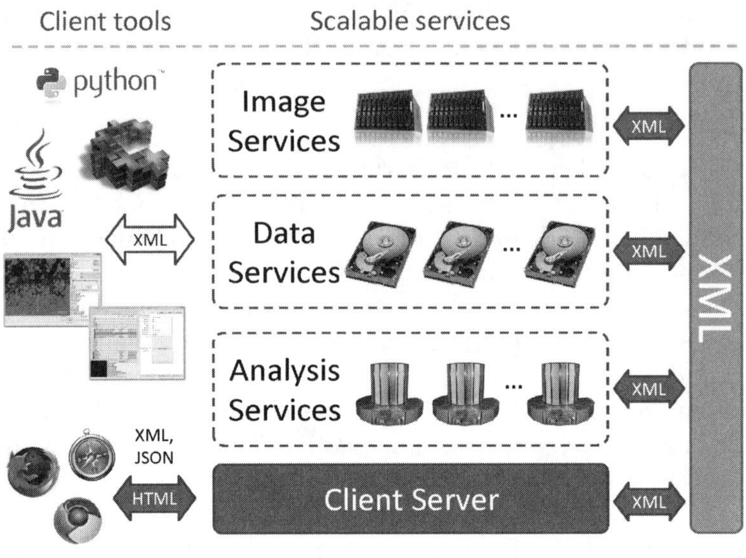

Figure 13.4 Scalable Web architecture (SOA): Each component of Bisque may act as an independent Web service. Image servers store and manipulate images. The data servers provide metadata storage and querying. Other components include the analysis servers (analysis execution) and the client/aggregation server. The web client application interacts with the component servers.

5D images, an inline image tagger, and a dynamic module execution manager that prepares Web input forms based on module specifications and tracks module execution within the system (Figure 13.7).

- *Data service:* The data service provides access to image metadata stored in a local DoughDB instance. Requests are delivered as XML documents to and from clients (either the Rich Client or other Web services). Data service providers exist for both our flexible data model and other systems (OME). The data server is responsible for storage and query requests. Data services can be used in parallel to increase performance and/or storage.
- *Image service:* The image service provides image storage and manipulation services. Any number of image servers can be employed for increased storage allowing the image collection to grow as needed. The image server also performs image manipulations like thumbnail generation, formatting, and slice preparation (for 3D, 4D, and 5D images).
- *Client/aggregate service:* The aggregate service acts as an intermediary between the Web Client and remote data, image, and modules servers. It is responsible for aggregating and distributing data and requests, respectively, for the client. The Web service delivers data to the client browser as HTML and XML documents. This component allows seamless viewing of images and metadata even though metadata may be spread among several systems, permitting expansion when a single DoughDB/data service instance will not suffice.

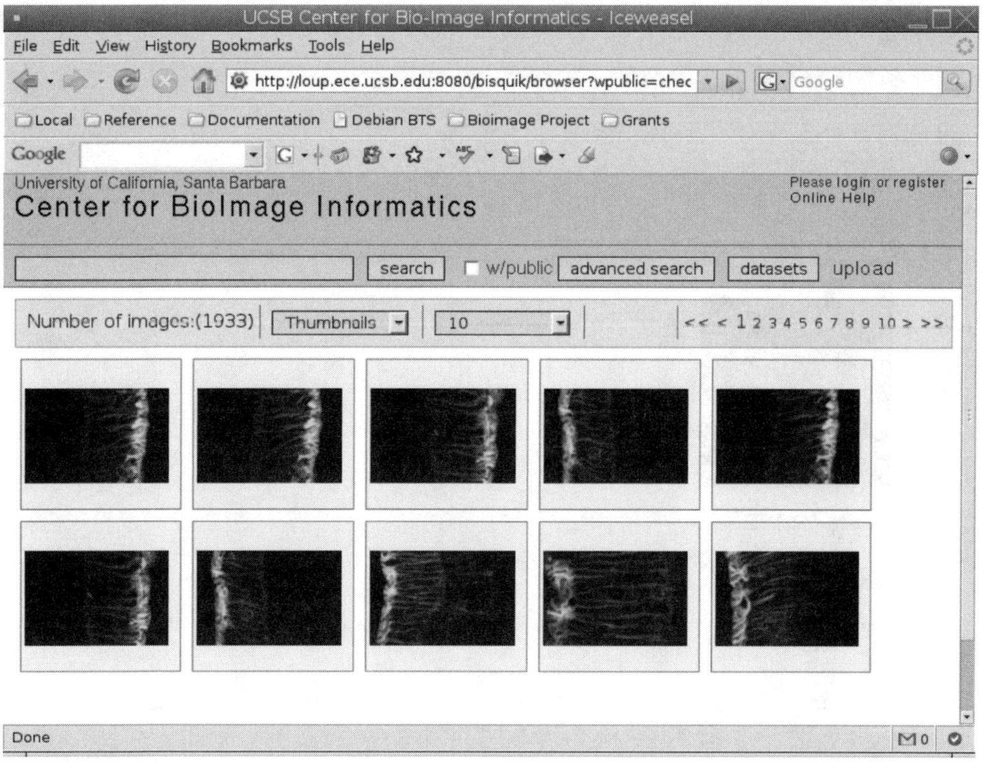

Figure 13.5 Bisque browser mock-up: A set of browsable images is shown with a portion of their tags. The search box can be used to search for tag expressions such as "antibody:vimentin."

We expect that any Bisque system may access several data servers, both local and remote servers (other labs). We have prototyped data servers for both Bisque style flexible databases and other image databases including our Bisque/OME database servers.

- *Module service:* The module service provides analysis registration and execution services. Requests are sent by the client or the Web service to the module service for execution. The module service interacts with the other servers (image, data) to collect resources and save results. Several engines (MATLAB, Python, Java) have been created and several more (C++, C#) are planned so that module writers can use their language of choice.

In addition to the core servers defined above, Bisque provides extensions based on the Web-service model. For example, the following extension servers have been defined:

- *Bisque/OME/PSLID bridge:* The Bisque/OME bridge server converts Bisque service requests into OME requests, effectively making any Bisque (or OME) server into a Bisque data/image server. This allows Bisque to effectively be used as a frontend to multiple Bisque/OME servers. This technique can be used to scale several backend image and data servers. In the future, we expect other bridge servers to be built for other image database systems as

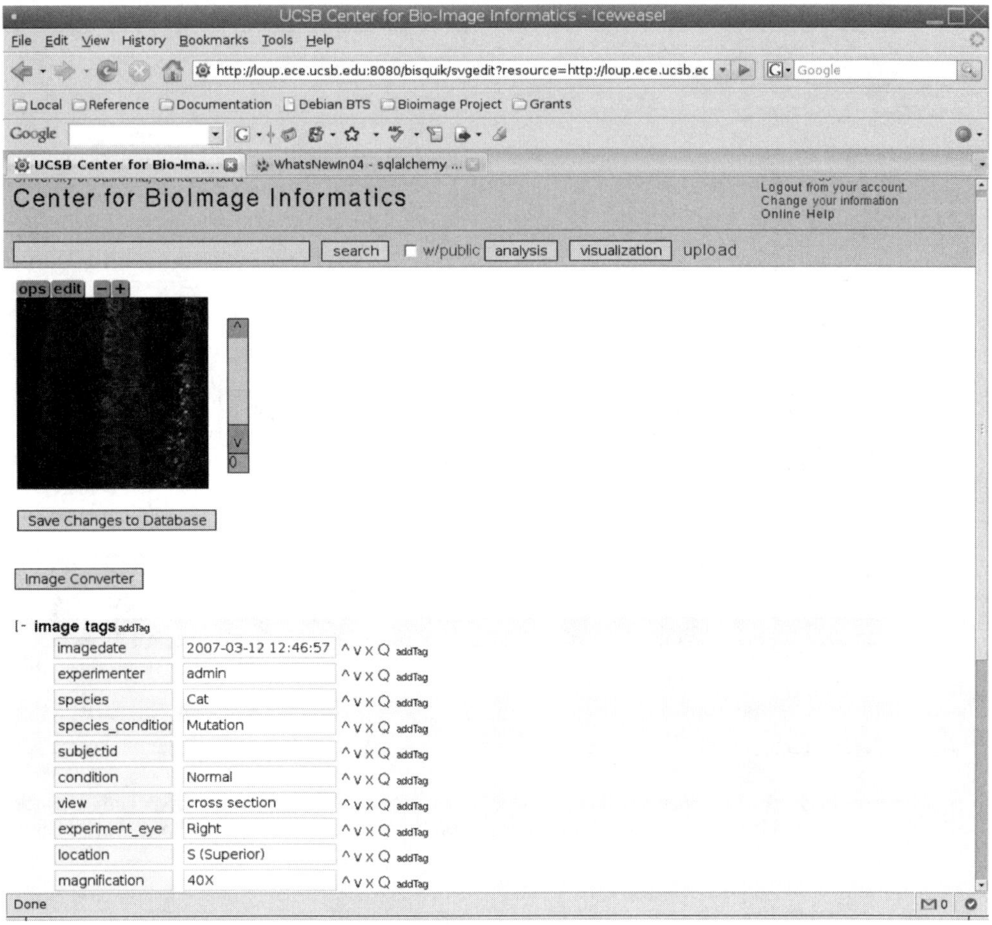

Figure 13.6 Bisque tagger prototype: Tags represent an expandable set of fields associated with the image. Tags can be added, deleted, and modified. Graphical markup will also be available with functions similar to the digital notebook.

needed. It should be noted that in this configuration, Bisque can be used to develop and test data models before migrating to Bisque/OME servers and fixed data models.

The core aggregate service allows users to locally tag and store local analysis results while operating on remote (bridged) images and data. The aggregate service combines the metadata from several sources including bridge servers.

- *Image similarity index:* The similarity search server permits scalable indexing and search by content for biological images. The search server maintains its own index and provides content-based search. The similarity index builds on our core servers using the data server and the image server. The data server defines basic image identifiers while the image server provides pixel data. The framework defines the interface the similarity-servers must adhere to, including initially building the index, extraction of relevant features, and searching. The framework does not specify the implementation of the index-

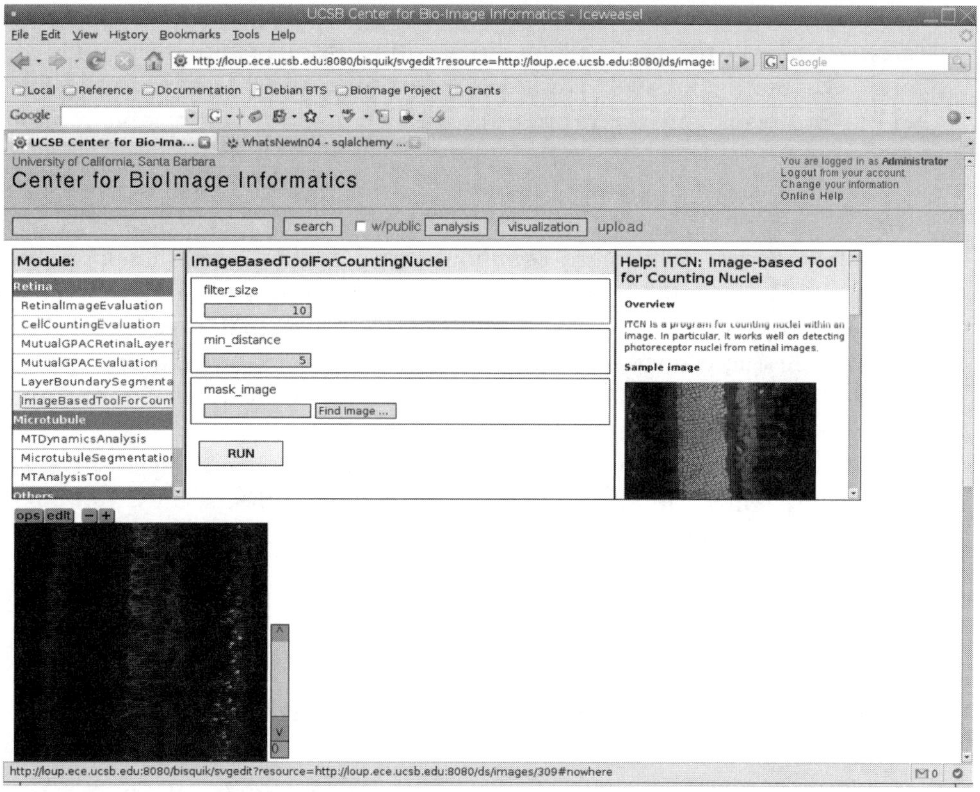

Figure 13.7 Bisque analysis prototype: The nuclei counting analysis generates an interface based on the module parameters. Parameters may be read from the image tags or from the user interface. Results of the analysis can be stored as tags for review or further analysis.

ing scheme, thus allowing different index structures and search algorithms to be integrated.

Most services use a RESTful architecture, which exposes resources through URIs instead of functional services such as XML-RPC and SOAP. In a restful architecture, resources are manipulated by using the standard HTTP verbs (GET, PUT, POST, DELETE, and HEAD). There are many benefits attributed to building Web services using the RESTful pattern, including scalability through Web caches, and the use of client side state. Based on our experience, we expect that all Bisque services and extensions will be built as component Web services allowing easy recombination and reutilization by virtue of their language and tool independence.

13.3.4 Analysis Framework

The Bisque system functionality is extensible in several different ways. The simplest mechanism for adding functionality is through the analysis modules. Analysis modules interact with other components of the Bisque system (i.e., the data server and the image server). Our framework allows an analysis to be written in any language with access to sockets and HTTP protocols. In several cases, there exists a language API to access all facilities within Bisque (initially MATLAB, Python, and Java, but the list will grow). However, since all services are accessible through HTTP protocols, any client program with Web access will be able to interact with our system.

Two styles of analysis modules are available. The first is termed "internal" for an analysis available through the Web interface and which uses our parallel execution engine. These are commonly used for analyses that have been found useful to a large number of users or that need significant computational resources. We also support an "external" analysis for those that need to operate outside the core Bisque framework. In essence, these analyses are external programs that utilize Bisque resources. It is often easier to initially create an external analysis and migrate it to an internal module as needed.

Both styles of analysis modules have access to the same Bisque services. In particular, analysis modules may create DoughDB objects, tags, and links through the data service, execute other modules, and store new images and masks through the image service. This key feature promises to make integrating analysis modules much simpler than current methods.

13.4 Analysis Architectures for Future Applications

Image management systems of the future will need to provide additional capabilities such as supporting the use of external computing resources for extensive computation-bound tasks and the use of libraries for modeling of complex systems. In this section, we develop the idea of probabilistic data management and analysis, which is becoming increasingly important for image databases.

The primary methods of scientific experimentation are becoming high throughput, producing vast amounts of data. However, much of the data being produced is not in the form of definitive measurements, but rather contains an inherent uncertainty from the acquisition process. The current state of scientific analysis does not usually take this uncertainty into account. In microscopy, for example, poor image acquisition, light pollution, and experimental error all produce images with poorly demarcated cell or tissue boundaries. The typical automated process of analysis on these images performs a threshold-based segmentation followed by feature extraction. While such analyses can provide accurate results, they are usually limited to high-quality, high-resolution images and only present one possible interpretation of the data [37]. No notion of uncertainty is presented; thus some of the original data is, in essence, lost.

Elements of uncertainty arise in all stages of the analysis work-flow including image acquisition, feature extraction, querying and management, and pattern discovery. Aspects of uncertainty must be explicitly modeled and accounted for in all stages, and their precise impact understood in a holistic manner. While the main context of this chapter is biological images, the ideas extend to a variety of applications in other image-based domains, such as environmental management, geographical information science, remote sensing, and interactive digital multimedia.

Recent work in the field of data management has seen a growing trend in the focus on uncertain data [19, 36]. Many sources inherently provide uncertain data due to technical limitations and acquisition conditions. As accuracy becomes important, we need to examine this data probabilistically to account for the uncertainty [2].

The accepted model for evaluation of probabilistic queries is the "possible-worlds" model [4, 11, 13, 19] in which a set of possible certain worlds is defined for the database objects, the query is evaluated in these certain worlds, and the result is aggregated to return the answer. In general, such evaluation is computationally challenging and makes a straightforward implementation all but infeasible for large databases. A number of strategies have been developed to find subclasses of queries that can be answered efficiently. This includes the isolation of safe query plans [11,12]. Other approaches combine uncertainty with lineage [5] and consider the decomposition of possible worlds into a manageable set of relations [3]. Aspects of completeness under various models of uncertainty have been considered [32].

Answering queries on probability density functions (pdf) has been well studied [6, 8] under the Gaussian and Uniform pdfs. These assumptions allow for interesting and efficient index structures, and can be appropriate for the uncertainty of many of the individual measurements. They are too restrictive for pdfs that occur as a result of summarization; however, the observations being summarized may be generated by different mechanisms. For instance, a summary of the density of bipolar cells in a detached cat retina will have two peaks, corresponding to parts of the retina that are injured and healthy, respectively. Fitting a model distribution, such as a Gaussian, to the data is only appropriate when the data are well understood, and in a scientific database, the most interesting data to query are precisely the ones that are not well understood. Others have considered the indexing of uncertain categorical data [35], the use of Monte Carlo simulations and state space search to answer top-k queries [29, 36], and skyline queries [10].

The realization that uncertainty plays an especially important role in biological data has been slow to emerge, yet progress is being made. Louie et al. focus on the integration of uncertain databases specific to the protein annotation domain [23]. Segal and Koller propose providing a model for hierarchical clustering of probabilistic gene expression data [33], and Rattray et al. provide a method for retaining the uncertainty inherent in microarray data analysis [28]. While these models and methods are clearly important, there is a general lack of work specific to the microscopy field. Ljosa and Singh present a model for probabilistic segmentation of retinal horizontal cells in microscopy images [22]. While this method certainly adds to the small body of probabilistic analyses, it does not provide a general framework to incorporate existing certainty-based analyses into uncertain databases and data mining.

Besides querying and mining, the computation of features is another important aspect of biological image analysis. A typical task here requires the isolation of a cell from a background and the computation of features such as soma size, branching pattern, and so on. Current work in the field of image analysis [17] focuses on segmentation, which is considered the first step of analysis. The general approach is outlined as a three step process: (1) segmentation of objects, (2) feature extraction, and (3) analysis. This leaves all analysis dependent on one specific interpretation of the image and gives no measure of the underlying data uncertainty. In an alternative approach, each pixel is classified as part of the cell or part of the background. More formally, let S be an image with N pixels and $X = X_s, s \in S$ be the set of corresponding random variables, where X_s is 0 or 1 depending on whether it is part of the background or cell. A possible world is an instantiation of all random variables, and thus defines a specific binary image. If images are 1,024 $\times$ 1,024, then we have $2^{1{,}024 \times 1{,}024}$ possible worlds. Exploring this entire space is clearly intractable, and we need a method to overcome this limitation.

One possible solution to the computational bottleneck is by sampling over the possible worlds. A simple approach is to assume pixel independence. In this method, we randomly classify each cell as part of the cell according to its probability. We repeat this operation for all pixels to obtain one possible world. Features can then be measured in this binary world. This process is then repeated until a sufficient sample size of worlds has been collected. A distribution of the features is obtained in this manner. A better and less error-prone approach is to assume that pixels are inherently dependent upon their surrounding neighbors. One can define a neighborhood system and assume that pixels are conditionally independent given their neighborhood [18]. The challenge then is to estimate the joint probability distribution. For this, one can employ Gibbs sampling, a common Markov Chain Monte Carlo (MCMC) technique [7]. In preliminary experimental results, this technique gave good results for a number of cellular features when compared with manual ground truth.

13.5 Concluding Remarks

Data-driven science has become a reality, and quantitative image analysis is becoming ever more important to many scientific endeavors. This chapter discusses new developments in data-driven image-based science, bridging the gap between image analysis and databases. Bisque extends traditional image databases with an extensible analysis integration framework. It is built using a distributed Web-based architecture to ensure scalability and ease of access for both users and analysis designers. The combination of a flexible schema architecture with ontologies (controlled vocabularies) promises easier data modeling for experts in many fields without the need to consult with database experts. Managing provenance of images and chains of analyses improves the development of new hypotheses and the reproducibility of conclusions. New techniques in managing uncertainty and probabilistic data have been explored, capturing nuances in data previously unseen. Bisque is a combination of emerging techniques and technologies providing an integrated package for scientists, image processing experts, and database researchers in order to explore this new domain.

References

[1] *Proceedings of the 23nd International Conference on Data Engineering, ICDE 2007*, April 15–20, 2007, Istanbul, Turkey. IEEE, 2007.

[2] C. Aggarwal. "On density based transforms for uncertain data mining." In *International Conference on Data Engineering*, 2007.

[3] L. Antova, C. Koch, and D. Olteanu. "10^{10^6} worlds and beyond: Efficient representation and processing of incomplete information." In *ICDE* [1], pp. 606–615.

[4] D. Barbará, H. Garcia-Molina, and D. Porter. "The management of probabilistic data." *IEEE Trans. Knowl. Data Eng.*, 4(5):487–502, 1992.

[5] O. Benjelloun, A. D. Sarma, A. Y. Halevy, and J. Widom. "Uldbs: Databases with uncertainty and lineage." In U. Dayal, K.-Y. Whang, D. B. Lomet, G. Alonso, G. M. Lohman, M. L. Kersten, S. K. Cha, and Y.-K. Kim, editors, *VLDB*, pp. 953–964. ACM, 2006.

[6] C. Böhm, A. Pryakhin, and M. Schubert. "The gauss-tree: Efficient object identification in databases of probabilistic feature vectors." In Liu et al. [20].

[7] G. Casella and E. I. George. "Explaining the gibbs sampler." *The American Statistician*, 46(3):167–174, 1992.

[8] R. Cheng, Y. Xia, S. Prabhakar, R. Shah, and J. S. Vitter. "Efficient indexing methods for probabilistic threshold queries over uncertain data." In Nascimento et al. [27], pp. 876–887.

[9] J. Coombs, D. V. der List, G. Y. Wang, and L. M. Chalupa. "Morphological properties of mouse retinal ganglion cells." *Neuroscience*, 140(1):123–136, 2006.

[10] X. Dai, M. Yiu, N. Mamoulis, Y. Tao, and M. Vaitis. "Probabilistic spatial queries on existentially uncertain data." In *Advances in Spatial and Temporal Databases*, Vol. 3633/2005, pp. 400–417. Springer, 2005.

[11] N. N. Dalvi and D. Suciu. "Efficient query evaluation on probabilistic databases." In Nascimento et al. [27], pp. 864–875.

[12] N. N. Dalvi and D. Suciu. "Management of probabilistic data: foundations and challenges." In L. Libkin, editor, *PODS*, pp. 1–12. ACM, 2007.

[13] D. Dey and S. Sarkar. "A probabilistic relational model and algebra." *ACM Trans. Database Syst.*, 21(3):339–369, 1996.

[14] I. Goldburg, et al. "Open microscopy environment." http://www.openmicroscopy.org/.

[15] J. Swedlow, et al. "Omero." http://www.openmicroscopy.org/.

[16] R. M., et al. "Protein subcellular localization image database." http://pslid.cbi.cmu.edu/public3/.

[17] Y. Fok, J. Chan, and R. Chin. "Automated analysis of nerve-cell images using active contour models." *IEEE Transactions on Medical Imaging*, 1996.

[18] S. Geman and D. Geman. "Stochastic relaxation, gibbs distributions, and the bayesian restoration of images." *PAMI*, 6(6), November 1984.

[19] L. V. S. Lakshmanan, N. Leone, R. B. Ross, and V. S. Subrahmanian. "Probview: A flexible probabilistic database system." *ACM Trans. Database Syst.*, 22(3):419–469, 1997.

[20] L. Liu, A. Reuter, K.-Y. Whang, and J. Zhang, editors. *Proceedings of the 22nd International Conference on Data Engineering, ICDE 2006, April 3–8, 2006, Atlanta, GA, USA*. IEEE Computer Society, 2006.

[21] V. Ljosa, A. Bhattacharya, and A. K. Singh. "Indexing spatially sensitive distance measures using multi-resolution lower bounds." In *Proceedings of the 10th International Conference on Extending Database Technology (EDBT)*, pp. 865–883, March 2006.

[22] V. Ljosa and A. Singh. "Probabilistic segmentation and analysis of horizontal cells." In Liu et al. [20], pp. 980–985.

[23] B. Louie, L. Detwiler, N. Dalvi, R. Shaker, P. Tarczy-Hornoch, and D. Suciu. "Incorporating uncertainty metrics into a general-purpose data integration system." In *Scientific and Statistical Database Management*, p. 199, 2007.

[24] W.-Y. Ma and B. S. Manjunath. "A texture thesaurus for browsing large aerial photographs." *J. American Society for Information Science*, 1998.

[25] P. C. Mahalanobis. "On the generalised distance in statistics." In *Proc. of the National Institute of Science of India*, 1936.

[26] M. E. Martone, S. Zhang, A. Gupta, X. Qian, H. He, D. Price, M. Wong, S. Santini, and M. H. Ellisman. "The cell-centered database: A database for mulstiscale structural and protein localization data from light and electron microscopy." *Neuroinformatics*, 1(3):379–396, 2003.

[27] M. A. Nascimento, M. T. Özsu, D. Kossmann, R. J. Miller, J. A. Blakeley, and K. B. Schiefer, editors. *(e)Proceedings of the Thirtieth International Conference on Very Large Data Bases, Toronto, Canada, August 31 - September 3, 2004*. Morgan Kaufmann, 2004.

[28] M. Rattray, X. Liu, G. Sanguinetti, M. Milo, and N. D. Lawrence. "Propagating uncertainty in microarray data analysis." *Brief Bioinformatics*, 2006.

[29] C. Re, N. N. Dalvi, and D. Suciu. "Efficient top-k query evaluation on probabilistic data." In *ICDE* [1], pp. 886–895.

[30] Y. Rubner, C. Tomasi, and L. J. Guibas. "The earth mover's distance as a metric for image retrieval." *Int.J. Comput. Vision*, 2000.

[31] Y. Rui, T. S. Huang, M. Ortega, and S. Mehrotra. "Relevance feedback: A power tool in interactive content-based image retrieval." *EEE Trans. Circuits and Systems for Video Technology*, 8(5), 1998.

[32] A. D. Sarma, O. Benjelloun, A. Y. Halevy, and J. Widom. "Working models for uncertain data." In Liu et al. [20].

[33] E. Segal and D. Koller. "Probabilistic hierarchical clustering for biological data." In *Proceedings of the Sixth Annual International Conference on Computational Biology*, pp. 273–280, 2002.

[34] A. K. Singh, B. Manjunath, and R. F. Murphy. "A distributed database for biomolecular images." *SIGMOD Record*, 33(2):65–71, June 2004.

[35] S. Singh, C. Mayfield, S. Prabhakar, R. Shah, and S. E. Hambrusch. "Indexing uncertain categorical data." In *ICDE* [1], pp. 616–625.

[36] M. A. Soliman, I. F. Ilyas, and K. C.-C. Chang. "Top-k query processing in uncertain databases." In *ICDE* [1], pp. 896–905.

[37] C. Weaver, P. Hof, S. Wearne, and W. Lindquist. "Automated algorithms for multiscale morphometry of neuronal dendrites." 2004.

[38] M. Werman, S. Peleg, and "A. Rosenfeld. A distance metric for multi-dimensional histograms." *Computer, Vision Graphics and Image Processing*, 32:328–336, 1985.

CHAPTER 14

High-Performance Computing Applications for Visualization of Large Microscopy Images

Rajvikram Singh, Abel W. Lin, Steven Peltier, Maryann E. Martone, and Mark H. Ellisman

Large data visualization problems are prevalent in microscopy and find some reprieve in high-performance computing (HPC). Clusters and multi-CPU architectures help in accelerating applications such as feature extraction, image processing, and analysis of large 2D and 3D datasets. Cluster driven tile-displays have recently become popular end points for large data exploration because of their high-resolution capability and scalability. Certain algorithms and strategies have played a key role in designing parallel applications for these high-resolution displays. Issues regarding performance tuning of graphics, processing, and networking subsystems have also become important factors in building efficient scientific visualization pipelines for microscopy data.

14.1 Mesoscale Problem: The Motivation

Researchers analyzing brain tissue samples at multiple resolutions are faced with a well-known problem when traversing scales spread across several orders of magnitudes: the higher the resolution, the smaller the scope. When an investigator is zoomed in on a sample at subcellular resolution he or she has lost context of where the region of interest (ROI) lies with reference to other ROIs. This gap between dimensional scales makes it difficult to understand how higher order structures are constructed from finer building blocks. Problems traversing scales are particularly acute in the dimensional range that is now called *mesoscale*, which is the dimensional range spanning hundreds of microns to nanometers. Structural elements within this range include subcellular structures, cell-cell interactions, and macromolecular constituents. Within the nervous system, the mesoscale encompasses those structures most associated with information processing (i.e., synaptic complexes, subcellular microdomains, and the fine structures of axons and dendrites). Understanding mesoscopic structures within the brain presents a unique challenge because of the extended nature of nerve cells, the cellular and molecular complexity of nervous tissue, and the intricate arrangement of cellular processes. Although the nervous system is perhaps the most extreme in terms of mesoscopic complexity, we are limited in our ability to understand even a well-ordered tissue such as muscle across large expanses in fine detail.

The mesoscale gap arises in part from the requirement to use multiple imaging technologies to examine a specimen across scales. Each technology requires different expertise, specimen preparation techniques, and contrast mechanisms, and also requires a severe reduction in the amount of tissue. For example, if the pipeline begins with an entire brain, the end results in one small block of tissue, $< 0.5\ mm^3$. These requirements make it difficult for individual researchers to bridge scales, both because single researchers may not be familiar with a given technology and because there is significant loss of context as the scope decreases with increasing resolution of imaging technologies. Bridging techniques such as multiphoton microscopy and electron tomography, correlated microscopy is a key methodology for acquiring the necessary multiscale data in order to fill in the resolution gaps between gross structural imaging and protein structure: data which is central to bridging the mesoscale gap and to the elucidation of the nervous system.

From a computer science perspective, the mesoscale presents a number of challenges. The two major ones are large data sizes and ultra-high resolution content. The typical size of a dataset collected by a microscope, capable of acquiring ultrawide field mosaics, ranges from a couple of gigabytes to a few terabytes on disk. The content resolution ranges from a few hundred megapixels to hundreds of gigavoxels. The largest CCD sensors for electron microscopes are now approaching 64 megapixels. Figure 14.1 shows an image acquired from a 64 megapixel sensor on an electronic microscope. With image formats of $8{,}000 \times 8{,}000$ pixels, these systems can be used to acquire wide-field mosaics (2D, 3D, and 4D) that exceed

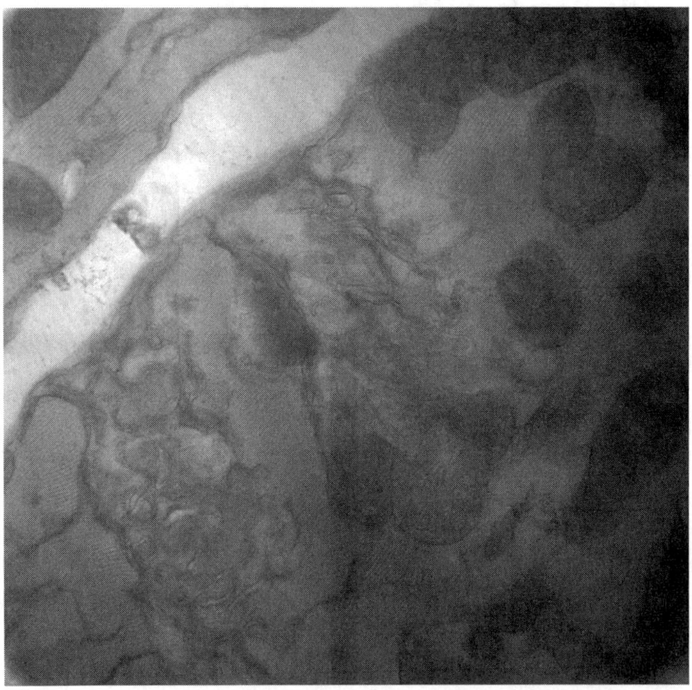

Figure 14.1 A section of the mouse heart as seen under National Center for Microscopy and Imaging Research's 8K × 8K CCD camera. The camera has a collective resolution of 64 megapixels and the specimen stage can be moved in X and Y to collect tiles. These tiles can then be stitched together form ultrawide field of view mosaic.

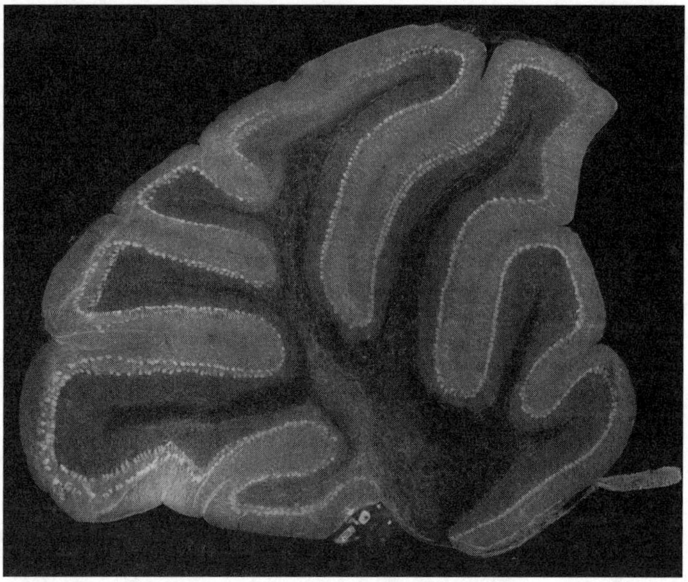

Figure 14.2 The figure shows a scaled down version of an 18,000 × 16,000 pixel montage created by data acquired from a light microscope. This "small" dataset has a resolution of 288 megapixels and is ~70 times the resolution of the best LCD monitor available in 2008.

hundreds of gigabytes of content [1]. Figure 14.2 shows one such moderate-sized montage constructed from hundred of smaller tiles.

For most analysis applications, researchers want to juxtapose many such datasets next to each other for visual comparison and analysis. Due to the large memory footprints of these datasets, it is not possible to load them entirely in the video memory or even in the RAM of typical workstations. To add to the complexity, since many research groups are collaboratively working on these datasets, they are usually stored offsite, typically at a data center. In order to move a 100-gigabyte dataset over a university fast Ethernet network it typically takes over 2 hours. Moving large data between the stages of the visualization pipeline not only requires efficient storage and query engines but networks that are several magnitudes faster by current standards.

14.2 High-Performance Computing for Visualization

The sheer size of the datasets generated by the field of microscopy has challenged computer scientists to come up with unique approaches for each phase of the visualization pipeline. Figure 14.3 shows a typical scientific visualization pipeline. The computational and rendering components of the pipeline are usually clusters or supercomputers hosted by universities and research organizations. Often these clusters are geographically located several hundreds of miles apart and at times in different countries since microscopy is a very collaboration-rich field. However, the final result of the visualization has to be delivered back to the researchers who initiated the process. To compound the problem, visualization applications

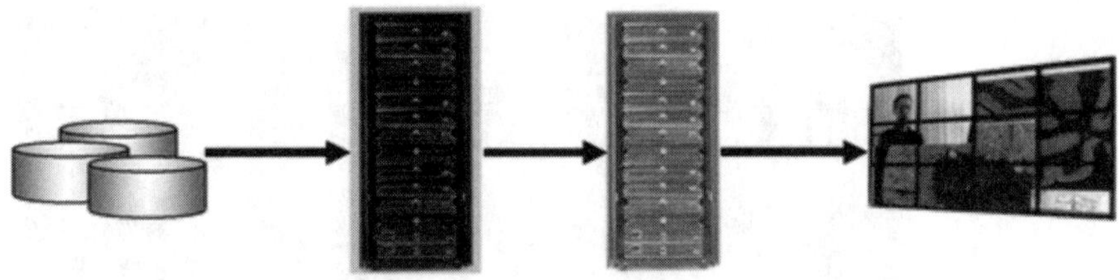

Figure 14.3 A typical scientific visualization pipeline.

are expected to be interactive in nature. This means that the popular model of queued job-submission in high-performance computing does not apply anymore since the users are expecting an instantaneous feedback. The causality of a user's mouse interaction perpetuates data to be pulled from data centers and delivered to the cluster-farms for computation and rendering. The result of the rendering, which is usually graphical primitives or pixels, is then delivered to the display at the user's end. All this needs to happen within a period of few hundred milliseconds to seconds in order for the visualization to be interactive. The following section describes how computational grids and high-speed optical networks are used for solving such data and compute intensive tasks.

14.2.1 Data Acquisition

Filling in the mesoscale information gap requires data from a number of different imaging instruments. Researchers can now collect correlated, multiscale data from electron and laser scanning light microscopes to create interconnected 2D, 3D, and 4D geometries to study structure function relationships within key cellular and tissue subsystems.

Over the past decade, this data generation has accelerated at an exponential rate, and scientific imaging instruments (e.g., electron and light microscopes) have been automated to now deliver large datasets, some exceeding 1 terabyte. From a resource perspective, such immense data sizes require seamless access to computational, data management, and visualization resources that scale beyond what can be effectively deployed by individual research groups [11, 21].

14.2.2 Computation

Visualization in microscopy involves some very computationally intensive processes such as volume rendering and tomographic reconstructions of large datasets. These tasks are time consuming because of the sheer number of calculations involved. For example, tomographic reconstruction of an 8K $\times$ 8K dataset with approximately 60 angles can take up to 4 days on a multi-CPU state-of-the-art computer. Researchers have been drawn towards high-performance computing in hope of finding some reprieve in clusters and computation grids [12, 13]. A significant amount of effort has been directed towards getting biological codes to run on parallel architectures in order to cut down the turnaround time. For shared computational resources, users have to use the job-submission model where a task is assigned to CPUs by a scheduler. The assignment is governed by existing load on the machines and the number of jobs submitted. Typically it is not possible to estimate a worst case turnaround time for a job and users might have to wait from anywhere between, say, a few hours to a few days. Ideally, resources such as CPU cycles, memory, and network bandwidth should be dedicated for a job so an upper bound can be estimated on the time needed. In the best case users would like to get a visual feedback of a lower resolution model of the data so that they have the option of adjusting the parameters in real-time instead of waiting for the task to finish computing the large data.

Besides faster computation it is also important to streamline this process for ease of use since the end users are typically not aware with the intricacies of grid computing. Web interfaces and the concept of a workflow are often employed to enrich the user experience and to guide them through the visualization pipeline [14–19]. This also ensures that the users are abstracted from the idiosyncrasies of the underlying computational architecture while providing the flexibility of future expansion of the resources.

14.2.3 Data Storage and Management

Modern microscopes are capable of generating several gigabytes of data on a daily basis. This translates to terabytes of raw data every year that needs to be made accessible to different collaborators for analysis and then archived for long-term storage. Researchers typically store the raw data generated by instruments on fast local disk arrays until it has been pruned and processed for analysis. It is at this point that the data is uploaded to data centers where it can be collaboratively viewed and analyzed by several groups. Specialized data centers have cropped up tending to the large data storage and serving needs of research groups across the globe. These centers not only provide scalable arrays of fast spinning disks but are accessible over fast optical networks. Efficient query of relevant data is a key component in the visualization pipeline and requires dedicated resources such as those provided by these data centers. It is not possible to keep all data on spinning media, and older, less-relevant data are usually relegated to slower, high-density media such as tapes where it is still accessible by the users.

As microscopes increase in resolution and data acquisition capabilities, data growth is not expected to slow down in the near future. A robust and scalable storage resource strategy is required for the archival and management of these large datasets [11, 21]. Researchers are constantly investigating new algorithms, architectures, and media to help them in managing the growing storage needs of the community [20].

14.2.4 Moving Large Data with Optical Networks

In its January 2001 edition, *Scientific American* published an article about the growing demand for network bandwidth, reporting that the growth rate for network bandwidth far exceeded the growth rate of processors as predicted by

Moore's law [2]. In just the past few years it has become possible for research organizations and universities to buy dedicated lambdas or optical light paths. With the necessary networking hardware, it is possible to connect rare resources like expensive instruments, supercomputers, and clusters over these high-speed networks. This trend was one of the main reasons for inception of the OptIPuter project [3]. The OptIPuter was a National Science Foundation project that was aimed at developing an advanced distributed computing infrastructure for collaborative data exploration in the fields of biological and earth sciences. The project helped set up several light paths between universities across countries including the United States, Canada, the United Kingdom, The Netherlands, and Japan, to name a few. The project also aimed at experimenting with the intriguing idea of user controlled light paths (UCLPs) [4, 5] where users for the first time get to schedule dedicated network resources and fiber between sites for experiments. This potentially allows a group of researchers to custom tailor a scientific visualization pipeline with dedicated bandwidth of several gigabits per second for moving large data. This approach would allow exploration of large data collaboratively by research groups separated by large geographical distances. Once the experiment is over, the resources would be freed and reconfigured for a different purpose. A significant amount of research has been directed towards developing new network protocols and performance tuning existing Transmission Control Protocol (TCP) and User Datagram Protocol (UDP) stacks inside operating systems to utilize these new networks more efficiently.

14.2.5 Challenges of Visualizing Large Data Interactively

As mentioned earlier, it is more challenging to model systems and software for interactively exploring large datasets since the queued job-submission model prevalent in high performance computing does not provide the desired quality of service (QoS) guarantees. The queued best-effort approach suffers from a serious drawback where results of a task might be returned after several minutes to a few hours. Since the users are expecting instantaneous feedback every time they interact with the software, the pipeline cannot possibly process the entire dataset at interactive rates. Parsing several gigabytes of data is intensive work even for the fastest parallel computers, and resource allocation has to be done in a dedicated manner to guarantee a certain QoS. One commonly employed scientific visualization scheme for handling large data involves mip-mapping [6]. A large dataset is sent through a preprocessing stage where multiple resolutions of the data are created and stored in a tree-like structure. To aid in fast query or recovery of this data, it is also indexed or chopped into smaller pieces. During the actual rendering phase, the users interact with a low-resolution version of the dataset. The software will automatically start filling in the high-resolution detail once the user has chosen a region of interest. Also depending on the zoom level and the region explored, only the data that is actually needed by the graphical hardware is read from storage and rendered on the screen. Thus effectively at any given point, the software is only dealing with a small manageable portion of the large raw data.

Apart from the problem of handling large datasets, the rendered output can span several hundreds of megapixels. Tile-displays have proven to be an effective solution for solving the need for large pixel real estates [7]. Figure 14.4 shows one such display being used by researchers for visualizing high-resolution content. The collective resolution of these displays can run up to a few hundred megapixels. They are usually run by a cluster where a single node drives one or more tiles and all the nodes are interconnected by gigabit Ethernet or equivalent hardware. The software responsible for managing the displays is inherently distributed in nature and can synchronize the visualization across tiles. Ideally we would like to use the entire display as one giant desktop where multiple applications can simultaneously reside. Datasets can be placed adjacent to each other for visual comparison using these applications. These windowed applications can either run locally on one or more local nodes or remotely across the network on a cluster.

Tile-displays at times span entire room lengths and can be cumbersome to interact with because of the large physical space they cover. Because of their large size, the paradigm of a keyboard and a 2D mouse on a fixed workstation does not apply very well. Users sometimes want to analyze the high-resolution content up close and sometimes want to step back to do a visual comparison of multiple datasets. Multiple users want to interact with different sections of the display simultaneously and want their own mouse pointer and keyboard. This is different from the typical one-desktop-one-user paradigm used by desktop operating systems. Thus, the displays bring with them a host of unique human-computer interface problems that require unique solutions. Computer scientists have experimented with alternate input devices such as wireless 3D mice and wireless tablet PCs which can be held by the user as they walk around the room and help provide user input to the displays. Groups are also working on camera-based face, gaze, and hand-tracking systems which will one day help do away with the need for carrying any physical input device on the person.

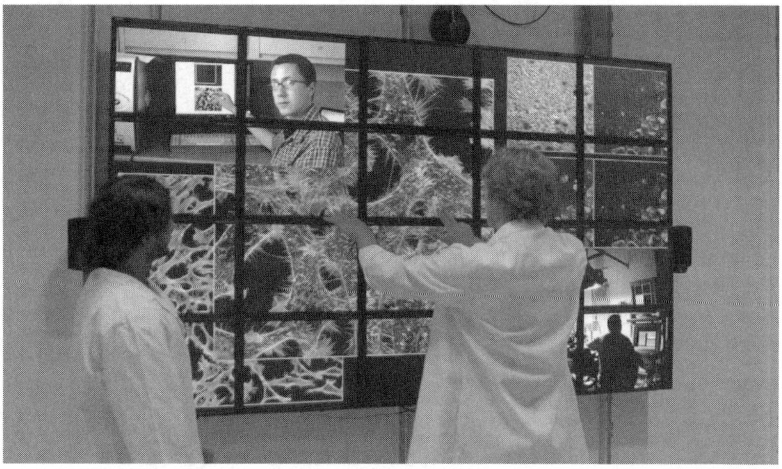

Figure 14.4 Researchers using NCMIR's 40-megapixel display wall for conducting collaborative data exploration experiments. The display is managed by Scalable Adaptive Graphics Environment (SAGE), which allows multiple graphical applications to use the tile-display like a shared desktop space.

14.3 Visualizing Large 2D Image Data

Large montages are generated by biologists on a daily basis and are created by stitching together contiguous overlapping tiles of images acquired by the microscopes in the X-Y plane. Final image resolutions vary from a few hundred megapixels to several gigapixels and storage on disk varies from a few hundred megabytes to terabytes. These datasets are perfect cases for using the mip-mapping approach [6] to achieve maximum interactivity. The software loads appropriate detail at different zoom levels by paging image data to video memory in a view-dependent manner. As noted earlier, the datasets cannot be used directly in their raw format and have to be sent through a preprocessing step in order to generate a hierarchical multiresolution structure. The tree is arranged in a way where lower or higher resolution images of a region can be accessed by traversing up or down the levels. Figure 14.5 depicts such a quad-tree where every node has four children and the child nodes are at higher resolution than their parents.

For the sake of simplicity, splitting is done with square tiles where the sides of the squares are power of 2. The edge cases can be buffered with null or black pixels to maintain power of 2 dimensions. The other advantage of using this approach is that OpenGL textures are managed as squares with power of 2 edges. Thus it is easy to do the mapping between files on disk and textures in graphics hardware memory and the load times are consistent. It also helps memory management by minimizing fragmentation. The other big advantage of splitting levels into smaller tiles is when different cluster nodes are trying to load their portion of the data from the shared network file system it drastically minimizes contention for access to the same portion of data. Smaller files are more easily cached in memory by the file system and can thus be transferred over the network more efficiently.

MagicCarpet [22] is software that supports mip-mapped based visualization of large 2D datasets (see Figure 14.6). It allows interactive viewing of multiple and possibly time-dependent datasets and even vector data. It has a simple intuitive user interface which allows users to zoom, pan, and flip through images. It has been observed that even though the mip-map creation needs to be done only once for

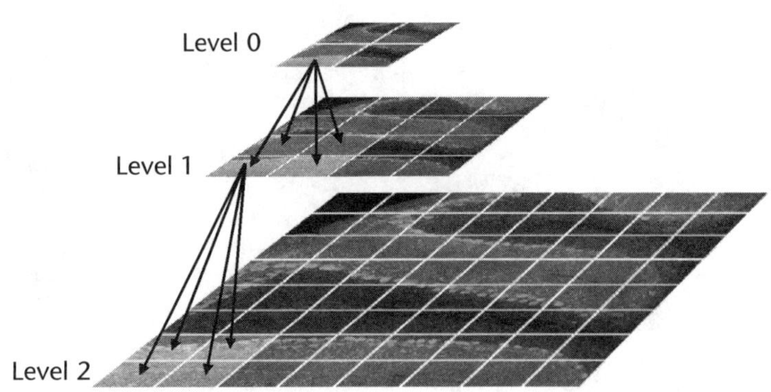

Figure 14.5 The figure shows a quad-tree hierarchy of images. Every parent node has four child tiles. All tiles are uniform squares with edge dimension which are power of two.

Figure 14.6 MagicCarpet running on the LambdaTable. The LambdaTable is a high-resolution, cluster-driven display table that supports multiple simultaneous users. (Image courtesy: Electronic Visualization Laboratory, University of Illinois at Chicago.)

every dataset, users find it inconvenient since the preprocessing can take several minutes to a few hours. This is a long time period especially in collaborative workspaces where users want to simply present the data to their collaborators during meetings. It is also unintuitive from the user interface perspective for the data to go through this step before it can be seen on the displays. Users typically get impatient and sometimes assume that the software has become unresponsive on account of it taking a long time.

However, the task is embarrassingly parallel [23] and can be sped up by several factors by running the preprocessing code on multiple CPUs. The OptIPuter project has made it possible for create a grid of computing resources over high-speed optical networks and the preprocessing step can be offloaded to clusters and supercomputers residing on this fast backplane. This type of grid computing approach is widely popular in the scientific community since it allows groups to share resources beyond the means of most research organizations. We notice a significant speedup by running the preprocessing code on this distributed framework. The amount of speedup depends on the number of CPUs used and resources available but modest tests have shown that figures greater than 10X are easily achievable.

14.4 Visualizing Large 3D Volume Data

Large 3D volumes are a result of montages collected along the Z axis and are best described as a stack of 2D images. These volumes can easily occupy several terabytes of disk space and require techniques like *ray tracing* or *ray casting* [24] for rendering. Ray casting is a popular image-order method that generates realistic renderings by casting viewing rays from each point on the image plane through the data [24]. Samples are taken at discrete intervals along the ray. These samples are used to generate pixels on an image plane. Thus the 3D volume space is projected on a 2D plane. The density of the rays and sampling frequency used decide the resolution of the final image rendered. Some of these lighting models are very complicated, and achieving interactive frame rates even with volumes that can be fitted into the graphics memory can get challenging very fast. (See Figure 14.7.)

There are other similar volume rendering methods such as 3D texture mapping [25] and several possible optimizations for making computation more efficient and even hardware support is built inside most modern GPUs, but the sheer number of calculations involved increases exponentially with volume size. Most techniques also require that the data be contained entirely in RAM if implemented on the CPU, or in texture memory if implemented using graphics hardware [25]. Modified approaches are used when data size exceeds the amount of available memory. The three primary approaches for dealing with large volume data involve data bricking/paging, use of compression, and parallel processing. Data bricking is analogous to the mip-mapping discussed in the previous section where raw data is sent through a preprocessing stage to generate a multiresolution hierarchical data structure. In the case of 3D data, this structure is an octree where every node has eight children as shown in Figure 14.8. As in the case of 2D mip-mapping, individual levels are divided into smaller manageable bricks. Depending on the view frustum requested by the user, bricks are loaded in to the graphics memory or discarded. Modern graphics hardware already supports a notion of paging akin to virtual memory where least recently used memory blocks are purged and overwritten with data from the main memory to aid in fast graphics pipeline processing.

Octreemizer [26] is one such system that uses an octree data structure coupled with a multilevel paging system and predictive cache to roam through large volumetric data. The multilevel cache operates between video memory and main

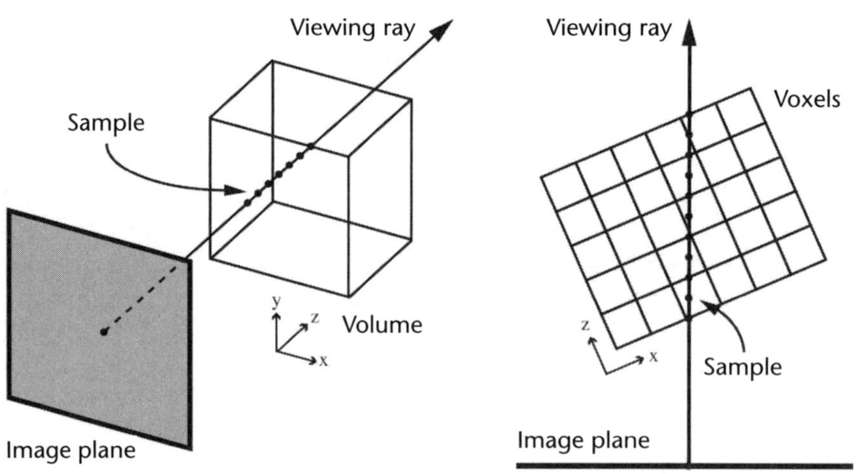

Figure 14.7 Volume rendering using ray casting. A ray starting at a point on the image plane is cast through the volume to evaluate the optical model. Samples are taken at evenly spaced intervals along the ray.

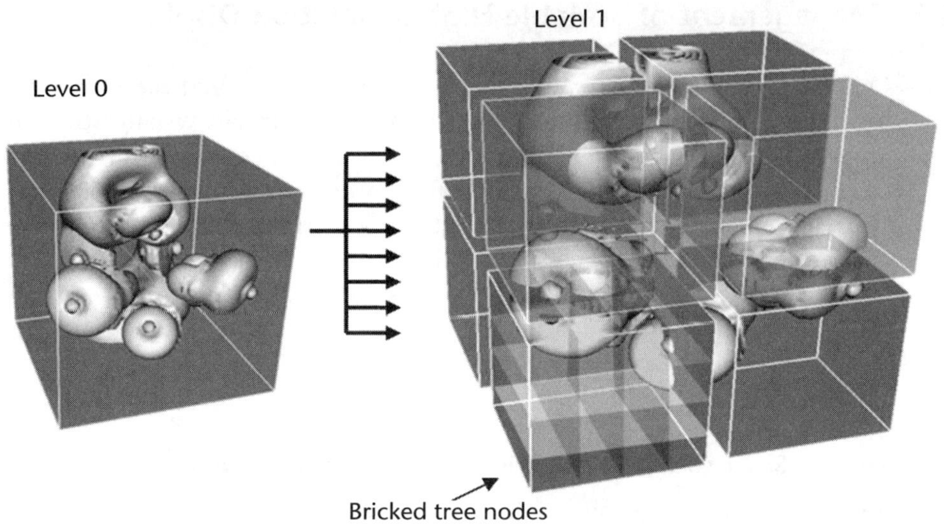

Figure 14.8 Volume data arranged as an octree where every level has eight children. Each node in the tree is further bricked into smaller pieces to aid in efficient loading times and memory management.

memory, and between main memory and disk. Using a least recently used (LRU) replacement strategy, the predictive cache fetches data based on the direction of user movement.

Parallel rendering of large volumes with multiple CPUs are primarily done using either an image-order approach or an object-order approach. Both approaches allow the task of rendering to be distributed to multiple CPUs simultaneously and then require an additional step to merge the individual results into a final image. Image-order rendering requires looking back from the final image and deciding on a portion of the image to be generated by each processor. Since this division is disjoint for every processor, they can each work on their sections of the data and render tiles. The compositing step is simple since all the tiles have to be stitched together for the final image. However, care needs to be taken to distribute the tasks efficiently since a processor with mostly black or empty voxels can finish its task earlier than others and will then sit idle while other processors are computing. Object-order rendering usually splits the data between processors in a fixed predetermined manner without worrying about the view frustum of every frame that is generated. During the execution, every CPU is supposed to calculate a projection based on the user-determined view port. However, the final compositing step is more complex in this case since it requires blending of all the individual projections in the right order.

Parallel computation of data is one requirement for handling large volumes interactively. The other requirement is display of the high resolution rendered output. Again cluster-driven tile-displays prove to be an invaluable asset in providing the necessary pixel real-estate to view the results across several hundreds of megapixels. The Volume Rendering Application (VRA) [27] is an image-order based parallel volume rendering system that can render on high-resolution display walls. It employs an octree-based multiresolution data structure to page in the correct bricks and resolution based on the view frustum.

14.5 Management of Scalable High-Resolution Displays

One basic requirement for working with these room-sized tile-displays is that they be treated as one contiguous desktop where multiple visualization applications and/or datasets can be rendered simultaneously. These applications can be a heterogeneous mixture of simple, single CPU programs to more complex parallel codes that run on clusters and supercomputers. Sometimes these applications execute on remote machines and/or clusters and need to stream their high-resolution output to the displays at the user end over fast networks. Examples of such compute-intensive applications are large 2D mosaic viewers and volume rendering software that are used by researchers for viewing and analyzing biological data.

14.5.1 SAGE (Scalable Adaptive Graphics Environment)

SAGE provides a graphics streaming architecture for supporting collaborative scientific visualization environments with potentially hundreds of megapixels of contiguous display resolution [8]. It was specifically designed for high-resolution cluster-driven scalable tile-displays and provides a giant contiguous "desktop" space. Applications that run on SAGE-enabled displays transport their rendered pixel frame buffers over the network [9, 10]. Multiple applications can be shown on these big desktops as windows that can be resized or moved by the users. Large datasets arranged next to each other for visual comparison are an important tool for the biologists since a lot of the analysis is still done manually. (See Figure 14.9.)

Figure 14.9 SAGE driven high-resolution displays allow users to arrange large datasets next to each other for visual comparison. Collaboration tools in the environment also support HD video conferencing.

The network-centric architecture of SAGE allows users to simultaneously run various compute intensive applications, such as 3D volume rendering, 2D montage viewers, and video streaming on local or remote computers. The environment also supports collaboration where the pixels from applications can be displayed at multiple sites simultaneously using network multicast or broadcast. Since the resolution of most graphical applications that run on these displays is very high, streaming requires a lot of bandwidth. For example, an uncompressed 30-fps HDTV stream with a resolution of $1,920 \times 1,080$ requires $\sim$1.5 Gbps. The OptIPuter infrastructure plays a crucial role in enabling this distributed computing architecture. Rendering and compute clusters can access the high-resolution displays over fast optical networks and stream their pixel frame buffers.

SAGE's streaming architecture is designed so that the output of arbitrary $M \times N$ pixel rendering cluster nodes can be streamed to $Q \times R$ pixel display screens [10], allowing user-definable layouts on the display. The dynamic pixel routing capability lets users freely move and resize each application's imagery over tiled displays in run-time, tightly synchronizing multiple component streams to form a single virtual stream.

14.5.2 COVISE (Collaborative Visualization and Simulation Environment)

COVISE [28] was originally been developed at the High Performance Computing Center Stuttgart (HLRS), and has been commercialized by the Stuttgart based VISENSO GmbH. It is a toolkit to integrate several stages of a scientific or technical application such as grid-generation, simulation, data import, post-processing, and visualization. Each step is implemented as a module. Using a visual user interface, these modules can be connected to a data flow network.

Each of the computational and I/O modules in this workflow can reside on a different computer. This allows distributing the work load among different machines. For instance, the pre- and post-processing modules can run on a visualization server, while the simulation runs on a remote supercomputer. The display modules can run on the workstation of a user, or on a visualization cluster driving a multiscreen visualization environment.

COVISE's virtual reality rendering module OpenCOVER can run on a variety of interactive displays and environments. Figure 14.10 shows COVISE managing the 15-tile rear-projected StarCAVE VR environment. It can even be used on a single computer with a mouse, but then the user cannot take advantage of its immersive capabilities. OpenCOVER is ideally run on tracked stereo environment, using 3D pointing devices. OpenCOVER uses the OpenSceneGraph API for its 3D rendering, which is an object-oriented framework on top of OpenGL. OpenCOVER has the ability to link multiple virtual environments together over the Internet, allowing for collaborative work between users in different buildings of a campus, or even on different continents. OpenCOVER is an open interface, in that the application programmer can write plug-in modules in C++ to create customized virtual reality applications, using COVISE's support of a large variety of virtual reality input and output devices, as well as interaction handling and network communication algorithms.

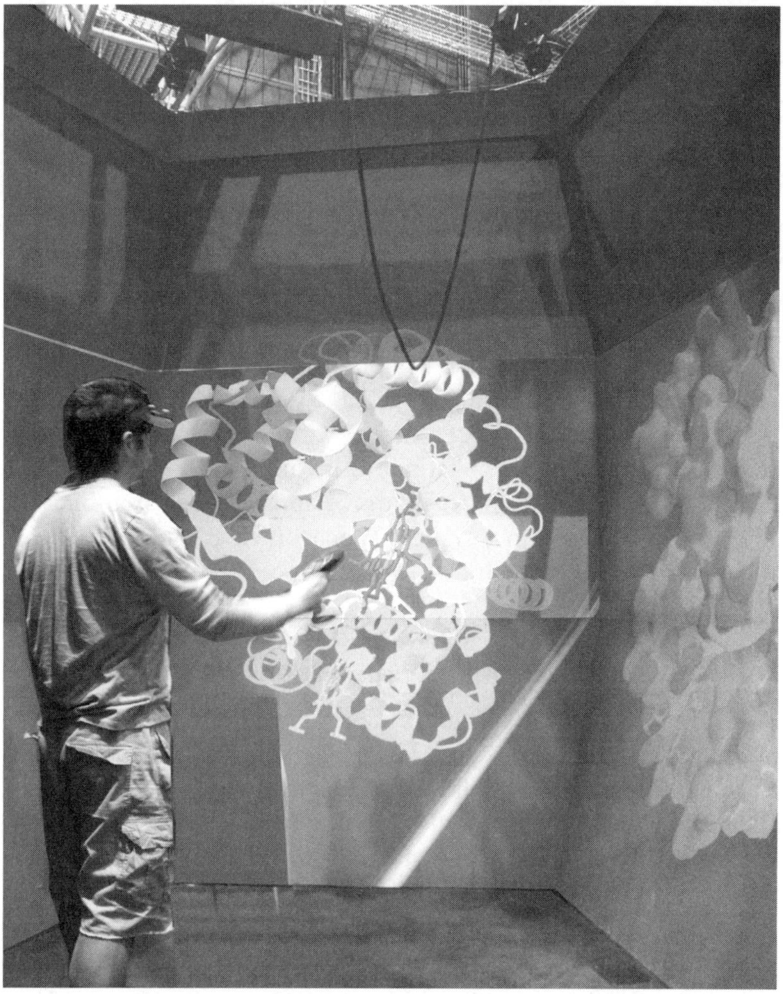

Figure 14.10 COVISE running on the 15-tile StarCAVE at the California Institute for Telecommunications and Information Technology at the University of California San Diego.

14.6 Virtual Reality Environments

Virtual reality display environments have historically proven invaluable for scientific visualization. Large 3D data exploration becomes more intuitive when the users are immersed in the dataset and forced to orient themselves with respect to regions of interest.

14.6.1 CAVE (Cave Automatic Virtual Environment)

The CAVE [29] is room-sized virtual reality apparatus that was invented at the Electronic Visualization Laboratory at the University of Illinois at Chicago. It is made up of high-resolution rear-projection displays and the first prototypes were cube-shaped rooms of dimensions 10 × 10 × 10 ft. Projection is done on the walls and the floor to enable an immersive VR experience. Users wear 3D glasses to see stereo and the system supports multiple simultaneous users. The primary viewer wears a headgear and is head-tracked so that the correct perspective view is generated for his/her eyes. The user can navigate through the 3D scene using a handheld controller like a joystick. The joystick position and orientation in space is tracked too so the user can potentially reach into the data and click on a point of interest. The first versions were run by expensive multi-CPU machines specialized for graphics processing and later versions are run by Linux clusters using commodity graphics and processing hardware.

Since its invention in 1991, many CAVE installations have appeared at universities and research organizations across the globe and proven to be invaluable for scientific visualization and data exploration. As seen in Figure 14.10, complex protein and biological structures can be understood better because of the high pixel

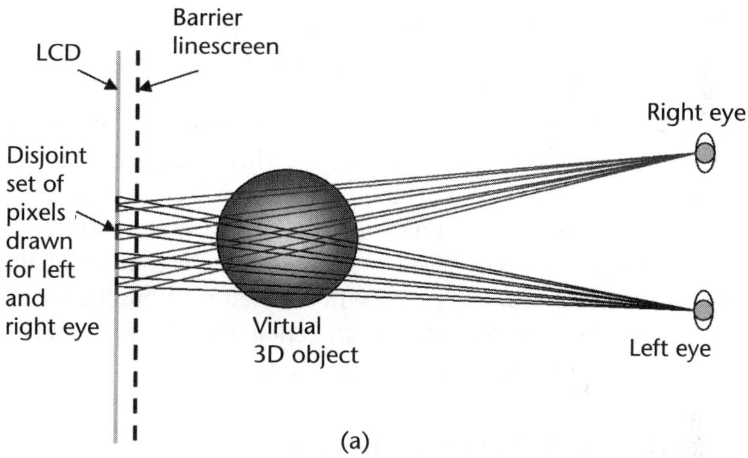

(a)

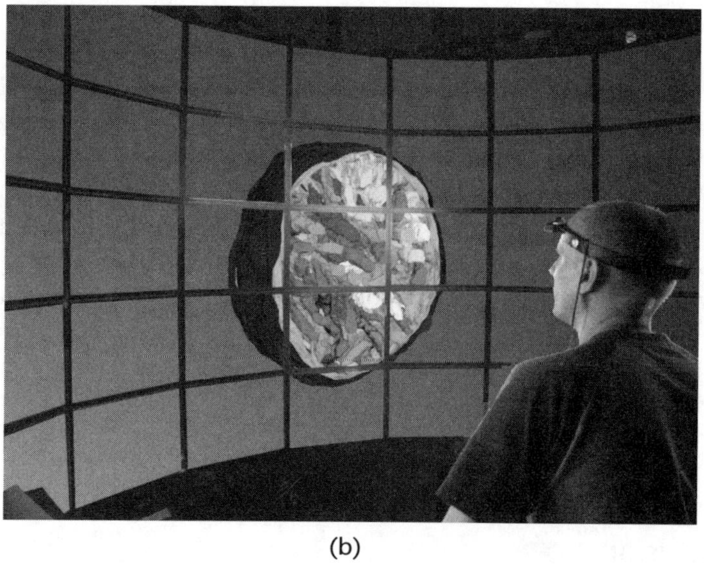

(b)

Figure 14.11 (a) How the barrier linescreen technology is used for generating pixel sets which are only visible to one eye at a time. Two such sets enable stereo vision. (b) A user interacting with the stereo display without using glasses.

count of the projected screens. It is even possible to achieve higher resolution by tiling projectors per wall.

14.6.2 Varrier

The Varrier [30, 31] is an auto-stereoscopic scalable virtual reality display that allows users to view stereo images without the need to wear any glasses. Unlike most stereo systems, which are based on projected displays that use active or passive stereo techniques, it uses the barrier stereography technique [30, 31] to generate images for the left and right eye. It can use LCD displays and so it is possible to build a high-resolution tile-display version of this auto-stereoscopic device. In the barrier method, a virtual barrier screen is created and placed in the virtual world in front of the projection plane. An off-axis perspective projection of this barrier screen, combined with the rest of the virtual world, is projected from at least two viewpoints corresponding to the eye positions of the head-tracked viewer. Figure 14.11 illustrates this principle.

The user does not have to deal with glasses but the system still uses a sensor based head-tracking system which lets the system know where the user's eyes are in 3D space. The projections for the eyes are then drawn behind the barrier linescreen accordingly. Like for the CAVE, the users also have a tracked joystick that allows users to pick and poke at 3D data. Work is well underway to develop a fast neural networks and camera-based head-tracking system that will help do away with the sensor headband. Future systems will also employ high-resolution cameras to do real-time hand gesture recognition to replace the joystick. The ultimate goal is to have a completely encumbrance free VR system.

14.7 Future of Large Data Visualization

The popular belief of using high-performance computing to group several CPUs together to achieve the power of future machines plays a key role in solving the large data-exploration problems of today. Multicore, multi-CPU parallel architectures seem to have an answer for handling large data visualization and we will see more of these machines appearing at research organizations across the globe. One recent exciting development has been the advent of programmable GPUs. These inexpensive processors currently host up to 128 cores and have dedicated fast memory and bus bandwidth. Driven by the video game industry, these graphics cards are available as commodity hardware and can be clubbed together for scalability. It is possible to envision a cluster farm of GPUs running scientific visualization codes for interactive data exploration. Due to the inexpensive hardware, they will also enable smaller research organizations to build their own graphics supercomputer without having to depend on expensive shared resources such as those hosted by supercomputing centers.

14.8 Conclusion

The problems generated by mesoscale in microscopy have challenged high-performance computing by bringing together large multimodal, multiscale data collected from a variety of instruments. However, the problems need a good solution in visualization to help in our understanding of the data. Scientific visualization has always proven to be a challenge for the best and fastest computing machines as scientific datasets, such as those generated by microscopy, get bigger every year. It is apparent that computers will be playing catch-up with the data for several years to come. Though the problem looks overwhelming at first glance, with the advent of fast and inexpensive optical networks and graphics and computing hardware, it is possible to tailor high-performance scientific visualization pipelines to help alleviate large data exploration problems to a great degree. Efficient parallel data handling algorithms and high-resolution scalable displays also play a key role in visualizing these datasets.

References

[1] Price, D.L., Chow, S.K., MacLean, N.A.B., Hakozaki, H., Peltier, S., Martone, M.E., and Ellisman, M.H., "High-Resolution Large-Scale Mosaic Imaging Using Multiphoton Microscopy to Characterize Transgenic Mouse Models of Human Neurological Disorders," *Neuroinformatics*, 4(1):65–80, 2006.

[2] http://www.sciamdigital.com/.

[3] http://www.optiputer.net/.

[4] He, E., Wang, X., Vishwanath, V., and Leigh, J., "AR-PIN/PDC: Flexible Advance Reservation of Intradomain and Interdomain Lightpaths," *IEEE GLOBECOM 2006*, June 31, 2006.

[5] Mambretti, J., "The Digital Communications Grid: Creating a New Architectural Foundation for 21st Century Digital Communications Based on Intelligent Lightpaths," in *Annual Review of Communications*, International Engineering Consortium, Vol. 58, 2005, pp. 307–314.

[6] Williams, L., "Pyramidal Parametrics," in *Proceedings of SIGGRAPH '83, Computer Graphics*, 17(3):1–11, July 1983.

[7] Park, K., Renambot, L., Leigh, J., and Johnson, A., "The Impact of Display-rich Environments for Enhancing Task Parallelism and Group Awareness in Advanced Collaboration Environments," Third Annual Workshop on Advanced Collaborative Environments, co-located at the Twelfth IEEE International Symposium on High Performance Distributed Computing (HPDC 12) and Global Grid Forum 8, Seattle, WA, June 22, 2003.

[8] http://www.evl.uic.edu/cavern/sage/.

[9] Jeong, B., Jagodic, R., Renambot, L., Singh, R., Johnson, A., and Leigh, J., "Scalable Graphics Architecture for High-Resolution Displays," *Proceedings of IEEE Information Visualization Workshop 2005*, Minneapolis, MN, October 23–25, 2005.

[10] Renambot, L., Jeong, B., Hur, H., Johnson, A., and Leigh, J., "Enabling High Resolution Collaborative Visualization in Display-Rich Virtual Organizations," *Future Generation of Computer Systems*, Vol. 25, 2009, pp. 161–168.

[11] Zhang, S., Price, D.L., Qian, X., Gupta, A., Wong, M., Ellisman, M.H., and Martone, M.E., "A Cell Centered Database (CCDB) for Multi-Scale Microscopy Data Management," *Microscopy and Microanalysis, Proceedings*, August 3–7, 2003, San Antonio, TX.

[12] Lin, A., Su, M., Kulungowski, A., Lathers, A., Mehta, G., Peltier, S., Deelman, E., and Ellisman, M., "Cyberinfrastructure for parallel tomographic reconstruction and analysis," *4th Intl. Congress on Electron Microscopy*, Nov. 5–8, 2006, San Diego, CA.

[13] Peltier, S.T., Lin, A., Lee, D., Smock, A., Lamont, S., Molina, T., Wong, M., Dai, L., Martone, M.E., and Ellisman, M.H., "The telescience portal for tomography applications," *J. Parallel Distributed Computing*, 63:539–550, 2003.

[14] Lathers, A., Su, M., Kulungowski, A., Lin, A.W., Mehta, G., Peltier, S.T., Deelman, E., and Ellisman, M.H., "Enabling Parallel Scientific Applications with Workflow Tools," *Proceedings of the 6th International Workshop on Challenges of Large Applications in Distributed Environments*, pp. 55–60, 2006.

[15] Molina, T., Yang, G., Lin, A.W., Peltier, S. and Ellisman, M.H., "A Generalized Service-Oriented Architecture for Remote Control of Scientific Imaging Instruments," *Proceedings of the 1st IEEE International Conference on e-Science and Grid Computing*, pp. 56–63, 2005.

[16] Lin, A.W., Dai, L., Mock, J., Peltier, S. and Ellisman, M.H., "The Telescience Tools: Version 2.0, Proceedings of The 1st IEEE International Conference on e-Science and Grid Computing, pp. 56–63, 2005.

[17] Lin, A.W., Dai, L., Ung, K., Peltier, S. and Ellisman, M.H., "The Telescience Project: Applications to Middleware Interaction Components," *Proceedings of the 18th IEEE International Symposium on Computer-Based Medical Systems*, pp. 543–548, 2005.

[18] Lee, D., Lin, A.W., Hutton, T., Akiyama, T., Shinji, S., Lin, F.P., Peltier, S. and Ellisman, M.H., "Global Telescience Featuring IPv6 at iGrid2002," *Future Generation of Computer Systems*, 19(6):1031–1039, 2003.

[19] Peltier, S.T., Lin, A.W., Lee, D., Mock, S., Lamont, S., Molina, T., Wong, M., Martone, M.E., and Ellisman, M.H., "The Telescience Portal for Advanced Tomography Applications," *Journal of Parallel and Distributed Applications, Special Edition on Computational Grids*, 63(5):539–550, 2003.

[20] Ludäscher B., Gupta, A., and Martone, M.E., "A Model-Based Mediator System for Scientific Data Management," in T. Critchlow and Z. Lacroix, (Eds.), *Bioinformatics: Managing Scientific Data*, Morgan Kaufmann, 2003.

[21] Martone, M.E., Zhang, S., Gupta, A., Qian, X., He, H., Price, D., Wong, M., Santini, S., and Ellisman, M.H., "The Cell-Centered Database: A database for multiscale structural and protein localization data from light and electron microscopy," *Neuroinformatics*, 1(3):379–396, 2003.

[22] http://www.evl.uic.edu/cavern/mc/index.html.

[23] http://en.wikipedia.org/wiki/Embarrassingly_parallel.

[24] http://en.wikipedia.org/wiki/Ray_tracing_%28graphics%29.

[25] http://portal.acm.org/citation.cfm?doid=378456.378484#.

[26] Plate, J., Tirtasana, M., Carmona, R., and Fröhlich, B., "A Hierarchical Approach for Interactive Roaming Through Very Large Volumes," *Joint Eurographics - IEEE TCVG Symposium on Visualization*, Barcelona, May 2002.

[27] http://www.evl.uic.edu/core.php?mod=4&type=3&indi=355.

[28] http://www.hlrs.de/organization/vis/covise/.

[29] Cruz-Neira, C., Sandin, D., DeFanti, T., Kenyon, R., and Hart, J., "The CAVE®: Audio Visual Experience Automatic Virtual Environment," *Communications of the ACM*, Vol. 35, No. 6, pp. 65–72, 1992.

[30] Sandin, D., Margolis, T., Ge, J., Girado, J., Peterka, T., and DeFanti, T., "The Varrier Autostereoscopic Virtual Reality Display," *ACM Transactions on Graphics, Proceedings of ACM SIGGRAPH 2005*, July 30–August 4, 2005.

[31] Peterka, T., D. Sandin, D., Ge, J., Girado, J., Kooima, R., Leigh, J., Johnson, A., Thiebaux, M., and DeFanti, T., "Personal Varrier: Autostereoscopic Virtual Reality for Distributed Scientific Visualization," *Future Generation Computing Systems*, October 1–31, 2006.

About the Editors

A. Ravishankar Rao is a research staff member at the Computational Biology Center within the IBM T.J. Watson Research Center, in Yorktown Heights, New York. His research interests include image processing, computer vision, and computational neuroscience. He is currently working on solving bio-imaging problems using high-performance computing, and the mathematical modeling of neural circuits, with the goal of understanding computation in the brain.

He chaired the SPIE Conference on Machine Vision and Applications from 1996-1998, and has served on the program committees of several conferences on imaging and vision. He served as the tutorials chair for the IS&T PICS Conference from 1999-2001. He is an associate editor of the journals *Pattern Recognition and Machine Vision and Applications*. He received his B.Tech degree in electrical engineering from the Indian Institute of Technology (Kanpur) in 1984, and his M.S. and Ph.D. degrees in computer engineering from the University of Michigan, Ann Arbor, in 1986 and 1989, respectively. His research has resulted in nineteen patents and over fifty publications. He has published a book entitled *A Taxonomy for Texture Description and Identification.* He co-teaches a course on global brain modeling at Columbia University. He was named a master inventor at IBM Research in 2004, in recognition of his sustained invention activity resulting in valuable patents within IBMs portfolio.

Guillermo A. Cecchi is a research staff member at the Computational Biology Center within the IBM T.J. Watson Research Center, in Yorktown Heights, New York. He received his M.Sc. in physics from the University of La Plata, Argentina, in 1991, and his Ph.D. in physics and biology from The Rockefeller University in 1999. He studied imaging in psychiatry as a postdoctorate fellow at Cornell University in 2000. He has researched diverse aspects of theoretical biology, including Brownian transport, molecular computation, spike reliability in neurons, song production and representation in songbirds, statistics of natural images and visual perception, statistics of natural language, and brain imaging. Since joining IBM Research in 2001, he has been working on computational models of brain function. Currently, his research is focused on high performance computing for brain imaging, and network theory approaches to analysis and modeling of brain function. He is also a member of the Systems Biology Center of New York, and a visiting scientist at The Rockefeller University.

List of Contributors

Louise C. Abbott
Department of Veterinary Integrative Biosciences
Texas A&M University
3112 TAMU
College Station, TX 77843-3112
e-mail: labbott@cvm.tamu.edu

Paul S. Albert
Biometric Research Branch
National Cancer Institute
National Institutes of Health
Bethesda, MD
e-mail: pa65b@nih.gov

Robert F. Bonner
Laboratory of Integrative and Medical Biophysics
National Institute for Child Health
and Human Development
National Institutes of Health
Bethesda, MD
e-mail: bonerr@mail.nih.gov

Filiz Bunyak
Department of Computer Science
University of Missouri
Columbia, MO 65211
e-mail: bunyak@missouri.edu

Guillermo A. Cecchi
IBM Research
PO Box 218
Yorktown Heights, NY 10598
e-mail: gcecchi@us.ibm.com

Yoonsuck Choe
Department of Computer Science
Texas A&M University
3112 TAMU
College Station, TX 77843-3112
e-mail: choe@tamu.edu

Rodrigo F. Chuaqui
Pathogenetics Unit
Laboratory of Pathology and Urologic Oncology
Branch Center for Cancer Research
National Cancer Institute
National Institutes of Health
Bethesda, MD
e-mail: chuaquir@mail.nih.gov

Lee Cooper
Ohio State University
205 Dreese Laboratory
2015 Neil Avenue
Columbus, OH 43210
e-mail: cooperl@ece.osu.edu

Mark H. Ellisman
University of California, San Diego
9500 Gilman Drive
M/C 0608
La Jolla, CA 92093
e-mail: mark@ncmir.ucsd.edu

Michael R. Emmert-Buck
Pathogenetics Unit
Laboratory of Pathology and Urologic Oncology
Branch Center for Cancer Research
National Cancer Institute
National Institutes of Health
Bethesda, MD
e-mail: buckm@mail.nih.gov

Heidi S. Erickson
Pathogenetics Unit
Laboratory of Pathology and Urologic Oncology
Branch Center for Cancer Research
National Cancer Institute
National Institutes of Health
Bethesda, MD
e-mail: HSErickson@mdaderson.gov

Yan Feng
Novartis Institutes for BioMedical Research
250 Massachusetts Avenue
Cambridge, MA 02139
e-mail: yan.feng@novartis.com

Rahul Garg
IBM Research
PO Box 218
Yorktown Heights, NY 10598
e-mail: grahul@us.ibm.com

Jason Goffeney
Department of Computer Science
University of Missouri
Columbia, MO 65211
e-mail: jgoffeney@gmail.com

Donhyeop Han
Department of Computer Science
Texas A&M University
3112 TAMU
College Station, TX 77843-3112
e-mail: dhan1@neo.tamu.edu

Jeffrey C. Hanson
Pathogenetics Unit
Laboratory of Pathology and Urologic Oncology
Branch Center for Cancer Research
National Cancer Institute
National Institutes of Health
Bethesda, MD
e-mail: hansoje@mail.nih.gov

Kun Huang
Ohio State University
3190 Graves Hall
333 W. 10th Avenue
Columbus, OH 43210
e-mail: khuang@bmi.osu.edu

Pei-San Huang
Department of Veterinary Integrative Biosciences
Texas A&M University
3112 TAMU
College Station, TX 77843-3112
e-mail: phuang@cvm.tamu.edu

Kirk Jordan
IBM Deep Computing
Systems and Technology Group
One Rogers St.
Cambridge, MA 02142
e-mail: kjordan@us.ibm.com

Ehud Kaplan
The Mount Sinai School of Medicine
New York, NY 10029
e-mail: ehud.kaplan@mssm.edu

John Keyser
Department of Computer Science
Texas A&M University
3112 TAMU
College Station, TX 77843-3112
e-mail: keyser@cs.tamu.edu

Kristian Kvilekval
University of California Santa Barbara
Santa Barbara, CA 93116
e-mail: kris@cs.ucsb.edu

Jaerock Kwon
Department of Computer Science
Texas A&M University
3112 TAMU
College Station, TX 77843-3112
e-mail: jrkwon@tamu.edu

Abel W. Lin
University of California, San Diego
9500 Gilman Drive
M/C 0446
La Jolla, CA 92093
e-mail: awlin@ncmir.ucsd.edu

Zigurts K. Majumdar
Laboratory of Integrative and Medical Biophysics
National Institute for Child Health
and Human Development
National Institutes of Health
Bethesda, MD
e-mail: majumdarz@mail.nih.gov

Maryann E. Martone
University of California, San Diego
9500 Gilman Drive
M/C 0446
La Jolla, CA 92093
e-mail: maryann@ncmir.ucsd.edu

David Mayerich
Department of Computer Science
Texas A&M University
3112 TAMU
College Station, TX 77843-3112
e-mail: mayerichd@neo.tamu.edu

Bruce H. McCormick
Department of Computer Science
Texas A&M University
3112 TAMU
College Station, TX 77843-3112
e-mail: mccormic@cs.tamu.edu

Zeki Melek
Department of Computer Science
Texas A&M University
3112 TAMU
College Station, TX 77843-3112
e-mail: z0m8905@cs.tamu.edu

Sumana Mukherjee
Pathogenetics Unit
Laboratory of Pathology and Urologic Oncology Branch
Center for Cancer Research
National Cancer Institute
National Institutes of Health
Bethesda, MD
e-mail: mukherjs@mail.nih.gov

Sumit Nath
GE Global Research
Imaging Technologies
John F. Welch Technology Center
122 EPIP Phase 2, Whitefield Road
Bangalore, Karnataka 560 066
India
e-mail: Sumit.Nath@ge.com

Kannappan Palaniappan
Department of Computer Science
University of Missouri
Columbia, MO 65211
e-mail: palaniappank@missouri.edu

Steven T. Peltier
University of California, San Diego
9500 Gilman Drive
M/C 0446
La Jolla, CA 92093
e-mail: peltier@ncmir.ucsd.edu

Thomas J. Pohida
Computational Bioscience
and Engineering Laboratory
Division of Computational Bioscience
Center for Information Technology
e-mail: pohidat@exchange.nih.gov

A. Ravishankar Rao
IBM Research
PO Box 218
Yorktown Heights, NY 10598
e-mail: ravirao@us.ibm.com

Jaiḿe Rodriguez-Canales
Pathogenetics Unit
Laboratory of Pathology and Urologic Oncology Branch
Center for Cancer Research
National Cancer Institute
National Institutes of Health
Bethesda, MD
e-mail: rodrigja@mail.nih.gov

Antonio Ruiz
University of Malaga
29071 Malaga, Spain
e-mail: aruiz@ac.uma.es

Jurgen P. Schulze
University of California, San Diego
9500 Gilman Drive # 0436
La Jolla, CA 92093-0436
e-mail: jschulze@soe.ucsd.edu

Ambuj Singh
University of California Santa Barbara
Santa Barbara, CA 93116
e-mail: singh@cs.ucsb.edu

Rajvikram Singh
University of California, San Diego
9500 Gilman Drive
M/C 0446
La Jolla, CA 92093
e-mail: raj@ncmir.ucsd.edu

Michael A. Tangrea
Tumor Angiogenesis Section
Surgery Branch
National Cancer Institute
National Institutes of Health
Bethesda, MD
e-mail: tangream@mail.nih.gov

Charles Y. Tao
Novartis Institutes for BioMedical Research
250 Massachusetts Avenue
Cambridge, MA 02139
e-mail: tao212@gmail.com

Manuel Ujaldon
University of Malaga
29071 Malaga, Spain
e-mail: ujaldon@ac.uma.es

John Wagner
Computational Biology Center
IBM TJ Watson Research Center
1101 Kitchawan Road
P.O. Box 218
Yorktown Heights, NY 10598
e-mail: wagnerjo@us.ibm.com

Youping Xiao
The Mount Sinai School of Medicine
New York, NY 10029
e-mail: youping.xiao@mssm.edu

Elad Yom-Tov
IBM Haifa Research Lab
165 Aba Hushi St.
Haifa 31905, Israel
e-mail: yomtov@il.ibm.com

Daniel W. Young
Novartis Institutes for BioMedical Research
250 Massachusetts Avenue
Cambridge, MA 02139
e-mail: daniel.w.young@gmail.com

Index

N